Offshore Structures

Volume II • Strength and Safety for Structural Design

Günther Clauss, Eike Lehmann
and Carsten Östergaard

Translated by M. J. Shields, FIInfSc, MITI

Offshore Structures

Volume II
Strength and Safety for Structural Design

With 176 Figures

Springer-Verlag
London Berlin Heidelberg New York
Paris Tokyo Hong Kong
Barcelona Budapest

Gunther Clauss, Prof. Dr.-Ing.
Technische Universität Berlin, Institut für Schiffs- und Meerestechnik,
Salzufer 17–19, 10587 Berlin 10, Germany

Eike Lehmann, Prof. Dr.-Ing.
Arbeitsbereich Schiffstechnische Konstruktionen und Berechnungen, Technische
Universität Hamburg/Harburg, Lavenbruch Ost, 21071 Hamburg 90, Germany

Carsten Östergaard, Dr.-Ing.
Germanischer Lloyd AG, Vorsetzen 32, D-20459 Hamburg 11, Germany

Translator
M. J. Shields, FIInfSc, MITI
Literary and Technical Language Services, 199 The Long Shoot, Nuneaton,
Warks CV11 6JQ, UK

Cover illustrations·
Ch. 2, Fig. 3 (Vol. I). Compliant piled tower. Ch. 7, Fig. 8. Safety domains in an $A-W$
(stress strength) safety format.

ISBN-13: 978-1-4471-2000-1   e-ISBN-13: 978-1-4471-1998-2
DOI: 10.1007/ 978-1-4471-1998-2

British Library Cataloguing in Publication Data
Clauss, Gunther
   Offshore Structures. –Vol 2· Strength
   and Safety for Structural Design
   I. Title II. Shields, M. J
   627.98
ISBN-13: 978-1-4471-2000-1

Library of Congress Cataloging-in-Publication Data
(Revised for volume 2)
Clauss, Günther
   Offshore structures.
   Rev. translation of: Meerestechnische Konstruktionen.
   Includes bibliographical references and indexes
   Contents. v. 1. Conceptual design and hydromechanics
—v. 2. Strength and safety for structural design
   1. Offshore structures – Design and construction.
2. Ocean engineering. I. Lehmann, E (Eike).
II. Östergaard, C (Carsten) III. Title.
TC1165.C5313   1992   627'.98   91-37296

First published in German as *Meerestechnische Konstruktionen* by Springer-Verlag Berlin
Heidelberg New York © 1988
Softcover reprint of the hardcover 1st edition 1988

Typeset by Thomson Press (India) Ltd., New Delhi

69/3830-543210 Printed on acid-free paper

Contents

Preface to Volumes I and II

As always with new technology, research and development needs a certain time for specialist knowledge arising from individual, arbitrary cases of practical or engineering activity to develop into a clear field of knowledge that can be placed in a general context. We are convinced that this point has now been reached in the field of analysis and evaluation of offshore structures, and we hope to produce the proof in this edition. Our main aim is to present the basics of the technology as broadly as possible, so as to facilitate technical communication between specialists in different and diverse disciplines relevant to the development of offshore structures.

In order to maintain this broad approach, which is also applicable to many other marine structures (including ships), we have deliberately but with considerable regret omitted some special fields that could have been covered in depth on the basis of current knowledge. It seemed more important, in terms of explaining the analysis and evaluation of offshore structures, that the reader should develop practical solutions directly from as many concrete examples as possible, or be referred to further special literature. For these, we have turned especially to design and development engineers who are trained or in training situations, and who are familiar with the basics of the mathematical and physical approach to technical problems.

In this way, the book serves as an introduction for students and practising engineers, or as a source of information for experts in this field, or for specialists in adjacent fields who wish to obtain an overall picture of where their own work on individual problems of offshore structures stands in relation to the field as a whole. We know that the text will at times be difficult for the reader, but we also believe that these difficulties will be repaid at the very next encounter with the practical problems of offshore structures.

The development of a structure basically requires an iterative process which can be roughly defined in terms of three stages – concept specification, analysis, and evaluation. The term analysis signifies the mathematical modelling and expression of the behaviour of a structure under arbitrary boundary conditions, while the term evaluation describes the determination of particular boundary conditions, the fulfillment of which can be accepted for data used for analysis or for the results obtained by analysis.

In support of this process, and in order to explain comprehensively the practically relevant basics for the development of offshore structures, we have arranged these books such that after an introductory survey of the field of marine science and ocean engineering (Chapter 1), alternative concepts in offshore structures are laid down and explained (Chapter 2). We then proceed to cover as extensively as possible the main points in the analysis of offshore structures – hydromechanical analysis (Chapter 3) and structural analysis (Chapter 4) – in the framework of the special requirements of this field. In design practice, such analyses are bound up in a stochastic evaluation concept, which divides into the evaluation of the various environmental conditions under which offshore structures operate (Chapter 5), and the evaluation of the structures themselves (Chapter 6). From this basis we go on to discuss new developments and peculiarities of dimensioning practice according to Regulations (Chapter 7). In all chapters our objective is to keep the material as self-contained and as comprehensible as possible for the reader, who is then referred to further literature.

The themes of this book, integrating different classical and modern fields of engineering science, arise from lectures we have given on the basics of hydromechanics, and on structural analysis and mechanical reliability evaluation, as well as from our long design practice in offshore structures and their components. The work in manuscript was divided according to our individual fields of interest: overall control of Chapters 1–3 went to G. Clauss, while Chapter 4 and Sections 7.1, 7.2 and 7.5, together with Appendix A2, were the responsibility of E. Lehmann. In Chapters 5 and 6, along with Sections 7.3 and 7.4 plus Appendix A1, C. Östergaard had overall control. From these basics, we proceeded from individual sections to an integrated whole. Because of field-specific conventions which could not simply be ignored, it was impossible to use all symbols consistently: in some cases the same symbol means different things in different disciplines, so that the nomenclature and basic methodology had to be separate for each chapter. This is particularly useful for readers who wish to use the books as a reference work on methods of rational analysis and evaluation of offshore structures within specific problem areas.

The books arose from the initiative of Springer-Verlag, who proved to be an understanding partner and support in all questions of form and production during the development of the manuscript. Basic to the text is an extensive treasure-trove of experience which we had access to at the Technical University of Berlin, the Technical University of Hamburg-Harburg, and Germanischer Lloyd, as well as through published research and development work and participation in projects of the offshore construction and shipbuilding industries. For many reasons we must be grateful for the facilities of the above universities and of Germanischer Lloyd used in the development of these books. We would especially like to thank the following colleagues and fellow-workers who have contributed towards various parts of this work:

At the Institute of Naval Architecture and Ocean Engineering of the Technical University of Berlin it is especially necessary to thank Mr H. Höhne, who with skill and endurance prepared the illustrations and diagrams for the first three chapters, overcoming with great patience the often contradictory requirements regarding layout. Corrections and completion of the manuscripts of the first three chapters lay in the hands of Messrs T. Riekert, L. Birk, and J. Heeg, who in an outstanding example of cooperation and commitment produced the final text.

In the Department of Marine Structures and Structural Analysis of the Technical University of Hamburg-Harburg, we were especially grateful to Messrs G. Nickel and W. Prediger for typing of the manuscript and preparation of diagrams for Chapter 4 and parts of Chapter 7.

In the Hydromechanics and Reliability Department of Germanischer Lloyd, Dr. T. E. Schellin and Dr. T. Jiang discussed special hydrodynamical questions of Chapters 3, 5 and 6, and Dr. G. Schall of the Technical University, Munich made valuable contributions to the last part of Chapter 6. Mr K. Hamann of Germanischer Lloyd prepared most of the illustrations for Chapters 5 and 6, which were eventually adapted to this English edition by Mr H. Höhne of the Technical University Berlin.

This English edition in two volumes is the translation of the one volume German text of 1988, which was carefully reviewed and updated, and to which essentially new parts have been added in Chapters 2, 3, 6 and 7.

G. Clauss
E. Lehmann
C. Östergaard

Note on Terminology

In the process of translating these books it was necessary to consider not only the correspondence between technical terms in German and English but also terminology in English itself. This proved especially difficult in the terms 'marine technology' and 'ocean engineering', which are often used interchangeably in English, yet which can have quite different implications.

For example, the term 'ocean engineering' is used in the names of technical universities and institutions and for many committees, conferences, and journals which cover engineering in the marine environment, while 'marine technology' is often associated with instrumentation in marine science. Yet, by definition, 'marine technology' must cover non-engineering activities in the marine environment, while 'ocean engineering' cannot.

For this reason, we eventually decided to use the terms 'marine science' and 'marine technology' in exact correspondence with the terms 'science' and 'technology' in the broader sense. In other words, all applications of technology in the marine environment come under the umbrella of 'marine technology'. This includes 'ocean engineering', which refers specifically to marine-based engineering operations, and not to other areas of marine technology such as marine biotechnology, or navigation.

We hope this definition will find general acceptance in the field, as it seems to us to be both logical and consistent.

G. Clauss, E. Lehmann, C. Östergaard (authors)
M. J. Shields (translator)

4. Marine Structural Analysis

Structural analysis has been integrated into the design process of marine structures to a far greater extent than conventional shipbuilding. The effectiveness of the design of, for example, a semisubmersible or a tension leg platform, significantly depends on its design weight. Design weight, in turn, depends on the scantlings of the structural members obtained from structural analysis based on a rational approach. An example may clarify this.

Usually in the design of a semisubmersible structure (e.g. Aker H3 in Fig. 4.1a), the pontoons are joined together with horizontal braces. From the viewpoint of statics, this is a compact, three-dimensional, load-bearing structure which is very effective against horizontal forces. However, if we consider the incident of a blow-out at a drilling location, this type of horizontal bracing actually turns out to be disadvantageous. A structure without horizontal bracing (e.g. Chris Chenery in Fig. 4.1b) has important implications for its behaviour. In this arrangement there are, at section $A-A$, additional bending stresses of considerable magnitude arising solely because of the lack of horizontal bracing. These stresses must also be resisted by the structure. It is therefore apparent that considerations based on structural behaviour are just as indispensible at the initial design stages as they are at the final stages of design.

In general, the analysis of such a structure requires a mathematical model which is by no means less complex, and this is only possible by taking recourse to the finite element method. It is, however, important to study the fundamentals of static and dynamic structural analysis of basic structural systems so as to formulate a problem in general terms, and to select the suitable finite element code for it.

Such fundamentals are especially important in the setting up of an extensive model and assessing the results of the calculations.

Structural analysis *per se* includes the following problems:

— ultimate strength under static and dynamic loading,
— strength analysis of individual modules (e.g. ring-stiffened cylinders in semi-submersibles),
— strength analysis of local components (e.g. pipe joints),
— vibration characteristics of individual components, modules and/or the whole structure.

The methods used in marine structural analysis do not differ significantly from

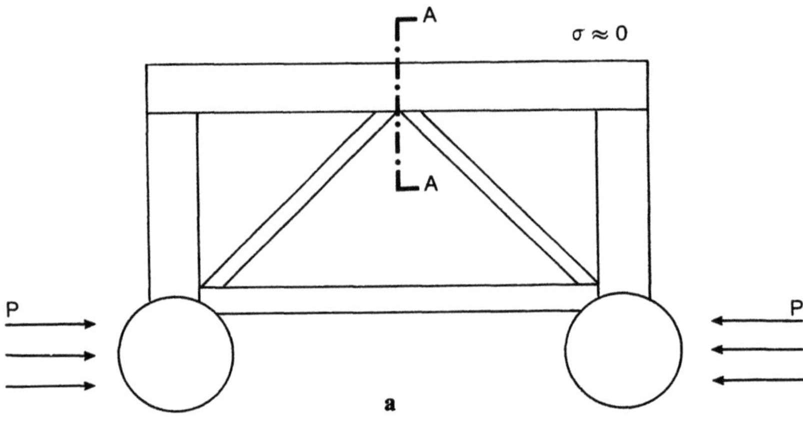

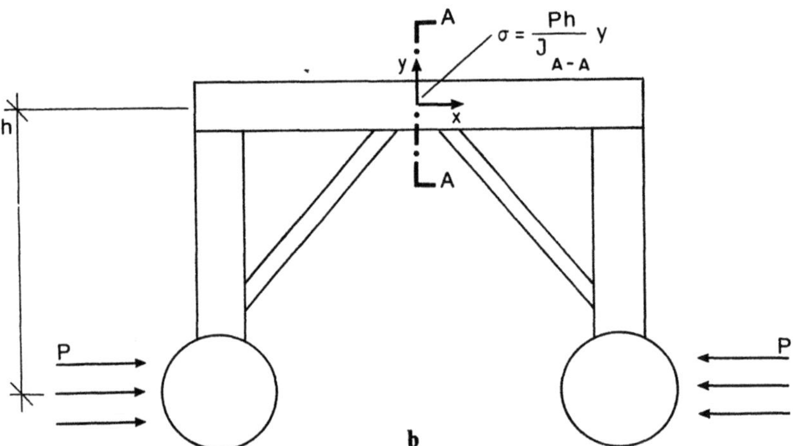

Fig. 4.1a,b. Various semisubmersible structures. a Aker H3, b Chris Chenery.

those used in ships, aircraft or steel structures, except for the assumptions on loadings and special conditions of construction.

4.1 Time-Independent Elastic Problems

In this section the more important fundamentals of structural analysis as applied to marine structures are set out, based on the general theory of elastic structures. When a sufficiently detailed coverage of a topic is infeasible due to a paucity of space, mention will be made of those aspects which aid a thorough appreciation of structural behaviour, and provide the basis for assessing numerical calculations (e.g. using finite element method). The various structural members can be classified as beams, membranes, plates and shells. Membrane structural elements such as deep webbed beams or flexible wide flanges are distinguished by the fact that the forces act only in the plane membrane, whereas plates can dissipate forces which act transversely to the plane of the plate, resulting in bending moments and shear forces. This applies for a linear approach. In several problems, however, a non-linear approach becomes essential, in which a strong coupling between the bending of the plate and the inplane stresses on the one hand, and coupling of the inplane stresses with the plate curvatures on the other, have to be taken into account. In addition to flat plate elements, there are also stiffened and unstiffened cylindrical shells, which are most frequently encountered as structural elements in marine structures. Their widespread use is due to the fact that shells have the property of transmitting the surface loads normal to its surface, not in the form of bending moments, but principally in the form of membrane forces, or in other words, forces in the shell surface. Since in most cases the membrane stiffness of a cylindrical shell is considerably greater than its bending stiffness, it gives the lightest structure, which is therefore most economical. This extremely favourable property is, however, mitigated by the action of bending moments arising from the local forces or support conditions. Because of this, we distinguish between membrane and bending theories of cylindrical shells. As pointed out earlier, it is impossible to present in complete detail all the basic elasticity theory as it is beyond the scope of this book and, in any case, there are several excellent texts in which further details may be found [1–3].

4.1.1 Frameworks

The internal stresses of an individual element of a structure can be determined by the so-called 'free-body' diagram. We isolate an infinitesimally small element from the component under investigation, and consider those forces and moments at its edges which hold this element in equilibrium. These forces are equivalent to the invisible internal stresses. We consider, for example, the horizontal bracing of the semisubmersible shown in Fig. 4.1a as a beam with two supports which are elastically restrained to counteract rotation (Fig. 4.2).

If we take at any position x a differential element of length dx and consider the internal forces which act on it, such that on the left-hand side of the element all forces and moments act in the negative direction, and on the right-hand side they act in a positive direction, then such an element must be in a static equilibrium. Obviously, the forces and moments on the positive side, will be reversed with respect to the negative side. For setting up equilibrium it is useful if these forces and moments are expressed as the sum of the forces and moments of the negative side and the change of the forces and moments at a distance dx. The three equilibrium conditions are essentially first order differential equations with higher order terms disregarded

$$\frac{dN}{dx} = 0, \quad \frac{dQ}{dx} = -q(x), \quad \frac{dM}{dx} = Q. \tag{4.1}$$

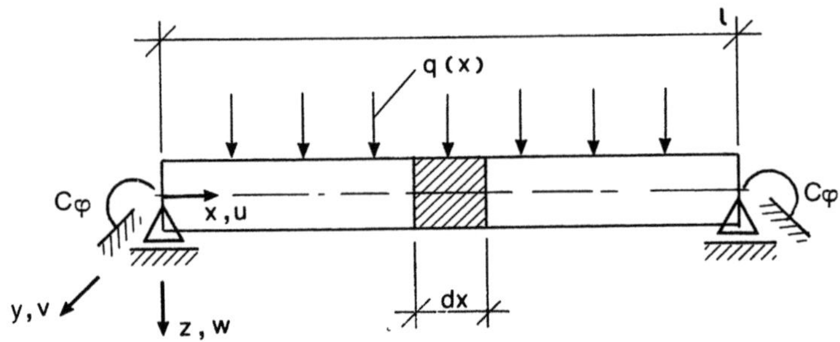

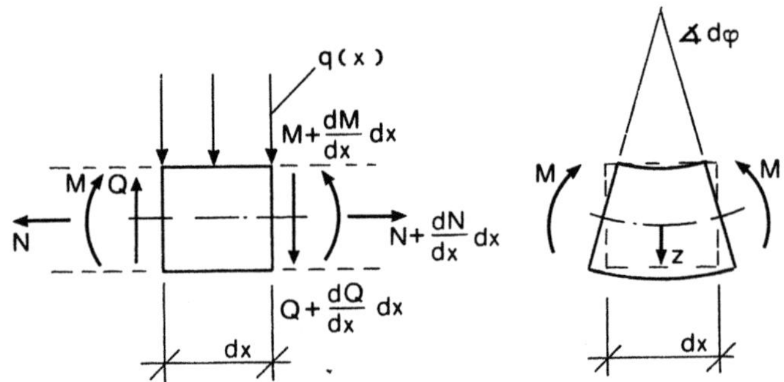

Fig. 4.2. Simply supported beam with elastic rotational constraints at supports.

By differentiation, the last two of these equations can be combined into a second order differential equation

$$\frac{d^2 M}{dx^2} = -q(x). \tag{4.2}$$

The solution is elementary, and has two constants of integration. For $q(x) = q$, we obtain

$$M = A_1 + A_2 x - \frac{qx^2}{2}.$$

The constants A_1 and A_2 are determined from boundary conditions which can have either the form of the solution itself (i.e. the bending moment), or its first derivative (i.e. the shear force) at $x = 0$ or $x = l$. However, for this example the boundary conditions are

$$M(x) = c_\varphi \varphi(x)|_{x=0, x=l}$$
$$w(x) = 0|_{x=0, x=l}, \tag{4.3}$$

which includes the conditions for deflections and rotations.

The relationship between deflection and the internal forces must now be derived, and so we consider the deformation of the beam element by these forces. The elongations or the strains of the individual fibres can be written, approximately with $\varphi = dw/dx$, as

$$\varepsilon = -z\frac{d^2w}{dx^2}. \tag{4.4}$$

The bending moment M can be expressed as the integral of the magnitudes of the axial stresses acting over a beam section A so that

$$M = \int_A \sigma z\, dA$$

By Hookes law, $\sigma = E\varepsilon$, and using the Bernoulli hypothesis (i.e. a linear distribution of normal stresses in the z-direction), we obtain

$$M = -E\int_A z^2\, dA \cdot \frac{d^2w}{dx^2}. \tag{4.5}$$

The integral in (4.5) is the area moment of inertia I, and hence

$$M = -EI\frac{d^2w}{dx^2}. \tag{4.6}$$

Differentiating this expression twice and using (4.2), the general differential equation for linear beam bending is obtained ·

$$EI\frac{d^4w}{dx^4} = q(x). \tag{4.7}$$

Its solution follows with $q = q(x)$

$$w(x) = C_1 + C_2 x + C_3 x^2 + C_4 x^3 + \frac{q}{EI}\frac{x^4}{24}. \tag{4.8}$$

Using the boundary conditions (4.3), the constants C_1 to C_4 can be obtained by substitution into (4.8) and solving the resulting system of linear equations. For bending we have

$$w(x) = \frac{ql^4}{24EI}\left[\frac{x}{l}\cdot\frac{1}{1+\bar{c}_\varphi} + \left(\frac{x}{l}\right)^2\frac{\bar{c}_\varphi}{1+\bar{c}_\varphi} - 2\left(\frac{x}{l}\right)^3 + \left(\frac{x}{l}\right)^4\right], \tag{4.9}$$

where the rotational constant c_φ has been made dimensionless with

$$\bar{c}_\varphi = \frac{c_\varphi l}{2EI}$$

The function of the bending moment is obtained by the second differentiation of (4.9) and substitution into (4.6)

$$M(x) = -\frac{ql^2}{12}\left[\frac{\bar{c}_\varphi}{1+\bar{c}_\varphi} - 6\frac{x}{l} + 6\left(\frac{x}{l}\right)^2\right]. \tag{4.10}$$

The shear force curve obtained by differentiation of (4.10) is independent of $\bar{c}_\varphi$ and

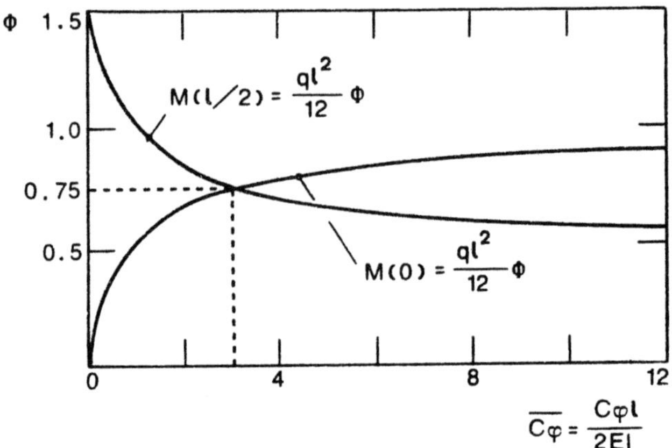

Fig. 4.3. Variation of slope and bending moment of simply supported beam with rotational constraint at supports under uniformly distributed loading.

thus the shear is independent of the rotational constraint

$$Q(x) = \frac{ql}{2}\left(1 - 2\frac{x}{l}\right). \qquad (4.11)$$

In Fig. 4.3, the moments at the mid span and at the supports are shown as functions of the rotational constant $\bar{c}_\varphi$. From this figure the following general observations may be made:

— The full support moment $M(0) = ql^2/12$ is reached approximately at a relatively greater spring stiffness $\bar{c}_\varphi$, which is very difficult to attain in a real structure.
— The optimum load-bearing capacity of the beam is reached when the support moment and the mid-span moment are equal, i.e. when $\bar{c}_\varphi = 3$ and we can write

$$M(0) = M\left(\frac{l}{2}\right) = \frac{ql^2}{16}.$$

— The sum of both moments is independent of $\bar{c}_\varphi$ and has a value of $ql^2/8$. The spring effect influences only the ratio of the two moments.

Very often, the load distribution $q(x)$ is not uniform, but, for example, triangular or trapezoidal. Figure 4.4 shows the distribution of moments for a clamped–clamped beam with both uniform and triangular load distributions.

It can be seen that the mid-span moment is identical in both cases, and that the maximum moment for triangular loading is not significantly greater than for uniform loading. It is thus sufficient in most practical cases, especially for lattice structures, to regard the loading as uniformly distributed.

The previous argument assumes that the beam is prismatic, i.e. the moment of inertia is constant along the length of the beam, which is sufficient for most practice cases. For the case of a beam with a variable section, we have, from (4.7),

$$EI(x)\frac{\mathrm{d}^4 w}{\mathrm{d}x^4} = q(x). \qquad (4.12)$$

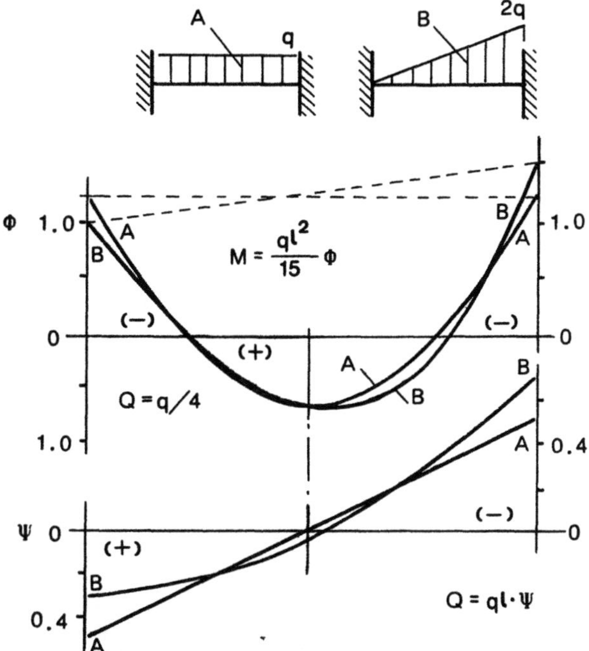

Fig. 4.4. Curves of bending moments and shear force of clamped–clamped beam under uniformly distributed load and triangular load.

For this case, the complete solution of (4.12) is more difficult. Depending on the distribution of the moment of inertia in the x-direction, we obtain a more or less complex, non-linear differential equation.

In practice we can, however, use an approximate solution. There are several known methods such as finite difference and integral equation methods, etc. An important approximate method will now be discussed which, in view of the computer-aided methods of Sect. 4.5, is of great significance. This approximate method is generally known as the principle of minimum potential energy in connection with the methods of Ritz [4] and Galerkin [5].

A beam on elastic foundations is considered for illustration. Such a problem serves as a model for structures resting on the sea bed, as shown in Fig. 4.5.

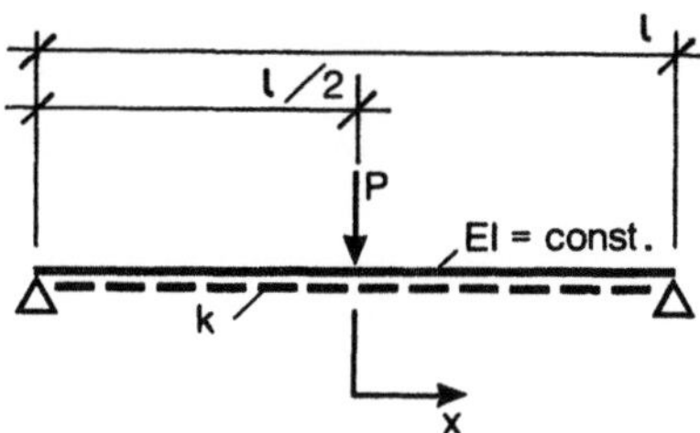

Fig. 4.5. Simply supported beam on elastic foundation with a central load.

The elastic potential is derived from the work done by internal forces during deformation, and the work done by the external loads. For the above problem we have

$$\Pi = \frac{1}{2}\int \sigma \cdot \varepsilon \, \mathrm{d}V + \frac{1}{2}\int kw(x)^2 \, \mathrm{d}x - P \cdot w(0), \tag{4.13}$$

where V is the volume of the beam.

The modulus of the foundation k is assumed to be constant, in other words, the foundation along the bottom is assumed to be equivalent to a set of closely spaced independent series of springs (the Winkler assumption, the dimension of k is N/mm^2). In general, this is not exactly fulfilled, but because of the rather approximate nature of data on the composition of the sea bed, such a simplified model can still be quite meaningful.

From (4.4) and the expression for Hookes law, we obtain, by integration over the section

$$\Pi = \frac{1}{2}\int_{-l/2}^{l/2} \left[EI\left(\frac{\mathrm{d}^2 w}{\mathrm{d}x^2}\right)^2 + kw^2 \right] \mathrm{d}x - P \cdot w(0). \tag{4.14}$$

The principle of minimum potential energy requires that this potential must be stationary in the neighbourhood of equilibrium. In other words, the variation of the potential must disappear, and hence

$$\delta\Pi = 0. \tag{4.15}$$

The correctness of expression (4.14) can be checked as follows. For a functional F we establish generally[†]

$$\delta\Pi = \delta \int_0^l F(w'', w', w, x)\mathrm{d}x$$

$$= \int_0^l \left(\frac{\partial F}{\partial w''}\delta w'' + \frac{\partial F}{\partial w'}\delta w' + \frac{\partial F}{\partial w}\delta w \right)\mathrm{d}x = 0.$$

After integration by parts we obtain

$$\delta\Pi = \int_0^l \left[\left(\frac{\partial F}{\partial w''}\right)'' - \left(\frac{\partial F}{\partial w'}\right)' + \frac{\partial F}{\partial w} \right]\delta w \, \mathrm{d}x = 0. \tag{4.16}$$

Because δw can be arbitrarily chosen, the bracketed expression in equation (4.16) must disappear. In our example the functional is given by

$$F = \frac{1}{2}\left[EI\left(\frac{\mathrm{d}^2 w}{\mathrm{d}x^2}\right)^2 + kw^2 \right]. \tag{4.17}$$

The Euler differential equation in the bracketed expression in (4.16) is obtained

---

[†] For convenience we instead take the Leibnitz expression of the differential dw/dx, the Lagrangian w' and w''.

using function (4.17):

$$\frac{\partial F}{\partial w''} = EI\frac{\mathrm{d}^2 w}{\mathrm{d}x^2}, \quad \frac{\partial F}{\partial w'} = 0, \quad \frac{\partial F}{\partial w} = kw,$$

$$EI\frac{\mathrm{d}^4 w}{\mathrm{d}x^4} + kw = 0, \tag{4.18}$$

or

$$\frac{\mathrm{d}^4 w}{\mathrm{d}x^4} + 4\alpha^4 w = 0$$

where

$$\alpha = \sqrt[4]{\frac{k}{4EI}}.$$

This is the homogeneous differential equation for a beam on an elastic foundation in our problem, which has the solution of the form

$$w(x) = e^{\alpha x}(C_1 \cos \alpha x + C_2 \sin \alpha x) + e^{-\alpha x}(C_3 \cos \alpha x + C_4 \sin \alpha x). \tag{4.19}$$

To determine the constants C_1–C_4 we need to use the following boundary conditions

$$\frac{\mathrm{d}w}{\mathrm{d}x} = 0\big|_{x=0}, \quad \frac{\mathrm{d}^3 w}{\mathrm{d}x^3} = -\frac{P}{2EI}\bigg|_{x=0},$$

$$w = 0\big|_{x=l/2}, \quad \frac{\mathrm{d}^2 w}{\mathrm{d}x^2} = 0\big|_{x=l/2}. \tag{4.20}$$

We obtain the maximum deflection at $x = 0$ as

$$\bar{w}_0 = \frac{P}{8EI\alpha^3} \cdot \frac{\sinh \alpha l - \sin \alpha l}{\cosh \alpha l + \cos \alpha l}.$$

The exact solution of this problem serves as a basis for the approximate solution to be obtained using the Rayleigh–Ritz method. We proceed by setting up an approximate solution for deflection in the form of a sum of basis functions $f(x)$

$$w(x) = \sum_{i=0}^{n} w_i \cdot f(x),$$

where these must fulfil geometrical boundary conditions, which are in our case deflection and rotation. The coefficients w_i must now be determined so that the elastic potential is a minimum. Since all events in nature occur at the smallest possible energy level, the exact solution also occurs at the minimum potential, and therefore all approximate solutions can only occur at a relative minimum which is greater than that of the exact solution. The rate of change of the elastic potential with respect to the individual coefficients w_i must vanish

$$\frac{\partial \Pi}{\partial w_i} = 0, \quad i = 1, 2, \ldots, n.$$

From this we obtain a system of n linear equations for the determination of the unknown coefficients w_i. Because this yields a linear system of equations, terms in w_i may at best be quadratic. We begin with an expression

$$w(x) = w_1 \cdot \cos \frac{\pi x}{l},$$

which by substitution in (4.14) results in

$$\Pi = \frac{1}{2} EI \int_{-l/2}^{l/2} \left(\frac{\pi}{l}\right)^4 w_1^2 \cos^2 \frac{\pi x}{l} \, dx + \frac{1}{2} k \int_{-l/2}^{l/2} w_1^2 \cos^2 \frac{\pi x}{l} \, dx - Pw_1$$

and the variation with respect to w_1 is

$$\frac{\partial \Pi}{\partial w_1} = w_1 EI \left(\frac{\pi}{l}\right)^4 \int_{-l/2}^{l/2} \cos^2 \frac{\pi x}{l} \, dx + w_1 k \int_{-l/2}^{l/2} \cos^2 \frac{\pi x}{l} \, dx - P = 0.$$

Since the integrals give $l/2$, we obtain

$$w_1 = \frac{2P}{l\left[EI\left(\frac{\pi}{l}\right)^4 + k\right]}. \tag{4.21}$$

As the error study shows in Fig. 4.6, in general an extraordinarily good result is achieved if we assume that the shear force is discontinuous and is approximated by a sine function. Only with very large $\alpha l > 2$ does the solution become notably more inaccurate. A better expression is

$$w(x) = w_1 \cos \frac{\pi x}{l} + w_2 \cos \frac{3\pi x}{l}.$$

With the orthogonality condition

$$\int \cos \frac{i\pi x}{l} \cos \frac{j\pi x}{l} \, dx = \begin{cases} 0 & \text{when} \quad i \neq j \\ \dfrac{l}{2} & \text{when} \quad i = j \end{cases} \tag{4.22}$$

which gives

$$\Pi = EI \left(\frac{\pi}{l}\right)^4 \frac{l}{4} (w_1^2 + 81w_2^2) + k \frac{l}{4}(w_1^2 + w_2^2) - P(w_1 + w_2)$$

$$\frac{\partial \Pi}{\partial w_1} = EI \left(\frac{\pi}{l}\right)^4 \frac{l}{2} w_1 + k \frac{l}{2} w_1 - P = 0 \tag{4.23}$$

$$\frac{\partial \Pi}{\partial w_2} = EI \left(\frac{\pi}{l}\right)^4 \frac{l}{2} 81w_2 + k \frac{l}{2} w_2 - P = 0$$

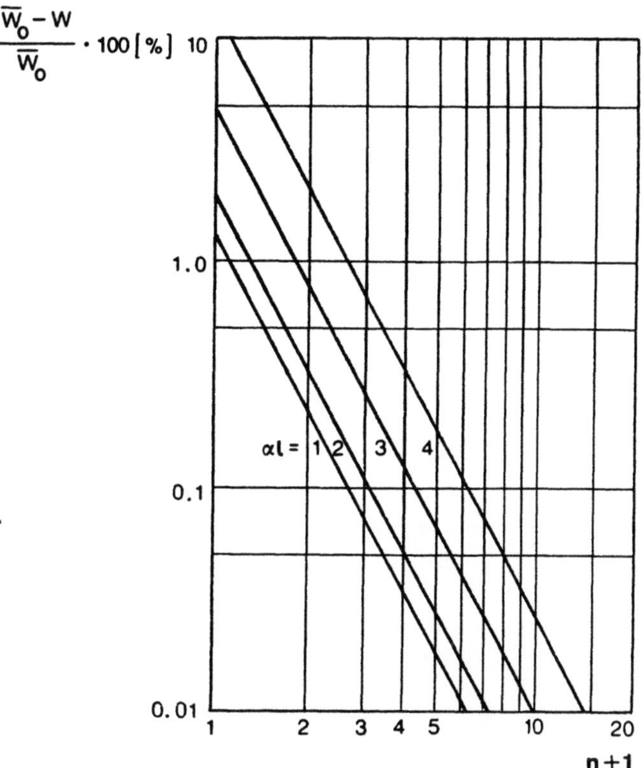

Fig. 4.6. Error study of the Ritz method for a simply supported beam with a central load on elastic foundation.

the system of equations is completely decoupled since all mixed integrals disappear. In (4.23), the mixed terms according to (4.22) are already dropped. We obtain the coefficient w_1 in agreement with (4.21) and w_2 as

$$w_2 = \frac{2P}{l\left[81EI\left(\dfrac{\pi}{l}\right)^4 + k\right]}.$$

Since the system of equations is decoupled, we can use the expression

$$w(x) = \sum_{i=1}^{n} w_i \cos\frac{(2i-1)\pi x}{l}.$$

By similar operations we obtain all coefficients as

$$w_i = \frac{2P}{l}\frac{1}{(2i-1)^4 EI\left(\dfrac{\pi}{l}\right)^4 + k}, \quad i = 1, 2, \ldots, n. \tag{4.24}$$

In Fig. 4.6 the errors are shown as functions of the number of terms of the deflection functions. The deflection under load P is given as an error magnitude.

From Fig. 4.6 some important conclusions can be made which are in general valid for the assessment of the numerical methods presented in Sect. 4.5:

— The error decreases monotonically with the increasing number of terms of the deflection functions.
— The reduction in error follows an exponential law, since the error curves are straight lines on log–log paper.
— The errors depend on the factor αl, i.e. with stiff bedding a significantly greater number of deflection functions are necessary.
— The value of the coefficients w_i are independent of the number of terms used in the expansion of the deflection function.

The Galerkin method arises from the energy principle in the formulation of the minimum of potential energy using the principle of virtual work. The variational problems which arise from this are solved using the property of orthogonality between the admissible functions for the displacement field and the resultant error function. For the error function the Euler differential equation of the respective problem is used (special case of the method of weighted residuals). For a beam with a constant moment of inertia and distributed load $q(x)$ we write, for the elastic potential,

$$\Pi = \frac{1}{2}\int_0^l EI\left(\frac{d^2 w}{dx^2}\right)^2 dx - \int_0^l q(x)w\,dx. \tag{4.25}$$

The variation of the elastic potential is

$$\delta\Pi = \frac{1}{2}\int_0^l EI\delta w''^2\,dx - \int_0^l q(x)\delta w\,dx = 0.$$

After integration by parts, we obtain

$$\int_0^l [EIw'''' - q(x)]\delta w\,dx + EIw''\delta w'|_0^l - EIw'''\delta w|_0^l = 0, \tag{4.26}$$

each term of which must identically be equal to 0. We also have

$$\int_0^l [EIw'''' - q(x)]\delta w\,dx = 0. \tag{4.27}$$

From (4.26) it is clear that the approximate solutions are only fulfilled in (4.27) when the boundary conditions are satisfied up to the third derivative of the approximation function, i.e. the boundary terms are effectively eliminated.

With the Galerkin method we obtain identical results from the same approximation function as with the Ritz method. However, while the Ritz method requires only the necessary geometrical boundary conditions, which in our case are the deflection and the rotation at the support, to be fulfilled the Galerkin method requires all boundary conditions to be fulfilled. This stronger requirement, in some cases, leads to more accurate solutions. In general, approximation functions which fulfil all boundary conditions are difficult to obtain. The Ritz method, however, is universally applicable.

We now move on to the formulation of typical two dimensional and structural elements used in marine structures.

4.1.2 Membranes

To determine the relationship between the stresses and deformations in a membrane element we consider an infinitesimally small element of it. The forces acting in the plane of the membrane per unit length are shown in Fig. 4.7. The condition of equilibrium of the element is expressed by

$$\frac{\partial n_x}{\partial x} + \frac{\partial n_{yx}}{\partial y} = 0, \quad \frac{\partial n_y}{\partial y} + \frac{\partial n_{xy}}{\partial x} = 0, \quad n_{xy} - n_{yx} = 0, \tag{4.28}$$

where body forces, such as the self weight, are not taken into account. The strain-displacement relations are given by

$$\varepsilon_x = \frac{\partial u}{\partial x}, \quad \varepsilon_y = \frac{\partial v}{\partial y}, \quad \gamma_{xy} = \frac{\partial v}{\partial x} + \frac{\partial u}{\partial y}. \tag{4.29}$$

Other strain (or stress) components are disregarded, since the thickness of the membrane is small in relation to its dimensions in the x- and y-directions, therefore the condition is that of the plane stress state. The differential equations (4.29) can be combined, resulting in

$$\frac{\partial^2 \varepsilon_x}{\partial y^2} - \frac{\partial^2 \gamma_{xy}}{\partial x\,\partial y} + \frac{\partial^2 \varepsilon_y}{\partial x^2} = 0, \tag{4.30}$$

which is called the compatibility condition of the plane stress state.

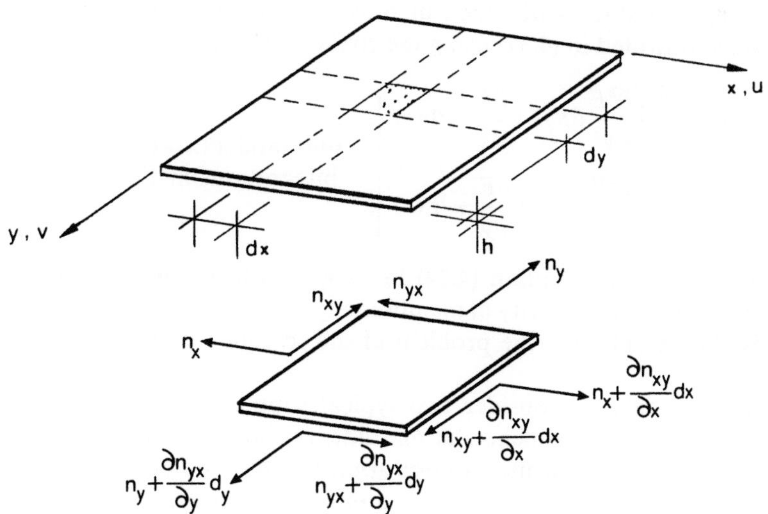

Fig. 4.7. Forces on an infinitesimal membrane (plane stress) element.

The stresses and strains are related by Hookes law

$$\varepsilon_x = \frac{1}{E}(\sigma_x - v\sigma_y),$$

$$\varepsilon_y = \frac{1}{E}(\sigma_y - v\sigma_x),$$

$$\gamma_{xy} = \frac{\tau}{G} = \frac{2(1+v)}{E}\tau_{xy}. \tag{4.31}$$

The substitution of Hookes law into the compatibility equation (4.30) leads to

$$2(1+v)\frac{\partial^2 \tau_{xy}}{\partial x \partial y} = \frac{\partial^2 \sigma_x}{\partial y^2} + \frac{\partial^2 \sigma_y}{\partial x^2} - v\left(\frac{\partial^2 \sigma_x}{\partial x^2} + \frac{\partial^2 \sigma_y}{\partial y^2}\right).$$

Since the stresses arising from the inplane forces per unit length are given by

$$\sigma_x = \frac{n_x}{h}, \quad \sigma_y = \frac{n_y}{h}, \quad \tau_{xy} = \frac{n_{xy}}{h}, \tag{4.32}$$

the equilibrium equations (4.28) can be obtained in terms of stresses. We now introduce the so-called Airy stress function F such that

$$\frac{\partial^2 F}{\partial y^2} = \sigma_x, \quad \frac{\partial^2 F}{\partial x^2} = \sigma_y, \quad -\frac{\partial^2 F}{\partial x \partial y} = \tau_{xy}. \tag{4.33}$$

By combining this with the equilibrium equation we obtain the homogeneous partial differential equation

$$\frac{\partial^4 F}{\partial x^4} + 2\frac{\partial^4 F}{\partial x^2 \partial y^2} + \frac{\partial^4 F}{\partial y^4} = 0 \quad \text{or} \quad \Delta\Delta F = 0. \tag{4.34}$$

The deformation state is obtained directly by integration of (4.29), where the ε_x and ε_y are substituted in (4.31) using the stress function F according to (4.33), thus

$$u = \frac{1}{E}\left[\int \frac{\partial^2 F}{\partial y^2}\,dx - v\frac{\partial F}{\partial x} + \Phi(y)\right],$$

$$v = \frac{1}{E}\left[\int \frac{\partial^2 F}{\partial x^2}\,dy - v\frac{\partial F}{\partial y} + \Psi(x)\right]. \tag{4.35}$$

($\Phi(y)$ and $\Psi(x)$ are integration functions.)

The Airy differential equation (4.34) in closed form can be used for only a few problems. Today, the majority of membrane type problems are solved by numerical methods. An exception is the problem of the so-called effective breadth or shear lag effects.

To illustrate this we consider two typical supports for a deck, or something similar, reinforced with stiffeners at one side, as shown in Figs. 4.8 and 4.9. While a linear distribution of normal stresses applies in the web of the beam, in the wide upper flanges the stress distribution is non-linear. Although we can today obtain the stress distribution with a finite-element model, in general, a simplified beam

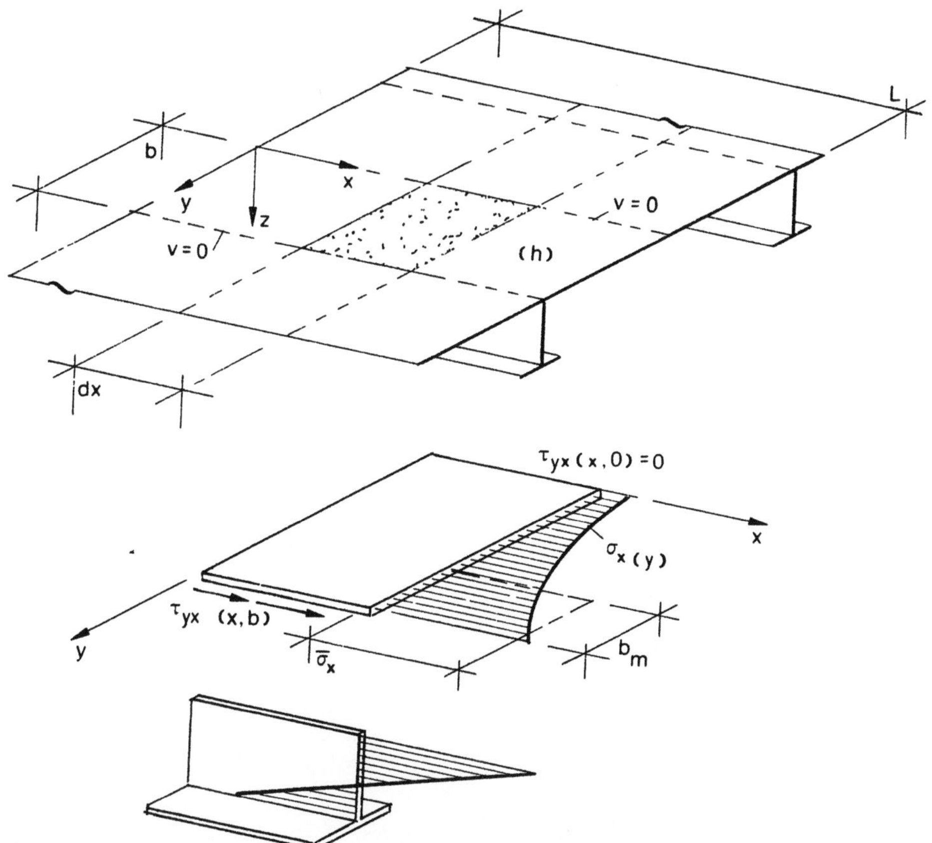

Fig. 4.8. Girder subject to bending with symmetrical, membrane type flange.

model is preferred. For this, the effective moment of inertia must be determined. This starts from the assumption that we need a fictitious flange of breadth b_m and thickness h that carries forces equal to that of the flange in question under constant stress distribution. Thus, for the assumed breadth b_m we have

$$\frac{b_m}{b} = \frac{\displaystyle\int_0^b \sigma_x(y)\,\mathrm{d}y}{b\bar{\sigma}_x} \tag{4.36}$$

assuming that no further stresses arise. In the case shown in Fig. 4.9, the flange must be cut off at $y = 0$, and the resulting shear stresses $\tau_{yx}(x, 0)$ must be considered in (4.36). We cannot go into further detail here, and so refer the reader elsewhere [6–9].

However, from the example in Fig. 4.8, we can at least indicate a solution. We therefore consider a membrane element of breadth b and length $\mathrm{d}x$. It is now necessary to formulate the membrane stress conditions to fulfil the Airy differential equation, taking into account the boundary conditions to substitute these into

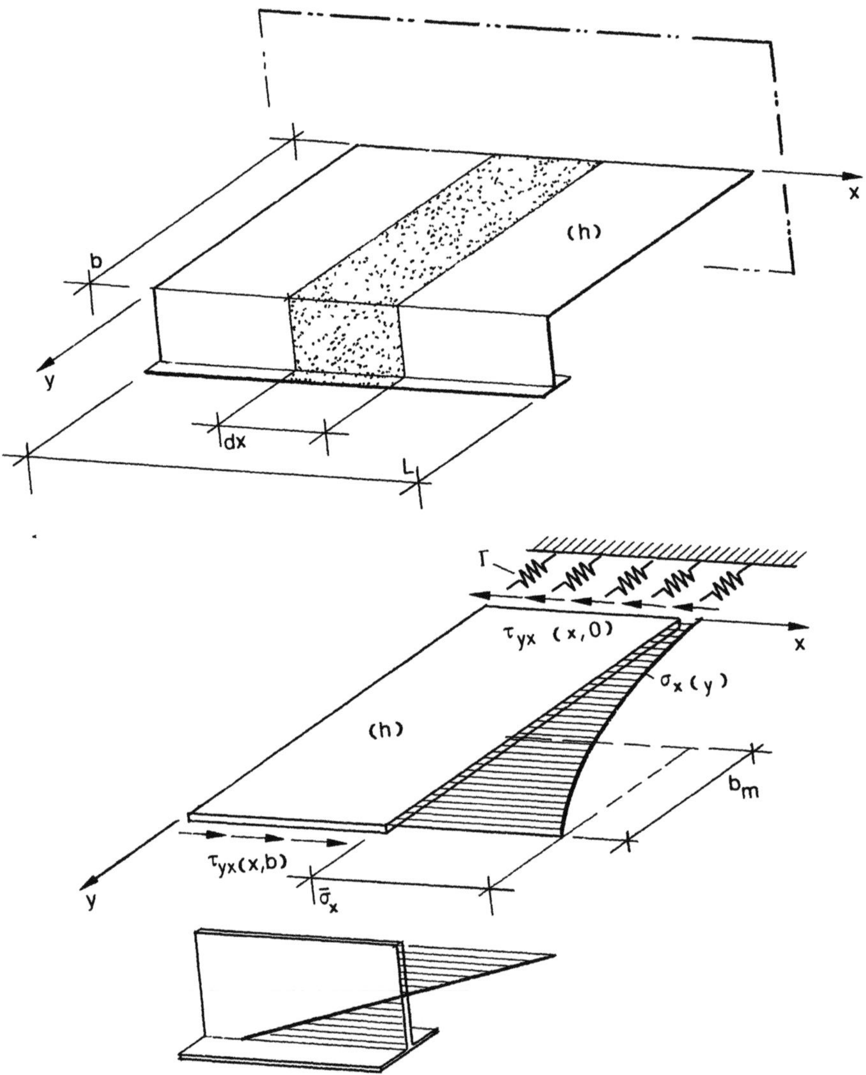

Fig. 4.9. Girder subject to bending with asymmetrical, membrane type flange

(4.36). Since it would be impossible to seek a general solution for the Airy differential equation, formulations corresponding to a harmonic series in the x-direction are used. When this is also carried out for the loading, we can calculate the separate effective breadth for each harmonic component, and superimpose this with respect to the harmonic load components. We begin with the expression

$$F(x, y) = \sum_{n=1}^{\infty} F_n,$$

where

$$F_n = f_n(y) \cdot \sin \frac{n\pi x}{L},$$

and obtain by (4.34) a series of n ordinary differential equations

$$\frac{d^4 f_n(y)}{dy^4} - 2\left(\frac{n\pi}{L}\right)^2 \frac{d^2 f_n(y)}{dy^2} + \left(\frac{n\pi}{yL}\right)^4 f_n(y) = 0, \tag{4.37}$$

with the solutions

$$f_n(y) = \left(A_n + C_n \frac{n\pi y}{L}\right)\cosh\frac{n\pi y}{L} + \left(B_n + D_n \frac{n\pi y}{L}\right)\sinh\frac{n\pi y}{L}. \tag{4.38}$$

In (4.36) we replace the stresses with the Airy stress function using (4.33), and obtain

$$\frac{b_m}{b} = \frac{\displaystyle\int_0^b \frac{\partial^2 F}{\partial y^2}\,dy}{b\bar{\sigma}_x}.$$

Three of the four constants in (4.38) can be determined using the boundary conditions, since they can be formulated in terms of the stress function or its derivatives. We still need to deal with the reference stress $\bar{\sigma}_x$. At the junction between a flange and web we can assume that the elongation of the flange

$$\varepsilon_x = \frac{1}{E}[\sigma_x(b) - v\sigma_y(b)]$$

is equal to the elongation of the web

$$\varepsilon_x = \frac{1}{E}\bar{\sigma}_x.$$

For the reference stress we then have

$$\bar{\sigma}_x = \sigma_x(b) - v\sigma_y(b).$$

The reference stress of the fictitious beam $\bar{\sigma}_x$ is for the case where the transverse stresses are smaller than the actual axial stress. In other words, the fictitious beam can have an effective flange breadth somewhat greater than in reality. We therefore obtain an expression for the effective breadth in relation to the stress function as

$$\frac{b_{mn}}{b} = \frac{\left.\dfrac{\partial F_n}{\partial y}\right|_0^b}{b\left(\dfrac{\partial^2 F_n}{\partial y^2} - v\dfrac{\partial^2 F_n}{\partial x^2}\right)_{y=b}}. \tag{4.39}$$

The unknown constants A_n cancel out, and thus need not be determined explicitly.

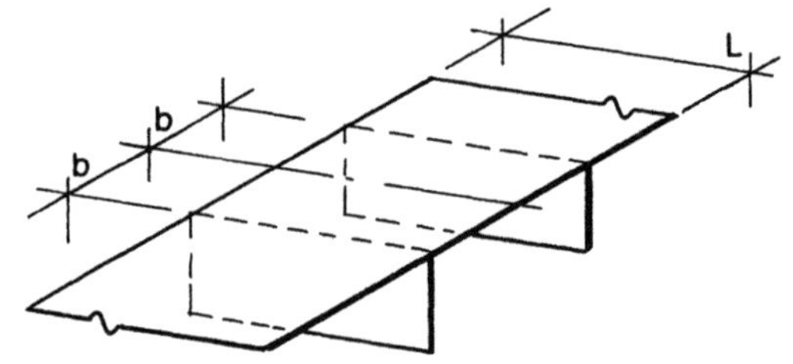

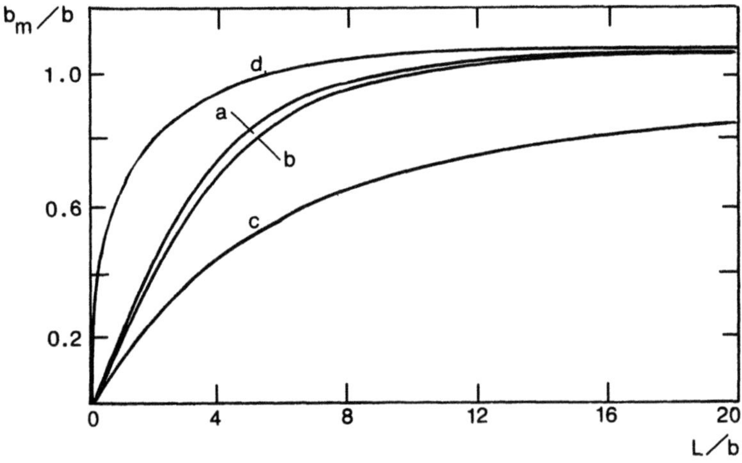

Fig. 4.10. Effective breadths of the stiffened panel for various loadings. a Uniformly distributed load, b sinusoidal load, c point load, d constant moment.

We finally obtain for the case under consideration

$$\frac{b_{mn}}{b} = \frac{4}{n\pi}\frac{L}{b}\frac{\cosh\dfrac{n\pi b}{L}-1}{(3-v)(1+v)\sinh\dfrac{n\pi b}{L}-(1+v)^2\dfrac{n\pi b}{L}}. \qquad (4.40)$$

The load must now be expressed as a Fourier series, i.e.

$$M(x) = \sum_{n=1}^{\infty} M_n \sin\frac{n\pi x}{L},$$

giving the reference stress

$$\bar{\sigma}_x = \sum_{n=1}^{\infty} \frac{M_n}{W_n}\sin\frac{n\pi x}{L},$$

where W_n is the section modulus of the beam. We therefore obtain, by summing

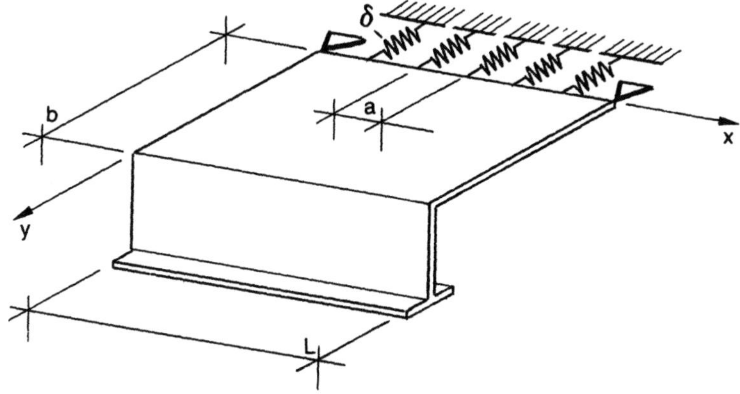

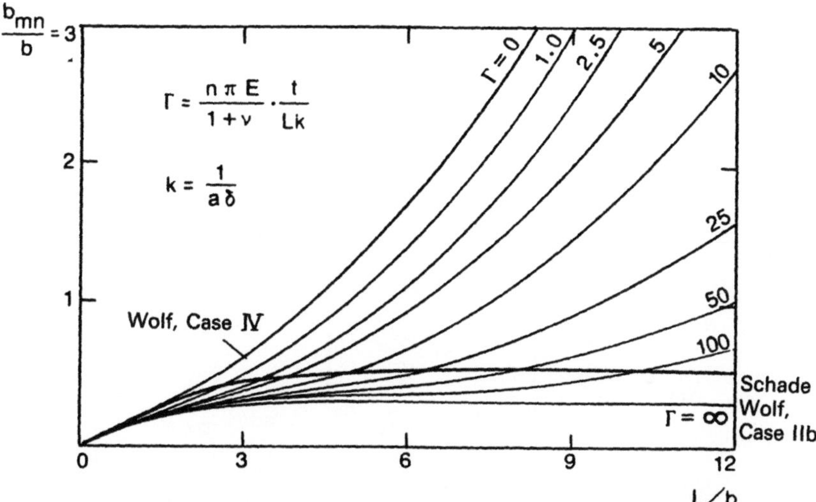

Fig. 4.11. Effective breadth of asymmetrical stiffened panel with sinusoidal loading. The parameter Γ accounts for the elastic lateral support (cf. solutions of Wolf [6] and Schade [7]).

over all values of n,

$$\frac{b_m}{b} = \frac{\displaystyle\sum_{n=1}^{\infty} \frac{b_{mn}}{b}\,\frac{M_n}{W_n}\sin\frac{n\pi x}{L}}{\displaystyle\sum_{n=1}^{\infty} \frac{M_n}{W_n}\sin\frac{n\pi x}{L}}. \tag{4.41}$$

In Figs. 4.10 and 4.11 the results are presented for some important cases. The length L refers to the distance between points of zero bending moment. The effective breadth shown is that at the midspan of the beam; the effective breadth can vary over the beam length. Using the fictitious moment of inertia in large 3-D frames such effects can be neglected partially.

4.1.3 Plates

Plate theory has a wider scope of application in marine structures for both isotropic and anisotropic orthogonal systems. The differential equation for the plate is obtained by considering an infinitesimal element cut out from the plate, as in Fig. 4.12, which shows an isotropic plate of constant thickness.

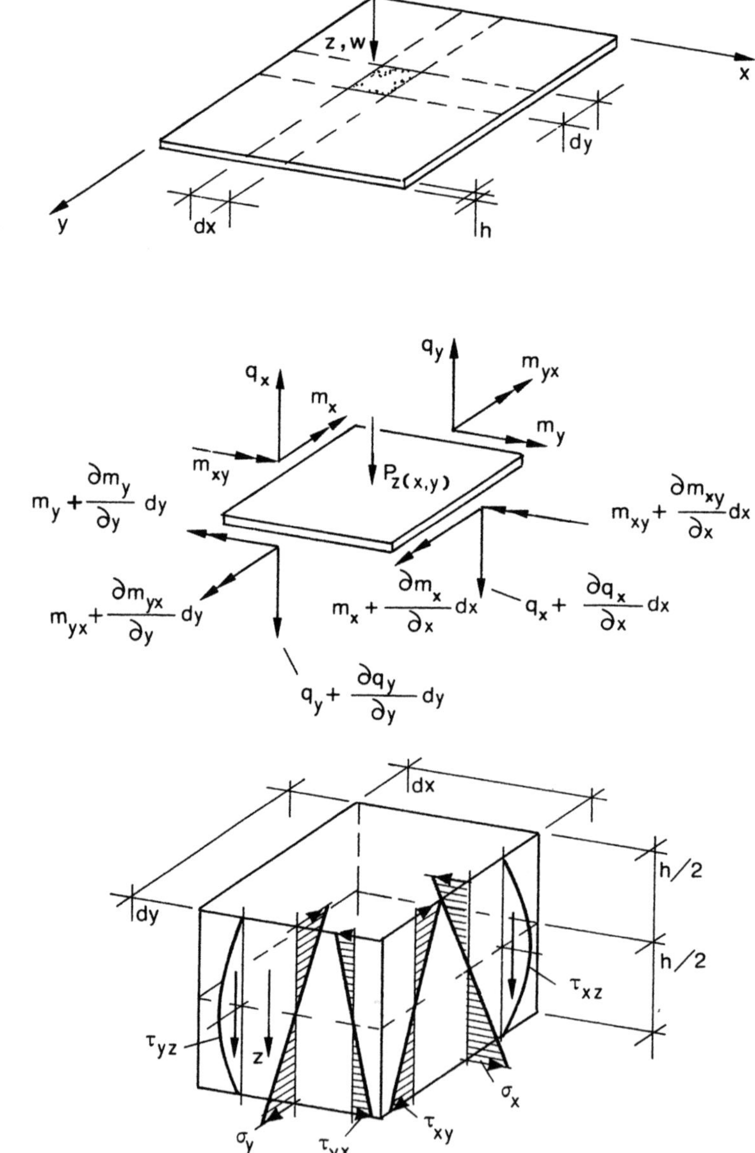

Fig. 4.12. Forces and moments in a plate element.

The general assumptions of plate theory require that the plate thickness is small relative to its other two dimensions. It then follows that the stresses and strains normal to the plane of the plate are negligible. Further, it means that all points on a normal to the middle plane of the plate during sagging lie on a straight line, perpendicular to the deformed neutral surface. This is the extended Bernoulli hypothesis that we have already used in beams. The bending must be small, i.e. the elongations and change of angle of the neutral surface are negligible. As with membrane theory, the changes in stress and displacement are only regarded as linear terms in the Taylor series expansion. In later sections we will refine some of these conditions to describe non-linear effects.

If we now turn our attention to Fig. 4.12, we see that the edge forces $m_x m_y$, etc., are defined for a unit length. We obtain these through integration of the stresses which, as with beam theory, are linearly distributed over the plate thickness. The forces and moments shown in Fig. 4.12 are in the direction of the stresses from which they result. The integration is carried out with $z = 0$ at the neutral surface each time.

Transverse forces

$$q_x = \int_{-h/2}^{h/2} \tau_{xz} dz, \quad q_y = \int_{-h/2}^{h/2} \tau_{yz} dz.$$

Bending moments

$$m_x = \int_{-h/2}^{h/2} \sigma_x z\, dz, \quad m_y = \int_{-h/2}^{h/2} \sigma_y z\, dz. \tag{4.42}$$

Torsional moments

$$m_{xy} = \int_{-h/2}^{h/2} \tau_{xy} z\, dz, \quad m_{yx} = \int_{-h/2}^{h/2} \tau_{yx} z\, dz.$$

The equilibrium conditions, namely that the sum of all forces in the z-direction and the sum of all moments be equal to zero, give

$$\frac{\partial q_x}{\partial x} + \frac{\partial q_y}{\partial y} = -p_z(x, y),$$

$$\frac{\partial m_y}{\partial y} + \frac{\partial m_{xy}}{\partial x} = q_y, \tag{4.43}$$

$$\frac{\partial m_x}{\partial x} + \frac{\partial m_{yx}}{\partial y} = q_x,$$

which on differentiation and summation give

$$\frac{\partial^2 m_x}{\partial x^2} + 2\frac{\partial^2 m_{xy}}{\partial x\,\partial y} + \frac{\partial^2 m_y}{\partial y^2} = -p_z(x, y). \tag{4.44}$$

The stresses in (4.42) can now be written in terms of the strains by using Hookes law

$$\sigma_x = \frac{E}{1-v^2}(\varepsilon_x + v\varepsilon_y), \quad \sigma_y = \frac{E}{1-v^2}(\varepsilon_y + v\varepsilon_x),$$

$$\tau_{xy} = \frac{E}{1-v^2}\frac{1-v}{2}\gamma_{xy}, \tag{4.45}$$

and the strains as in beam theory by the curvatures as

$$\varepsilon_x = -z\frac{\partial^2 w}{\partial x^2}, \quad \varepsilon_y = -z\frac{\partial^2 w}{\partial y^2}, \quad \gamma_{xy} = -2z\frac{\partial^2 w}{\partial x \partial y}, \tag{4.46}$$

so that we obtain the bending moments as

$$m_x = \int_{-h/2}^{h/2} \sigma_x z\,\mathrm{d}z = -D\left(\frac{\partial^2 w}{\partial x^2} + v\frac{\partial^2 w}{\partial y^2}\right),$$

$$m_y = \int_{-h/2}^{h/2} \sigma_y z\,\mathrm{d}z = -D\left(\frac{\partial^2 w}{\partial y^2} + v\frac{\partial^2 w}{\partial x^2}\right), \tag{4.47}$$

$$m_{xy} = \int_{-h/2}^{h/2} \tau_{xy} z\,\mathrm{d}z = -D(1-v)\frac{\partial^2 w}{\partial x \partial y},$$

where

$$D = \frac{Eh^3}{12(1-v^2)}$$

is the flexural rigidity of the plate. If we substitute these expressions in (4.44), we then obtain the Kirchhoff differential equation for plate bending

$$\frac{\partial^4 w}{\partial x^4} + 2\frac{\partial^4 w}{\partial x^2 \partial y^2} + \frac{\partial^4 w}{\partial y^4} = \frac{p_z(x,y)}{D}, \quad \Delta\Delta w = \frac{p_z(x,y)}{D}. \tag{4.48}$$

For a number of technically important problems, solutions of this equation can be successfully obtained. As an example, we will use the reinforced plate shown in Fig. 4.13. Due to the symmetry of the problem, only the hatched portion will be taken into account. The solution for (4.48) can be set out in the form of

$$w(x,y) = \sum_{m=1}^{\infty} f_m(y)\sin\frac{m\pi x}{l},$$

where, because of the nature of the sine function, free rotational supports on the edges $x = 0, x = l$ must be assumed. For the loading, a corresponding expression is obtained as

$$p_z(x,y) = \sum_{m=1}^{\infty} p_{zm}(y)\sin\frac{m\pi x}{l}.$$

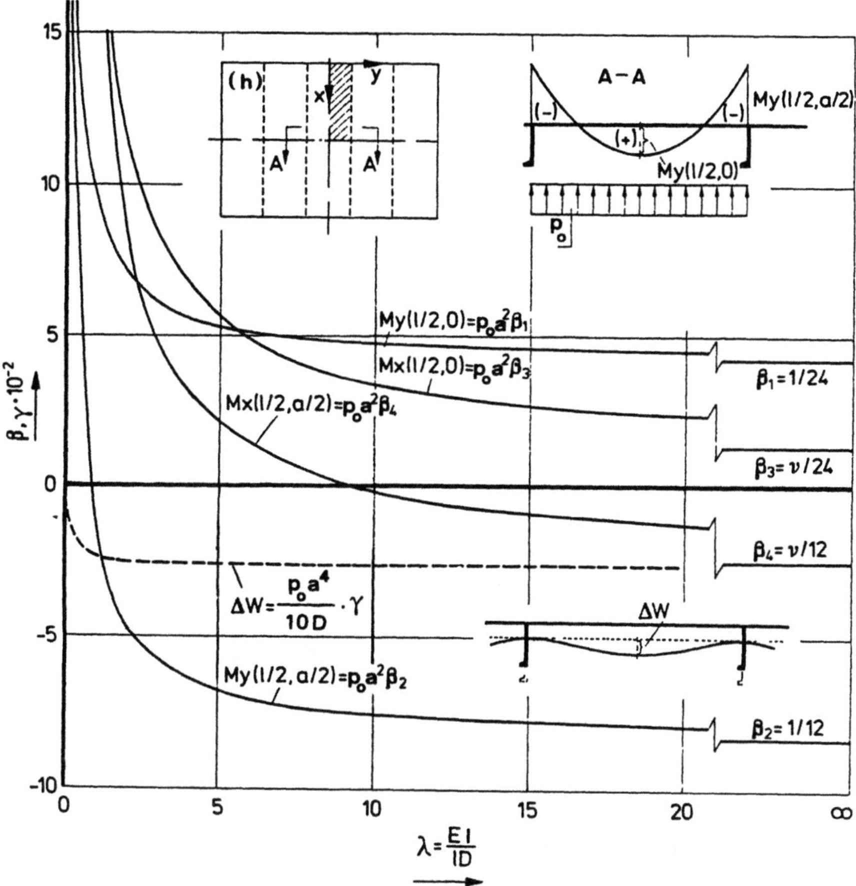

Fig. 4.13. Effect of reinforcement stiffness on bending moments and deflection of a plate

These two expressions, when substituted in the equation for plate bending, give

$$\sum_{m=1}^{\infty}\left[\left(\frac{m\pi}{l}\right)^4 f_{\mathrm{m}}(y) - 2\left(\frac{m\pi}{l}\right)^2\frac{\mathrm{d}^2 f_{\mathrm{m}}}{\mathrm{d}y^2} + \frac{\mathrm{d}^4 f_{\mathrm{m}}}{\mathrm{d}y^4}\right]\sin\frac{m\pi x}{l}$$

$$= \frac{1}{D}\sum_{m=1}^{\infty} p_{zm}(y)\sin\frac{m\pi x}{l}. \tag{4.49}$$

For the case of uniformly distributed loading we obtain, through Fourier analysis,

$$p_{zm}(y) = p_0\frac{4}{m\pi}.$$

In this way, we obtain a total of m linear ordinary fourth order differential equations

$$\left(\frac{m\pi}{l}\right)^4 f_{\mathrm{m}}(y) - 2\left(\frac{m\pi}{l}\right)^2\frac{\mathrm{d}^2 f_{\mathrm{m}}}{\mathrm{d}y^2} + \frac{\mathrm{d}^4 f_{\mathrm{m}}}{\mathrm{d}y^4} = \frac{p_0}{D}\frac{4}{m\pi}. \tag{4.50}$$

The solution of these differential equations is obtained by considering the symmetry of the problem, giving

$$f_m(y) = \frac{p_0 a^4}{D}\left(\frac{l}{a}\right)^4\left(\frac{4}{(m\pi)^5} + A_m \cosh\frac{m\pi y}{l} + B_m \frac{m\pi y}{l}\sinh\frac{m\pi y}{l}\right). \qquad (4.51)$$

The constants can be determined using the boundary conditions, which are

$$\frac{\partial w}{\partial y} = 0\Big|_{y=\pm a/2},$$

$$D\left[\frac{\partial^3 w}{\partial y^3} + (2-v)\frac{\partial^3 w}{\partial x^2\,\partial y}\right] = EI\frac{\partial^4 w}{\partial y^4}\Big|_{y=\pm a/2}, \qquad (4.52)$$

resulting in

$$B_m = \frac{4}{m^5\pi^5}\frac{1}{\left(\dfrac{2}{\lambda m\pi} - \alpha_m\right)\sinh\alpha_m + (1 + \alpha_m\coth\alpha_m)\cos\alpha_m},$$

and

$$A_m = -\frac{4}{m^5\pi^5}\frac{1 + \alpha_m\coth\alpha_m}{\left(\dfrac{2}{\lambda m\pi} - \alpha_m\right)\sinh\alpha_m + (1 + \alpha_m\coth\alpha_m)\coth\alpha_m},$$

where

$$\lambda = \frac{EI}{lD} \quad \text{and} \quad \alpha_m = \frac{m\pi a}{2l}.$$

The bending moments are determined using (4.47). In Fig. 4.13 these are shown as functions of the parameter λ. We know that for large values of λ the stiffeners act as fixed supports, and the moments approximate those of a plate supported on both sides. The curves shown in Fig. 4.13 are valid for typical stiffening in shipbuilding and marine structures in which the aspect ratio of the plate is relatively large. For many technically important problems, further solutions that are helpful even in this age of computerized structural analysis may be found in [1], [3], and elsewhere.

As with beam theory, we also make use of the energy method in plate theory. The elastic potential of a plate with dimensions $a\cdot l$ may be expressed as

$$\Pi = \int_0^a\int_0^l\left\{-p_z w + \frac{D}{2}\left[\left(\frac{\partial^2 w}{\partial x^2} + \frac{\partial^2 w}{\partial y^2}\right)^2\right.\right.$$

$$\left.\left. - 2(1-v)\left(\frac{\partial^2 w}{\partial x^2}\cdot\frac{\partial^2 w}{\partial y^2} - \left(\frac{\partial^2 w}{\partial x\,\partial y}\right)^2\right)\right]\right\}dx\,dy, \qquad (4.53)$$

where p_z can be a function of x and y. Apart from the term $p_z w$, the potential of the plate is formally identical with that of the membrane. If we interpret the integrand (4.53) as a functional F then

$$\Pi = \int_0^a \int_0^l F(x, y, w, w', w^{\boldsymbol{\cdot}}, w'', w^{\boldsymbol{\cdot\cdot}}, w'^{\boldsymbol{\cdot}})\,dx\,dy, \tag{4.54}$$

where $w' = dw/dx$, etc., and $w^{\boldsymbol{\cdot}} = dw/dy$, etc. Then we can verify, using the Euler differential equation

$$\frac{\partial F}{\partial w} - \frac{\partial F}{\partial w'} - \frac{\partial F}{\partial w^{\boldsymbol{\cdot}}} + \frac{\partial^2 F}{\partial w''} + \frac{\partial^2 F}{\partial w^{\boldsymbol{\cdot\cdot}}} + \frac{\partial^2 F}{\partial w'^{\boldsymbol{\cdot}}} = 0, \tag{4.55}$$

the correctness of the potential (4.53).

From the Euler differential equation of the functional F we can obtain the Kirchhoff differential equation. The Rayleigh–Ritz method used for beams can now be used for plates. This is of particular importance in view of the complexity of the above differential equation. It should also be noted that a remarkable simplification of the expression for the potential can be obtained by using Green's expression for plates supported at all sides

$$\Pi = \int_0^a \int_0^l \left[-p_z w + \frac{D}{2}\left(\frac{\partial^2 w}{\partial x^2} + \frac{\partial^2 w}{\partial y^2} \right)^2 \right] dx\,dy. \tag{4.56}$$

As an example of the use of the energy method, we take the plate with a uniformly distributed load clamped at all four edges (Fig. 4.14). For such a plate the following expression applies:

$$w(x, y) = \sum_{m=1}^{\infty} \sum_{n=1}^{\infty} w_{mn} \frac{1}{2}\left[1 - (-1)^m \cos\frac{2m\pi x}{a} \right].$$

$$\frac{1}{2}\left[1 - (-1)^n \cos\frac{2n\pi y}{l} \right] \quad \text{with} \quad m, n = 1, 3, 5\ldots \tag{4.57}$$

The expression fulfils all the boundary conditions. We choose the single term

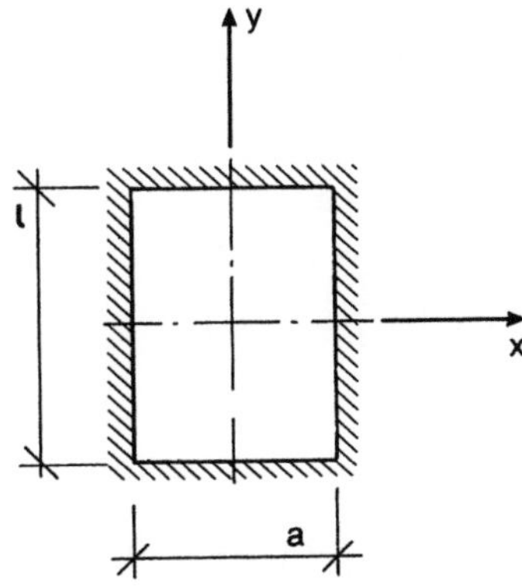

Fig. 4.14. Plate fixed on four sides.

expression $m = n = 1$

$$w(x, y) = w_{11} \frac{1}{4} \left(1 + \cos \frac{2\pi x}{a} \right) \left(1 + \cos \frac{2\pi y}{l} \right), \qquad (4.58)$$

and obtain from (4.56) after integration with $p_z = p_0$

$$\Pi = - p_0 w_{11} \frac{al}{4} + \frac{D\pi^4}{8} \frac{1}{al} w_{11}^2 \left[3\left(\frac{l}{a}\right)^2 + 3\left(\frac{a}{l}\right)^2 + 2 \right]. \qquad (4.59)$$

Since we are trying to find the value of w_{11} such that the potential is a minimum, we have

$$\frac{\partial \Pi}{\partial w_{11}} = 0.$$

We thus obtain the coefficient w_{11} as

$$w_{11} = \frac{p_0 (al)^2}{D\pi^4} \frac{1}{3\left(\frac{l}{a}\right)^2 + 3\left(\frac{a}{l}\right)^2 + 2}. \qquad (4.60)$$

We substitute this expression in (4.59) and obtain

$$\Pi_{\min} = - \frac{1}{2} p_0 w_{11} \frac{al}{4}.$$

In this way we obtain several important features. Minimum potential is always negative. External potential is always twice as large as inner potential, since external work is equal to the work done during deformation. This is independent of whether we substitute an approximate or an exact solution for the plate deflection. If we take $a = l$ (square plate), we obtain w_{11} from (4.60) as

$$w_{11} = 0.00128 \frac{p_0 a^4}{D}.$$

If we choose m and n very large, then the solution becomes exact. For the aspect ratio $a/l = 1$ we have

$$w_{\text{exact}} = 0.00128 \frac{p_0 a^4}{D}.$$

If we use a further approximation, i.e.

$$w = w_{11} \left[\left(\frac{x}{a}\right)^2 - \frac{1}{4} \right]^2 \left[\left(\frac{y}{l}\right)^2 - \frac{1}{4} \right]^2$$

then we obtain

$$w_{11} = \frac{p_0 a^4}{D} \frac{1}{7 + 7\left(\frac{a}{l}\right)^4 + 4\left(\frac{a}{l}\right)^2} \frac{49}{128 \cdot 16},$$

which gives for an aspect ratio of $a/l = 1$,

$$w_{11} = 0.00133 \frac{p_0 a^4}{D}.$$

It is interesting to note that the exact solution gives a smaller deflection. Each approximation can only give a potential greater than that given by the exact solution, which means a stiffer behaviour of the plate with smaller deflections. It is therefore only true for the case with an external individual force P, and therefore the associated potential

$$\Pi = -\frac{1}{2} w_{11} P.$$

For such a case an approximate solution will always lead to smaller bending. We take, for example, the case of a clamped square plate with uniformly distributed loading, as in [2]:

$$w_{\text{exact}} = 0.0056 \frac{pa^2}{D}$$

for which an approximation gives

$$w_{11} = 0.00513 \cdot \frac{pa^2}{D}.$$

Obviously, the potential can in general be approximated just as easily from below (i.e. from larger deflections) as from above (i.e. from smaller deflection). This is evident in Fig. 4.15. These observations are of considerable significance to the numerical approximation methods based on the principle of minimum potential energy, to be discussed later.

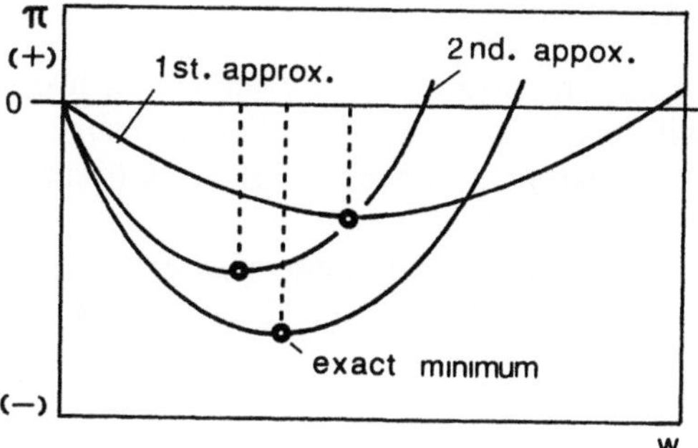

Fig. 4.15. Variation of elastic potential with maximum deflection showing various approximations and the exact solution.

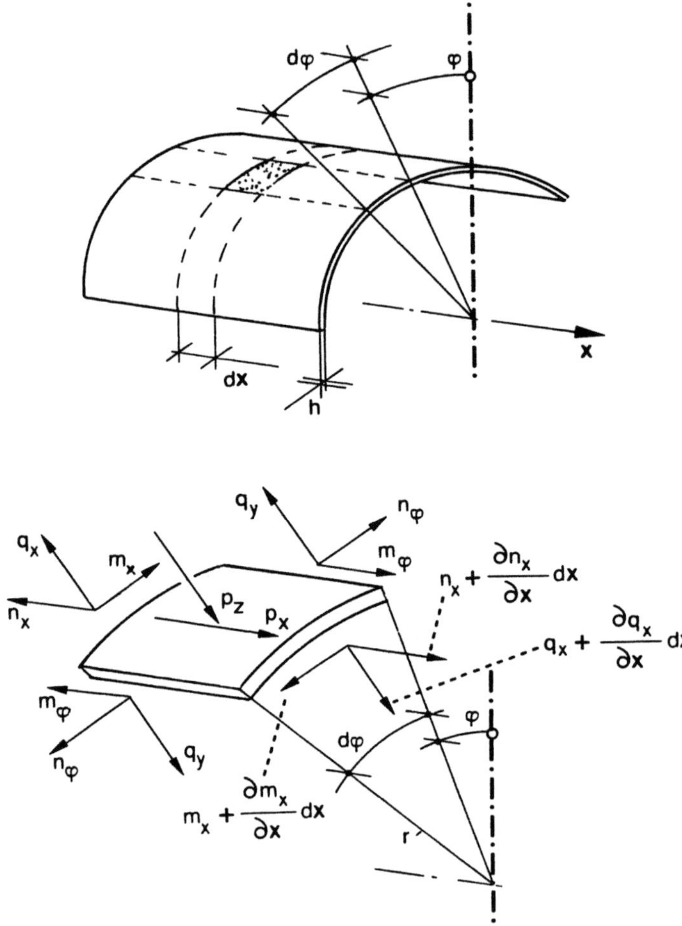

Fig. 4.16. Forces and moments in a cylindrical membrane shell element.

4.1.4 Cylindrical Shells

At this point it is not our intention to give even a brief insight into general shell theory. Instead, we will restrict ourselves to cylindrical shells, which are important in marine structures. By analogy to the membranes, we first examine a cylindrically curved membrane shell (Fig. 4.16).

Forces which are radial, axial and tangential to the surface are taken into consideration. The force and moment equilibrium are, for edge lengths dx and $r\mathrm{d}\varphi$, given by

$$\frac{\partial n_x}{\partial x} + \frac{\partial n_{\varphi x}}{r\partial \varphi} = -p_x, \quad \frac{\partial n_{x\varphi}}{\partial x} + \frac{\partial n_\varphi}{r\partial \varphi} = -p_y, \quad \frac{n_\varphi}{r} = -p_z. \tag{4.61}$$

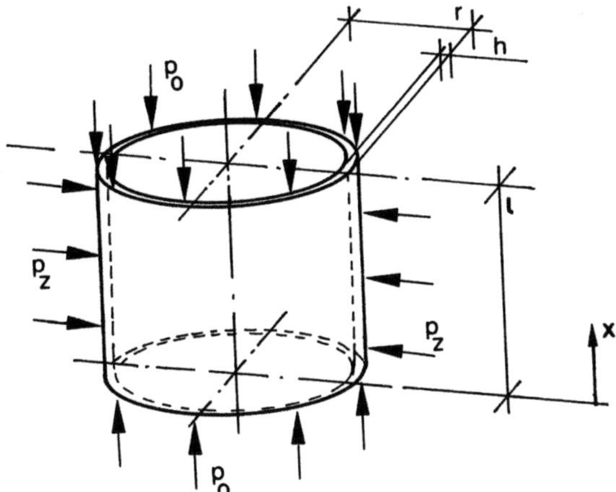

Fig. 4.17. Cylindrical shell under external and axial pressure.

Integration of (4.61) gives the forces $n_{x\varphi}$ and n_x

$$n_{x\varphi} = -\int\left(p_y + \frac{\partial n_\varphi}{r\partial\varphi}\right)dx + C_1(\varphi),$$

$$n_x = -\int\left(p_x + \frac{\partial n_{\varphi x}}{r\partial\varphi}\right)dx + C_2(\varphi). \tag{4.62}$$

Here, we consider the case of a cylinder under constant axial load and constant external pressure (Fig. 4.17), and thus

$$n_\varphi = -p_z r, \quad n_x = -p_0 h, \quad n_{x\varphi} = 0, \tag{4.63}$$

or, expressed in terms of stress,

$$\sigma_\varphi^m = -p_z\frac{r}{h}, \quad \sigma_x^m = -p_0, \quad \tau_{x\varphi}^m = 0. \tag{4.64}$$

The pressure forces compress the cylinder so that while the radius of the axially loaded cylinder tends to increase, the pressure p_z simultaneously tends to reduce the radius. The relationships between the strains and displacements are similar to those for membranes:

$$\varepsilon_x = \frac{\partial u}{\partial x}, \quad \varepsilon_\varphi = \frac{\partial v}{r\partial\varphi} + \frac{w}{r}, \quad \gamma_{x\varphi} = \frac{\partial v}{\partial x} + \frac{\partial u}{r\partial\varphi}. \tag{4.65}$$

Compared to the membrane, there is simply an extra term for hoop stress ε_φ taking into account the overall pressure. By rotational symmetry the first term is completely eliminated. We can again use Hookes law

$$\varepsilon_x = \frac{1}{Eh}(n_x - \nu n_\varphi),$$

$$\varepsilon_\varphi = \frac{1}{Eh}(n_\varphi - \nu n_x), \tag{4.66}$$

$$\gamma_{x\varphi} = \frac{2(1+\nu)}{Eh} n_{x\varphi}$$

to integrate the displacements equations:

$$u = \frac{1}{Eh}\int_0^x (n_x - \nu n_\varphi)\mathrm{d}x,$$

$$v = \frac{2(1+\nu)}{Eh}\int_0^x n_{x\varphi}\mathrm{d}x - \int_0^x \frac{\partial u}{r\partial\varphi}\mathrm{d}x, \tag{4.67}$$

$$w = \frac{r}{Eh}(n_\varphi - \nu n_x) - \frac{\partial v}{\partial\varphi}.$$

Carrying out the integrations we obtain

$$\frac{u}{x} = -\frac{1}{E}\left(p_0 - \nu\frac{r}{h}p_z\right),$$

$$\frac{w}{r} = -\frac{1}{E}\left(\frac{r}{h}p_z - \nu p_0\right). \tag{4.68}$$

This simple example shows great similarity between membrane shell theory and membrane theory of plates, but at the same time, it also brings out the limitations of membrane shell theory. This becomes evident if we look at a practical example of such a cylinder. In normal use this would be welded to other structural elements at its ends or, perhaps, horizontally stiffened through stringers (i.e. pipe deformation is radially constrained). In this case, the displacement w cannot be obtained using membrane shell theory. In other words, the boundary conditions encountered in practice do not allow membrane shell theory to represent the actual behaviour of a cylindrical shell. Further considerations are necessary; obviously, shell bending must be taken into account.

Let us therefore consider a cylindrical pipe (Fig. 4.18), and in the usual way cut out an infinitesimally small element and allow the equilibrium forces and moments to act upon it. Here we take into account the complete rotational symmetry of the problem, i.e. $P_y = 0$, and no variation of the forces n_φ and q_y in the peripheral direction. The equilibrium conditions are obtained from Fig. 4.18 as

$$\frac{\partial n_x}{\partial x} = -p_x, \quad \frac{\partial q_x}{\partial x} + \frac{n_\varphi}{r} = -p_z, \quad \frac{\partial m_x}{\partial x} = q_x. \tag{4.69}$$

The three forces n_x, n_φ, q_x and the moment m_x meet three equilibrium requirements.

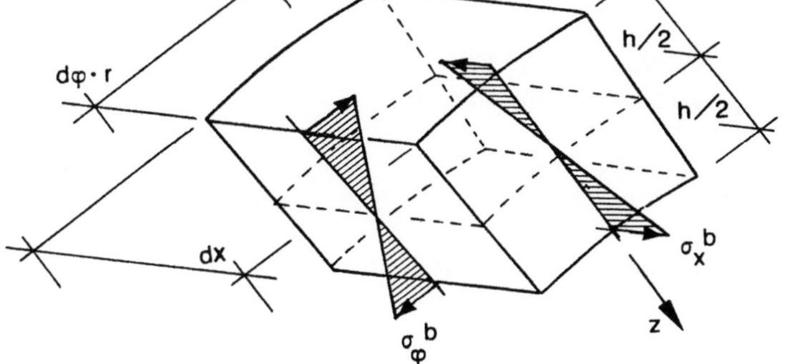

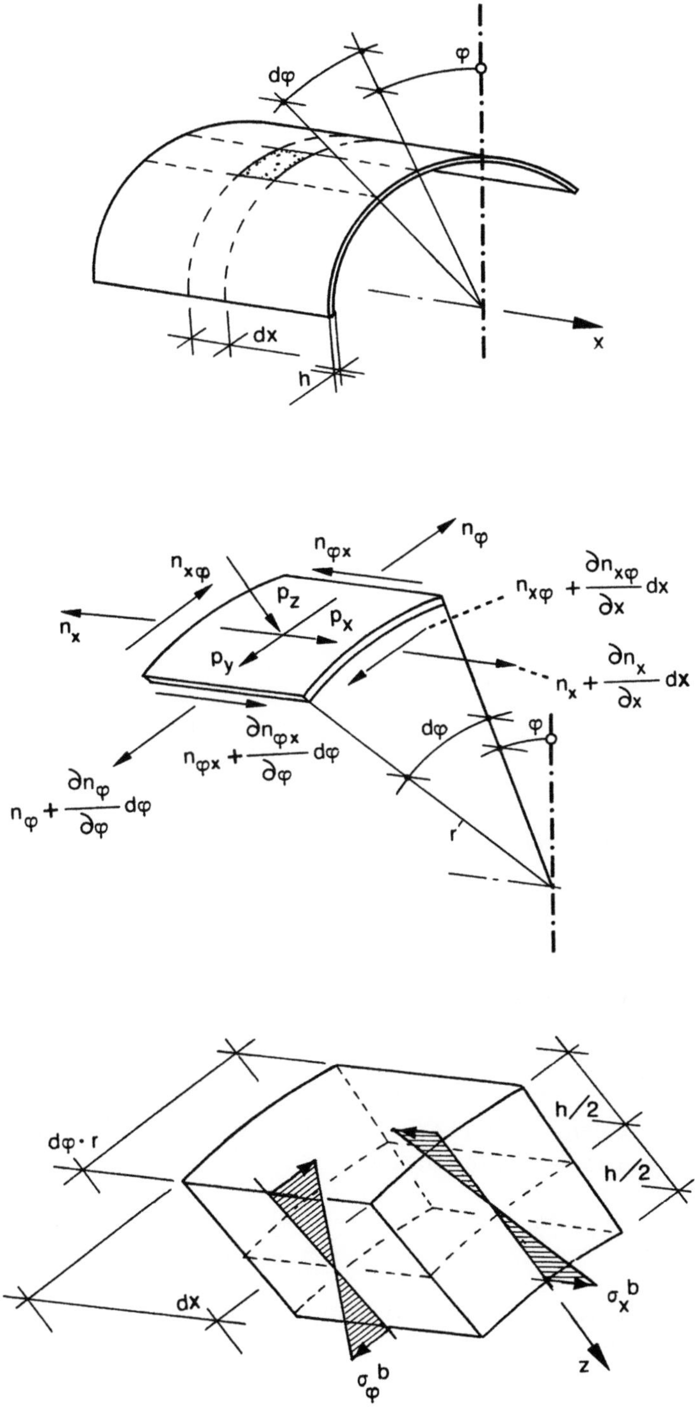

Fig. 4.18. Forces and moments in a cylindrical shell element in bending.

The fourth requirement we obtain from the bending deformation of the shell. The strains are given by

$$\varepsilon_x = \frac{\partial u}{\partial x} - z\frac{\partial^2 w}{\partial x^2}, \quad \varepsilon_\varphi = \frac{w}{r}. \tag{4.70}$$

The stresses (i.e. the membrane and bending stresses) are obtained from (4.66) using Hookes law as

$$\sigma_x = \frac{E}{1-v^2}(\varepsilon_x + v\varepsilon_\varphi), \quad \sigma_\varphi = \frac{E}{1-v^2}(\varepsilon_\varphi + v\varepsilon_x) \tag{4.71}$$

where use of $\sigma_x = n_x \cdot h$ and $\sigma_\varphi = n_\varphi \cdot h$ lead to

$$\sigma_x = \frac{E}{1-v^2}\left(\frac{\partial u}{\partial x} - z\frac{\partial^2 w}{\partial x^2} + v\frac{w}{r}\right),$$

$$\sigma_\varphi = \frac{E}{1-v^2}\left(\frac{w}{r} + v\frac{\partial u}{\partial x} - vz\frac{\partial^2 w}{\partial x^2}\right). \tag{4.72}$$

Obviously, since membrane stress components make no contribution to the bending moment, we obtain

$$m_x = -\int_{-h/2}^{h/2} \sigma_x z\,dz = D\frac{\partial^2 w}{\partial x^2},$$

$$m_\varphi = -\int_{-h/2}^{h/2} \sigma_\varphi z\,dz = vD\frac{\partial^2 w}{\partial x^2}. \tag{4.73}$$

The membrane stresses are given by

$$\sigma_x^m = \frac{B}{h}\left(\frac{\partial u}{\partial x} + v\frac{w}{r}\right), \quad \sigma_\varphi^m = \frac{B}{h}\left(\frac{w}{r} + v\frac{\partial u}{\partial x}\right), \tag{4.74}$$

where B is the inplane rigidity, given by

$$B = \frac{Eh}{1-v^2}.$$

If we substitute the displacements as in (4.68) into (4.74), we obtain (4.64) again. The bending stresses on the surface of the shell are

$$\sigma_x^b = \pm 6\frac{m_x}{h^2}, \quad \sigma_\varphi^b = \pm 6\frac{m_\varphi}{h^2}. \tag{4.75}$$

Now we have enough equations to determine the equilibrium forces and moments, and so we bring (4.74) to a single equation

$$n_\varphi = B(1-v^2)\frac{w}{r} + vn_x.$$

If we now substitute this into (4.69), we obtain, by inspection of (4.73),

$$D\frac{d^4w}{dx^4} + B\frac{1-v^2}{r^2}w = -p_z + \frac{v}{r}\int_0^x p_x\,dx. \tag{4.76}$$

The partial derivatives have been replaced here by ordinary ones, as integration needs to be carried out only with respect to a single coordinate. This is now exactly the differential equation which we formally obtained for a beam on an elastic foundation. We further examine the example in Fig. 4.17, and consider a horizontal stringer at half height. The length of the pipe is long enough so that the influence of the stringer is not felt by the upper or lower ends of the pipe. The differential equation (4.76), somewhat transformed, gives

$$\frac{d^4w}{dx^4} + 4\alpha^4 w = -\frac{p_z}{D} + \frac{v}{Dr}\int_0^x p_x\,dx, \tag{4.77}$$

where

$$\alpha^4 = \frac{3(1-v^2)}{r^2h^2}.$$

The homogeneous solution of differential equation (4.77) has already been obtained in (4.19). In the previous example we determined the deformation of the pipe according to membrane shell theory. We now need to take into account the effect of the stringer on these deformations. Since this is limited to the region of the stringer, the homogeneous solution suffices. Since, as we have said, at a large distance from the stringer its influence on the deformation is insignificant, both the coefficients C_1 and C_2 must also disappear. The rotation at $x = 0$ disappears likewise, so that $C_3 = C_4 = C$. The solution is now dependent on one unknown. The constant C is obtained as in (4.68). We now let the pipe, without the supporting effect of the stringers, be compressed by an amount

$$w(0) = -\frac{r}{E}\left(\frac{r}{h}p_z - vp_0\right).$$

Since the stringer prevents radial distortion, the pipe must expand by precisely this amount. The constant C is therefore $-w(0)$. We then obtain the displacement equation as

$$w(x) = C[e^{-\alpha x}(\cos \alpha x + \sin \alpha x) - 1]$$

and the rotation as

$$\frac{dw}{dx} = -2\alpha C\,e^{-\alpha x}\sin \alpha x.$$

The bending moment is obtained by repeated differentation

$$m_x(x) = -2D\alpha^2 C\,e^{-\alpha x}(\cos \alpha x - \sin \alpha x),$$

or

$$m_\varphi(x) = vm_x(x).$$

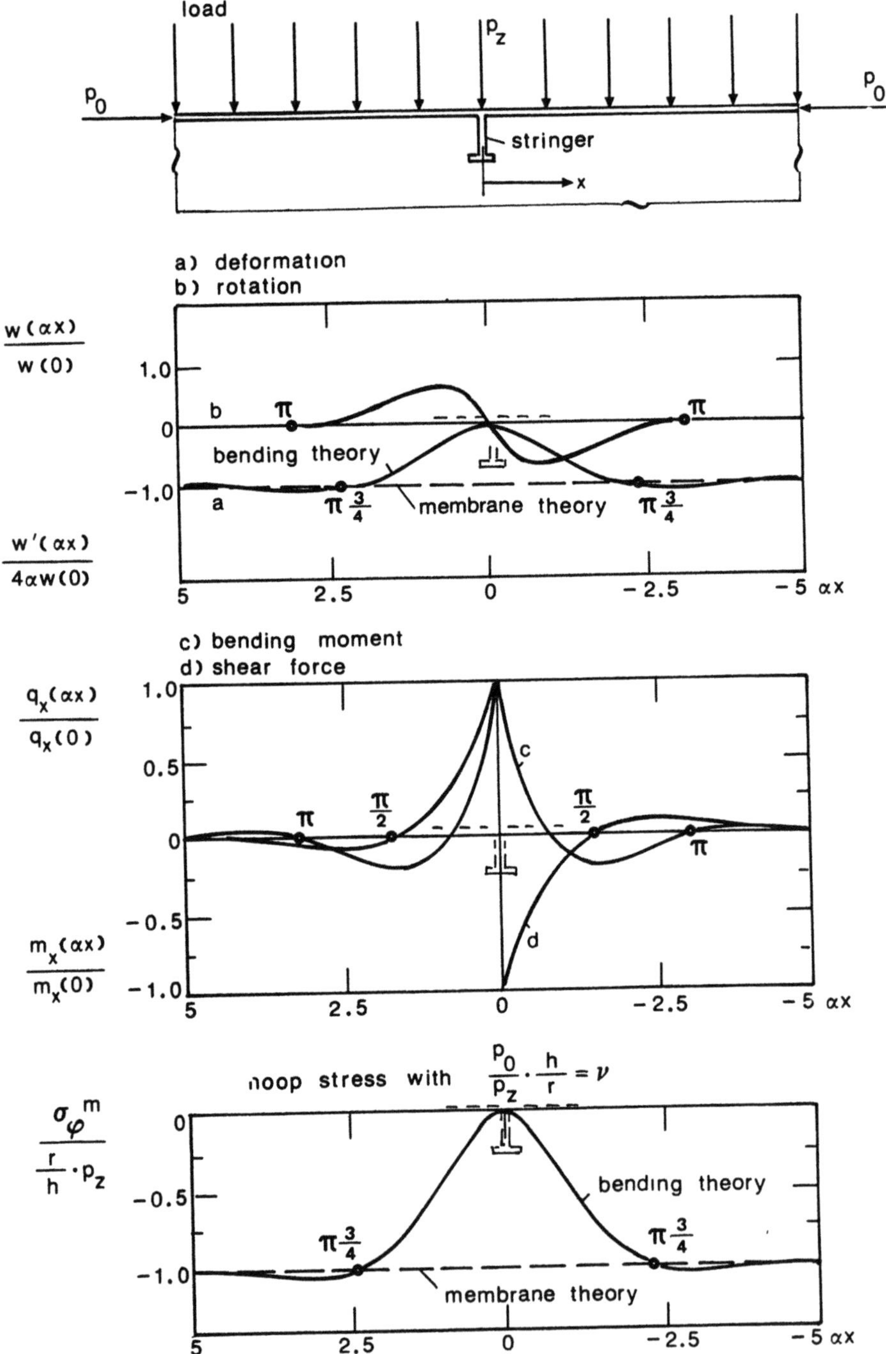

Fig. 4.19. Stresses and deflections in a pipe in the vicinity of a stringer.

Finally, we obtain the shear force as

$$q_x(x) = -4D\alpha^3 C\, e^{-\alpha x} \cos \alpha x.$$

These expressions are shown in Fig. 4.19, where the radial displacement according to membrane shell theory is shown dotted, so that the local effect of the stringer can be clearly seen. It can immediately be seen that, for example, at a distance of $\alpha x = 3/4\pi$ the local restraining effect of the stringer disappears. The rotation of the shell disappears at a distance of $\alpha x = \pi$, and the bending moment alters its sign at $\alpha x = \pi/4$ and becomes insignificantly small at $\alpha x = \pi$. The shear force changes sign at $\alpha x = \pi/2$, and decays fast beyond this point. We now obtain the total stress as

$$\sigma_x = \sigma_x^b + \sigma_x^m, \quad \sigma_\varphi = \sigma_\varphi^b + \sigma_\varphi^m, \tag{4.78}$$

where

$$\sigma_x^b = \mp \frac{12}{h^2} D\alpha w(0) e^{-\alpha x}(\cos \alpha x - \sin \alpha x),$$

$$\sigma_\varphi^b = v\sigma_x^b \tag{4.79}$$

are the bending components, and the membrane components are given by

$$\sigma_\varphi^m = -\frac{r}{h} p_z + \frac{1}{1-v^2}\left(\frac{r}{h} p_z - v p_0\right) e^{-\alpha x}(\cos \alpha x + \sin \alpha x),$$

$$\sigma_x^m = -p_0 + \frac{v}{1-v^2}\left(\frac{r}{h} p_z - v p_0\right) e^{-\alpha x}(\cos \alpha x + \sin \alpha x). \tag{4.80}$$

At large distances from the stringer, $\alpha x \to \infty$, i.e. the undisturbed membrane stresses prevail.

The complete theory with a number of applications can be found elsewhere [10], as well as in textbooks [1, 3]. Practical use of the above bending theory for pipes with misalignments or pipes with variable thickness is described in [11]. An excellent review of all these problems can be found in [12], as well as in the classic books [13] and [14].

4.2 Stability and Second-Order Stress Theory

In previous sections it has been the practice to establish the equilibrium of the undeformed elements. We have therefore been able to derive linear relationships between strains and displacements, as well as between stresses and strains. These are accurate enough in most cases; solutions on this basis always produce proportional increase of the stress and loading. Thus a doubling of the external load will lead to a doubling of the displacement or the stress and no theoretical limit in imposed on this linear behaviour. Let us now study the effects of second-order stress theory with simple examples. The results of computer calculations are considerably more difficult to assess in this context than in the linear theory, thus

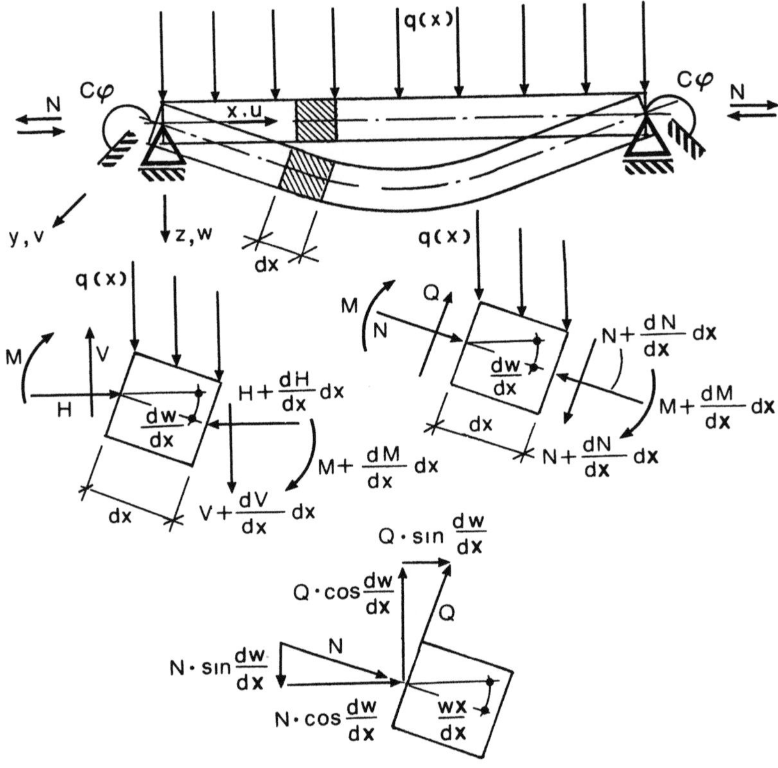

Fig. 4.20. Equilibrium of the deformed beam element.

a study of the basic relationships is especially important. The application of the second-order stress theory will be made to the case of a slender tube used in marine structures. We take a beam as in Fig. 4.2, cut out an element, and consider the equilibrium of the deformed element. While in the linear beam theory shear forces and axial forces are decoupled, in the second-order stress theory, they are coupled. From Fig. 4.20 we obtain the equilibrium conditions as

$$\sum P_x = H - \left(H + \frac{dH}{dx} dx \right) = 0,$$

$$\sum P_z = -V + \left(V + \frac{dV}{dx} dx \right) + q(x)dx = 0,$$

$$\sum M = -M + \left(M + \frac{dM}{dx} dx \right) - q(x)dx\frac{dx}{2}$$

$$-\left(V + \frac{dV}{dx} dx \right)dx - H\frac{dw}{dx} dx = 0. \tag{4.81}$$

The vertical and horizontal forces are obtained in terms of transverse and axial forces acting on the element as

$$V = Q \cos\frac{dw}{dx} - N \sin\frac{dw}{dx}, \quad H = Q \sin\frac{dw}{dx} + N \cos\frac{dw}{dx}. \tag{4.82}$$

For small angles the cosine is equal to 1 and the sine is equal to the angle in radians, and hence

$$V = Q - N\frac{dw}{dx}, \quad H = Q\frac{dw}{dx} + N. \tag{4.83}$$

If we substitute these in the equilibrium conditions and neglect the higher terms, we obtain

$$\frac{dN}{dx} = 0, \quad \frac{dQ}{dx} = -q(x) + N\frac{d^2w}{dx^2}, \quad \frac{dM}{dx} = Q. \tag{4.84}$$

Comparison with (4.1) shows that these equations differ only in the vertical equilibrium. It is not possible to convert this differential equation into a differential equation similar to (4.2), because now not only forces but also displacements are to be considered. Equation (4.6) still applies, as it gives the relationship between bending moment, stiffness EI, and the curvature of the beam. This is differentiated twice with respect to w to give

$$\frac{d^2M}{dx^2} = -EI\frac{d^4w}{dx^4}. \tag{4.85}$$

The equilibrium conditions of the deformed element are now substituted to give

$$EI\frac{d^4w}{dx^4} + N\frac{d^2w}{dx^2} = q(x). \tag{4.86}$$

When the axial force is tensile we have

$$EI\frac{d^4w}{dx^4} - N\frac{d^2w}{dx^2} = q(x).$$

In the usual way, we write

$$\frac{d^4w}{dx^4} + \alpha^2\frac{d^2w}{dx^2} = \frac{q(x)}{EI}, \tag{4.87}$$

where

$$\alpha^2 = \pm\frac{N}{EI}.$$

The solution for the case of compressive force with $q(x) = q$ is given by

$$w(x) = C_1 + C_2 x + C_3 \sin \alpha x + C_4 \cos \alpha x + \frac{ql^2}{2EI}\left(\frac{x}{\alpha l}\right)^2 \tag{4.88}$$

and for the tensile force by

$$w(x) = C_1 + C_2 x + C_3 \sinh \alpha x + C_4 \cosh \alpha x - \frac{ql^2}{2EI}\left(\frac{x}{\alpha l}\right)^2. \tag{4.89}$$

The boundary conditions are

$$w = \frac{d^2 w}{dx^2} = 0\big|_{x=0,\, x=l},$$

where $C_\varphi = 0$. Using these boundary conditions, coefficients C_1 to C_4 can be determined so that for bending we can write

$$w(x) = \frac{ql^4}{2EI(\alpha l)^4}\left[-2 - (\alpha l)^2 \frac{x}{l} + (\alpha l)^2\left(\frac{x}{l}\right)^2 + 2\,\mathrm{tg}\frac{\alpha l}{2}\sin \alpha x + 2\cos \alpha x \right],$$

and for the tensile force

$$w(x) = \frac{ql^4}{2EI}\left(\frac{1}{\alpha l}\right)^4\left[-2 + (\alpha l)^2 \frac{x}{l} - (\alpha l)^2\left(\frac{x}{l}\right)^2 - 2\,\mathrm{tgh}\frac{\alpha l}{2}\sinh \alpha x + 2\cosh \alpha x \right].$$

We now set the transverse loadings in relation to the longitudinal loadings to $ql = \Phi \cdot N$, and so we obtain the deflection at midspan

$$\frac{w}{l} = \Phi \frac{N}{N_{K_1}} \frac{\pi^2}{16} \frac{1}{\beta^4}\left[\frac{1}{\cos \beta} - 1 - \frac{\beta^2}{2} \right], \tag{4.90}$$

where

$$N_{K_1} = \frac{EI\pi^2}{l^2} \tag{4.91}$$

is the previously obtained Euler bifurcation load and $\beta = \alpha l/2$. In Fig. 4.21, bending deflection is expressed as a function of axial load for various levels of transverse loads. We realize that as the transverse loading disappears, i.e. $\Phi = 0$, the curve breaks down into vertical and horizontal lines, and the point of intersection is the bifurcation point. The equilibrium fails because the deflection is either zero or unknown. If such a condition exists in a frame system, we refer to it as a loss of stability in the Euler sense. Expressed mathematically, the eigenvalue problem exists. If the load–deflection curve is steady (i.e. $\Phi \neq 0$) then we refer to it as a second-order stress problem which, since small transverse loadings cannot usually be prevented, gives a considerably more realistic representation of a framed structure than the usual linear solution, as shown dotted in Fig. 4.21. This applies particularly in regions of high axial forces. We can also see from Fig. 4.21 that solutions from second-order stress theory, obtained using (4.87) for a number of problems in marine structures, give no useful information in the region of the Euler buckling load, since according to this theory the deflections increase to high levels, and even a deflection of $w/l \geqslant 0.5$ is not possible on geometrical grounds. The cause lies, of course, in expression (4.6), in which the linear relationship between bending moment and curvature is established.

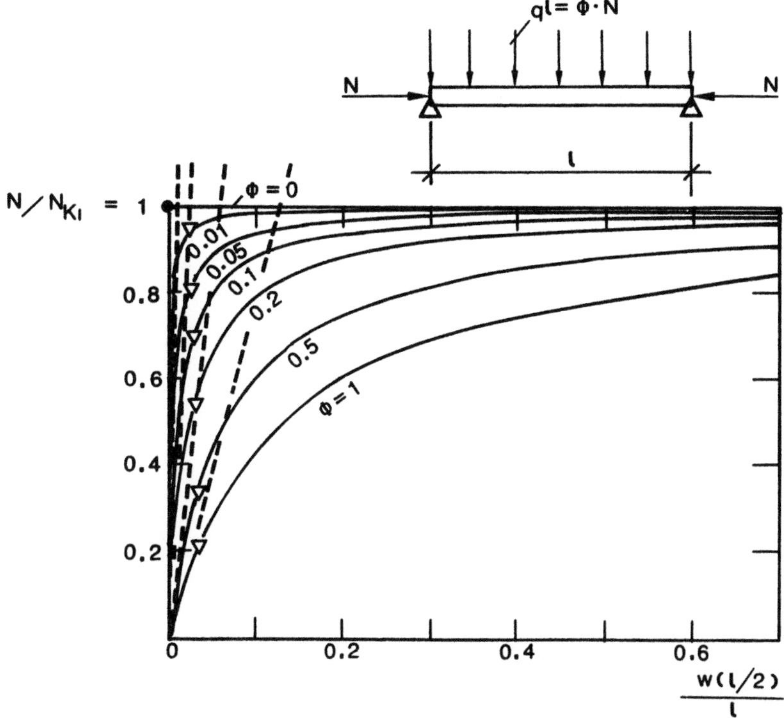

Fig. 4.21. Deflection of simply supported beam under transverse and axial loading. ∇ Load limit according to (4.185).

On the other hand, it is usual to have structural components in which deflection is small relative to the length of an individual beam, so that as a rule it lies between $N/N_{Ki} = 0$ to 0.5. However, the significance of the ideal Euler buckling bifurcation load can be determined. We can arrive at a conclusion that a homogeneous problem, for example a beam without transverse loading, is always described as a problem in which the bifurcation point or (mathematically) the eigenvalue has to be determined, whereas an inhomogeneous problem is a second-order stress problem. This is insufficient, as we illustrate in the following problem.

We consider, for example, one of the supports in a jacket subjected to external loading. Here it is often the case that the axial force N is eccentric by a distance e (see Fig. 4.22). In addition to transverse loadings, external moments from adjacent members must be considered as well as the moments resulting from axial forces. So as not to overcomplicate the problem, we take a beam that is loaded only with eccentric axial forces. Here, the homogeneous part of the differential equation (4.87) with the solution (4.88) $q = 0$, applies. The coefficients are obtained using boundary conditions which, for the curvature at $x = 0$ and $x = l$, include the external moment and the condition that the deflection disappears at the ends of

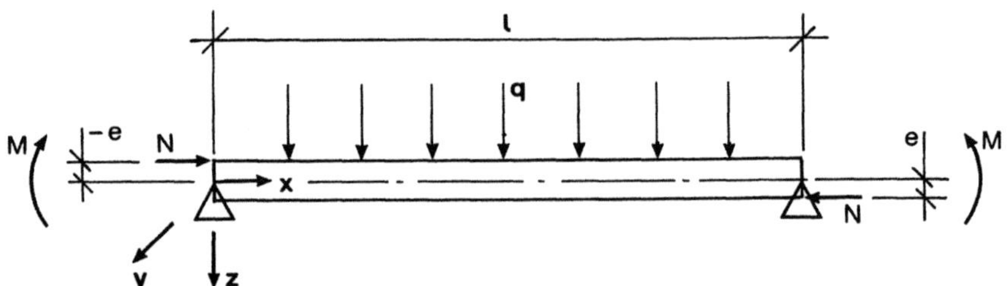

Fig. 4.22. Simply supported beam with eccentric loads in the axial direction.

the beam. We obtain for the symmetrical case a deflection at $l/4$

$$\frac{w}{e} = \frac{\sin\dfrac{3\alpha l}{4} + \sin\dfrac{\alpha l}{4}}{\sin \alpha l} - 1$$

and for the antisymmetrical case

$$\frac{w}{e} = \frac{\sin\dfrac{3\alpha l}{4} - \sin\dfrac{\alpha l}{4}}{\sin \alpha l} - \frac{1}{2}.$$

Figure 4.23 shows deflection dimensionalized with the lever e in relation to longitudinal force N dimensionalized with the Euler buckling load. We observe

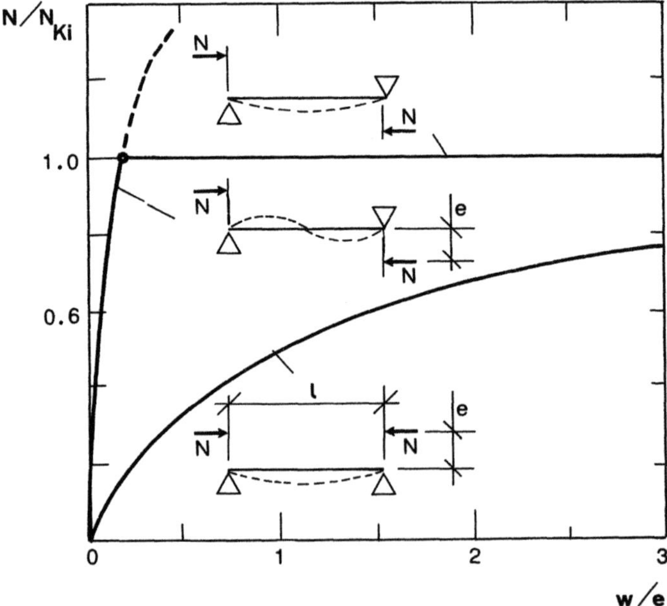

Fig. 4.23. Deflection of simply supported beam with eccentric loads in the axial direction.

that with a symmetrical bending line, a stress problem exists, and with an anti-symmetric moment application, a bifurcation point occurs. The criterion for whether a case of bifurcation exists, i.e. whether or not we have a case of stability, is now as in [15]: if the bending line includes the lowest buckling mode, then a stress problem exists. If the lowest buckling mode does not occur we have the antisymmetric case, which is dealt with as a stability problem.

We now consider the example of a symmetrically loaded beam with two supports. Taking moments as in (4.6), the bending moment at the middle is obtained as follows:

$$M = \frac{ql^2}{8}\left(\frac{2}{\alpha l}\right)^2\left[\frac{2}{\cos \alpha l/2} - 2\right]$$

or

$$M = M_0 f,$$

where $M_0 = ql^2/8$ is the bending moment without the effect of axial forces, and f is the so-called amplification factor. These factors are given in handbooks (e.g. [5, 16]). The expressions for f are often very complex, and so we have to try to find simpler expressions. Such a simplified expression is given by

$$f = \frac{1 + \Psi \dfrac{N}{N_{Ki}}}{1 - \dfrac{N}{N_{Ki}}}. \tag{4.92}$$

In Fig. 4.24 the factors Ψ are given for a number of practical cases. A numerical check shows that (4.92) is indeed a very good approximation.

Occasionally, multi-parameter eigenvalue problems occur. Here, the interactions between individual eigenvalues must be determined. An illustrative example is a tube loaded with both longitudinal forces and torsional moments. Such a problem arises, for example, in a drill string that is under both tensile and compressive longitudinal forces. We assume, for simplicity, a freely supported tube as in Fig. 4.25.

In its deformed configuration, the tube bends both in y- and z-directions under the effect of the torsional moment. We thus have two differential equations for the bending in each direction

$$EI\frac{d^2w}{dx^2} = -Nz + M_T\frac{dv}{dx},$$

$$EI\frac{d^2w}{dy^2} = -Ny - M_T\frac{dw}{dx}. \tag{4.93}$$

We can convert these equations into a single fourth-order differential equation by carrying out the appropriate differentiations. The result is

$$\frac{d^4w(x)}{dx^4} + \left[\frac{2N}{EI} + \left(\frac{M_T}{EI}\right)^2\right]\frac{d^2w}{dx^2} + \left(\frac{N}{EI}\right)^2 w(x) = 0. \tag{4.94}$$

CASE	Ψ	M_O
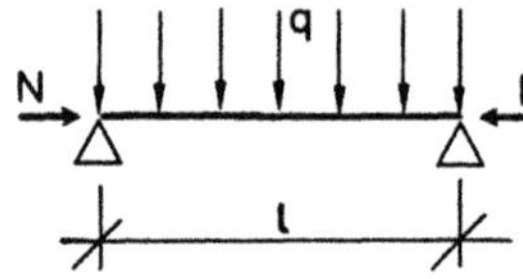 (simply supported beam, distributed load q, span l, axial forces N)	0	$\dfrac{ql^2}{8}$
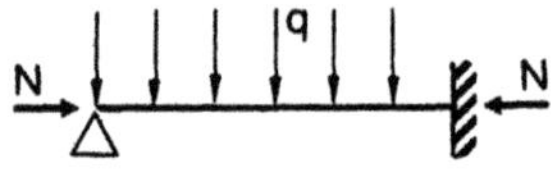 (propped cantilever, distributed load q, axial forces N)	-0.3	
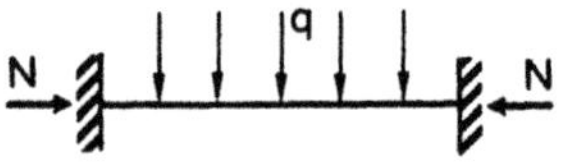 (fixed-fixed beam, distributed load q, axial forces N)	-0.4	$\dfrac{ql^2}{12}$
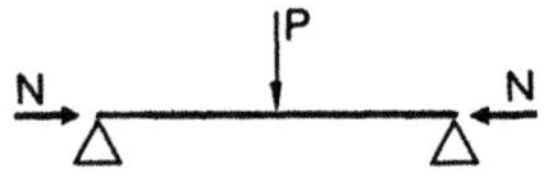 (simply supported beam, central point load P, axial forces N)	-0.2	$\dfrac{Pl}{4}$
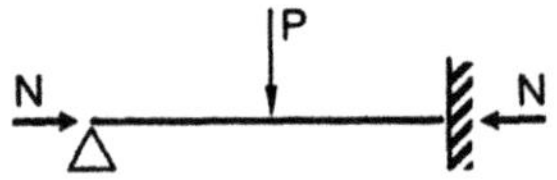 (propped cantilever, central point load P, axial forces N)	-0.4	$\dfrac{3Pl}{16}$
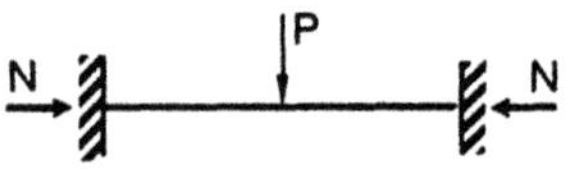 (fixed-fixed beam, central point load P, axial forces N)	-0.6	$\dfrac{Pl}{8}$
 (beam with initial deflection w_0, axial forces N)	0	$N \cdot w_0$
(beam with end moments M_A, M_B, axial forces N) $M_B/M_A < 0$	$\dfrac{-0.4\left(1+\dfrac{M_B}{M_A}\right)}{N/N_{KI}}$ $-1 < M_B/M_A < 0.5$	$M_A \quad M_A > M_B$ $M_B \quad M_B > M_A$
$M_B/M_A > 0$	$-\dfrac{0.6}{N/N_{KI}}$ $M_B/M_A > 0.5$	

Fig. 4.24. Amplification factors for effect of axial forces on maximum bending moment for variç load cases according to AISC [28], etc.

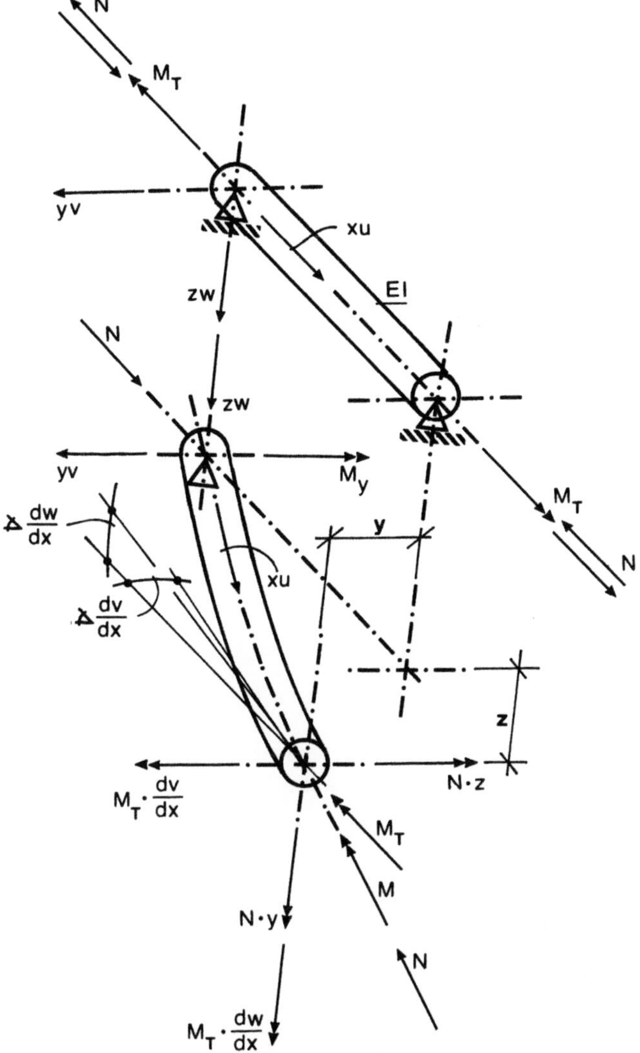

Fig. 4.25. Equilibrium of beam element subject to (a) axial force (b) torsional moment in the deformed state.

The solution of this differential equation is

$$w(x) = C_1 \sin \lambda_1 \frac{x}{l} + C_2 \cos \lambda_1 \frac{x}{l} + C_3 \sin \lambda_2 \frac{x}{l} + C_4 \cos \lambda_2 \frac{x}{l},$$

where

$$\left(\frac{\lambda_{1,2}}{l}\right)^2 = \left[\frac{N}{EI} + \frac{1}{2}\left(\frac{M_T}{EI}\right)^2\right] \pm \frac{M_T}{EI}\sqrt{\frac{N}{EI} + \frac{1}{4}\left(\frac{M_T}{EI}\right)^2}.$$

The constants C_1 to C_4 depend on the boundary conditions of a freely-supported tube which are

$$w(x) = 0|_{x=0, x=l},$$

$$\left[\frac{d^3w}{dx^3} + \frac{N}{EI} + \left(\frac{M_T}{EI}\right)^2 \frac{dw}{dx}\right] = 0|_{x=0, x=l}. \tag{4.95}$$

The characteristic equation obtained from the determinant of the homogeneous system of equations is

$$2\Phi_1 \Phi_2 (1 - \cos \lambda_1 \cos \lambda_2) - \sin \lambda_1 \sin \lambda_2 (\Phi_1^2 + \Phi_2^2) = 0, \tag{4.96}$$

where

$$\Phi_{1,2} = \frac{\lambda_{1,2}}{l}\left[\left(\frac{\lambda_{1,2}}{l}\right)^2 - \frac{N}{EI} - \left(\frac{M_T}{EI}\right)^2\right].$$

The solution of (4.96) is

$$\frac{1}{4}\left(\frac{M_T}{EI}\right)^2 + \frac{N}{EI} = \left(\frac{\pi}{l}\right)^2, \tag{4.97}$$

or

$$\frac{\lambda_{1,2}}{l} = \frac{\pi}{l} \pm \frac{1}{2}\frac{M_T}{EI}.$$

With the Euler buckling load and the critical torsional moment, and without taking into account the influence of the axial force

$$M_{TK_1} = 2\frac{EI\pi}{l}$$

we obtain the interaction curve

$$\left(\frac{M_T}{M_{TK_1}}\right)^2 + \frac{N}{N_{Ki}} = 1, \tag{4.98}$$

From this, shown in Fig. 4.26, it is clear that in the upper region of a drill string tensile forces increase the critical torque, $M_T/M_{TKi} \geqslant 1$, while compressive forces in the lower region tend to reduce it, $M_T/M_{TK_1} \leqslant 1$ (right half of figure).

Second-order stress theory, including the eigenvalue problem, can be applied in a similar manner to plate problems. For this purpose, we must represent the equilibrium of forces on the deformed plate element, remembering that, as with the beam, equilibrium moments in the deformed state must agree with those in the undeformed state. The equilibrium conditions (4.43) need only be expanded for the sum of all forces in the z-direction, as we can see from Fig. 4.27

$$\frac{\partial q_x}{\partial x} + \frac{\partial q_y}{\partial y} = -p_z(x, y) + n_x\frac{\partial^2 w}{\partial x^2} + n_y\frac{\partial^2 w}{\partial y^2} + 2n_{xy}\frac{\partial^2 w}{\partial x\,\partial y},$$

$$\frac{\partial m_y}{\partial y} + \frac{\partial m_{xy}}{\partial x} = q_y, \quad \frac{\partial m_x}{\partial x} + \frac{\partial m_{xy}}{\partial y} = q_x. \tag{4.99}$$

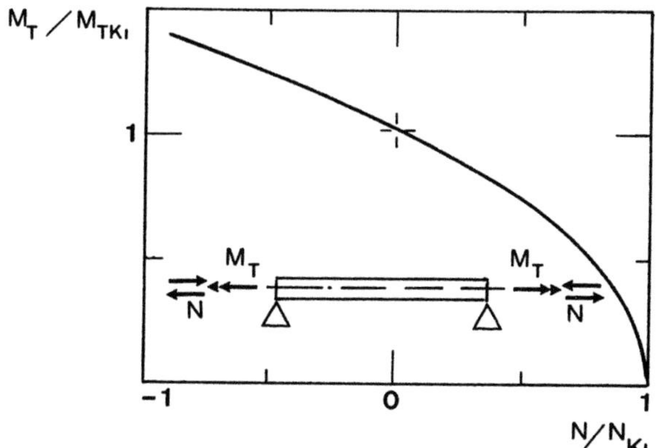

Fig. 4.26. Interaction curve between critical axial force and torsional moment.

After differentiation and substitution we obtain

$$\frac{\partial^2 m_x}{\partial x^2} + 2\frac{\partial^2 m_{xy}}{\partial x\,\partial y} + \frac{\partial^2 m_y}{\partial y^2} = -p_z(x, y) + n_x\frac{\partial^2 w}{\partial x^2} + n_y\frac{\partial^2 w}{\partial y^2} + 2n_{xy}\frac{\partial^2 w}{\partial x\,\partial y}.$$

$$(4.100)$$

Taking (4.47) into account, we finally obtain the plate differential equation

$$D\Delta\Delta w = n_x\frac{\partial^2 w}{\partial x^2} + n_y\frac{\partial^2 w}{\partial y^2} + 2n_{xy}\frac{\partial^2 w}{\partial x\,\partial y} + p_z(x, y). \tag{4.101}$$

A rigorous solution to this differential equation is possible for a freely supported plate with inplane loading and a constant transverse load p_0 (Fig. 4.28). For the deflection we choose a sine series expansion

$$w(x, y) = \sum_{m=1}^{\infty} \sum_{n=1}^{\infty} w_{mn} \sin\frac{m\pi x}{a} \sin\frac{n\pi y}{b},$$

and for the transverse loading $p_z(x, y) = p_0$ we use the expression

$$p_z(x, y) = \frac{16p_0}{\pi^2} \sum_{m=1}^{\infty} \sum_{n=1}^{\infty} \frac{1}{mn} \sin\frac{m\pi x}{a} \sin\frac{n\pi y}{b}.$$

If we insert these functions in (4.101), we obtain the unknown coefficients as

$$w_{mn} = \frac{16p_0}{\pi^2 mn\left\{D\left[\left(\dfrac{m\pi}{a}\right)^2 + \left(\dfrac{n\pi}{b}\right)^2\right]^2 + n_x\left(\dfrac{m\pi}{a}\right)^2 + n_y\left(\dfrac{n\pi}{b}\right)^2\right\}},$$

Hence the solution of the differential equation is

$$w(x, y) = \frac{16p_0}{D\pi^6} \sum_{m=1}^{\infty} \sum_{n=1}^{\infty} \frac{\sin\dfrac{m\pi x}{a}\sin\dfrac{n\pi y}{b}}{mn\left\{\left[\left(\dfrac{m}{a}\right)^2 + \left(\dfrac{n}{b}\right)^2\right]^2 + \dfrac{n_x}{\pi^2 D}\left(\dfrac{m}{a}\right)^2 + \dfrac{n_y}{\pi^2 D}\left(\dfrac{n}{b}\right)^2\right\}}.$$

$$m, n = 1, 3, 5\ldots \tag{4.102}$$

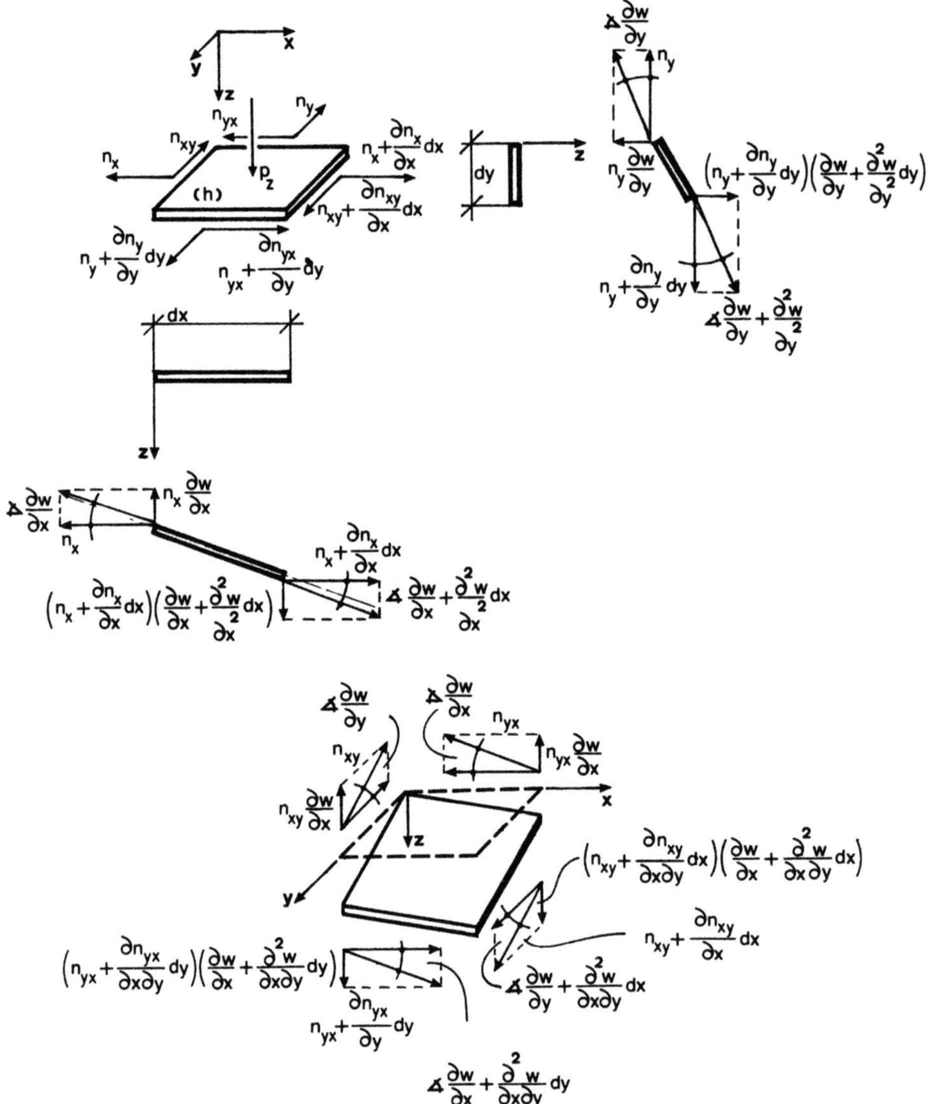

Fig. 4.27. Equilibrium of deformed membrane element.

The bending moments are

$$m_x = \frac{16p_0}{\pi^4} \sum_{m=1}^{\infty} \sum_{n=1}^{\infty} \frac{\left\{\left(\dfrac{m}{a}\right)^2 + v\left(\dfrac{n}{b}\right)^2\right\} \sin\dfrac{m\pi x}{a} \sin\dfrac{n\pi y}{b}}{mn\left\{\left[\left(\dfrac{m}{b}\right)^2 + \left(\dfrac{n}{b}\right)^2\right]^2 + \dfrac{n_x}{\pi^2 D}\left(\dfrac{m}{a}\right)^2 + \dfrac{n_y}{\pi^2 D}\left(\dfrac{n}{b}\right)^2\right\}},$$

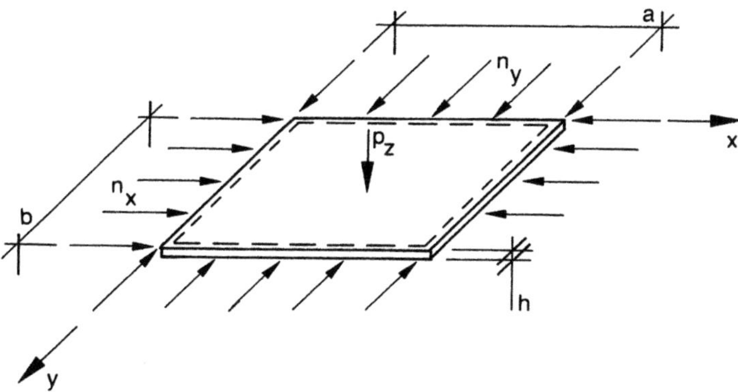

Fig. 4.28. Simply supported plate subject to inplane and transverse loading ($p_z = p_0 = $ const.).

$$m_y = \frac{16p_0}{\pi^4} \sum_{m=1}^{\infty} \sum_{n=1}^{\infty} \frac{\left\{\left(\frac{n}{b}\right)^2 + v\left(\frac{m}{a}\right)^2\right\}\sin\frac{m\pi x}{a}\sin\frac{n\pi y}{b}}{mn\left\{\left[\left(\frac{m}{a}\right)^2 + \left(\frac{n}{b}\right)^2\right]^2 + \frac{n_x}{\pi^2 D}\left(\frac{m}{a}\right)^2 + \frac{n_y}{\pi^2 D}\left(\frac{n}{b}\right)^2\right\}}.$$

We now distinguish between four cases:

Case	n_x	n_y
1	$+\sigma_x h$	$+\sigma_y h$
2	$-\sigma_x h$	$+\sigma_y h$
3	$+\sigma_x h$	$-\sigma_y h$
4	$-\sigma_x h$	$-\sigma_y h$

In Case 1, membrane stresses reduce deflections. The denominator in (4.102) is always non-zero, and positive.

In the three other cases, the denominator may become zero with appropriate choice of membrane stresses, and hence the deflection tends to infinity. To clarify the effect of membrane stresses on deflection, let us consider the freely supported plate subjected to unidirectional inplane and transverse loading, as in Fig. 4.29, where we should have

$$n_y = 0; \quad \alpha = \frac{a}{b} = 1; \quad \beta = \frac{a}{h} = 100; \quad \gamma = \pm\frac{n_x}{p_0 h} = 10^3; \quad \bar{p}_0 = \frac{Dh\pi^6}{16a^4}; \quad p_z = p_0.$$

The deflection is then obtained as

$$\frac{w(x,y)}{h} = \frac{p_0}{\bar{p}_0} \sum_{m=1}^{\infty} \sum_{n=1}^{\infty} \frac{\sin\frac{m\pi x}{a}\sin\frac{n\pi y}{b}}{mn\left\{[m^2 + (\alpha n)^2]^2 \pm \frac{p_0}{\bar{p}_0}m^2\frac{\gamma\pi^4}{\beta^2 16}\right\}}.$$

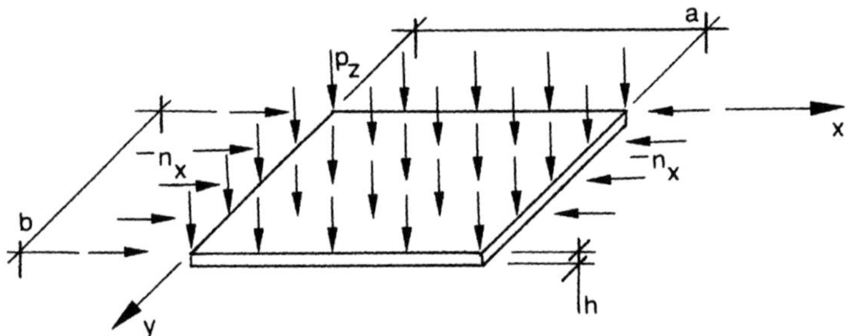

Fig. 4.29. Simply supported plate subject to uniaxial and transverse loading.

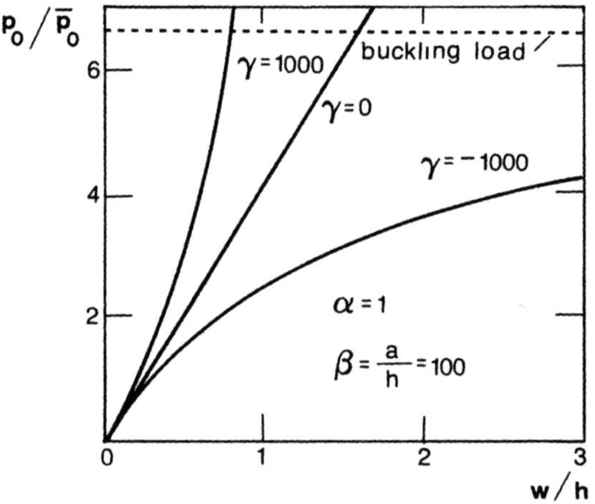

Fig. 4.30. Deflection at the centre of plate with an aspect ratio of $\alpha = 1$

Figure 4.30 shows the deflection at the centre of the plate with load $m = n = 1$. We can see that deflection increases sharply in case of compression, and at the critical buckling load the denominator becomes zero, i.e. deflection becomes infinity. Of course, such a calculation cannot be used in the region of the critical buckling load; the reasons for this have already been given. This is due to the fact that (4.102) does not apply in the region of the critical buckling loads, as the effect of large deformations must then be taken into account. It is of practical interest to study the behaviour of plates with larger aspect ratios, and we choose the factors as follows:

$$n_y = 0; \quad \alpha = \frac{a}{b} = 3; \quad \beta = \frac{a}{h} = 300; \quad \gamma = \frac{n_x}{p_0 h} = \pm 2 \cdot 10^6.$$

Figure 4.31 shows the curve of deflection at $(a/2, b/2)$. We can see that the plate buckles under inplane forces, taking into account that with an aspect ratio of $\alpha = 3$ we have $m = 3$ and $n = 1$. The behaviour depends, of course, on γ. If the transverse load is dominant, it cannot buckle, but if the inplane force is dominant, then a map through effect is to be expected.

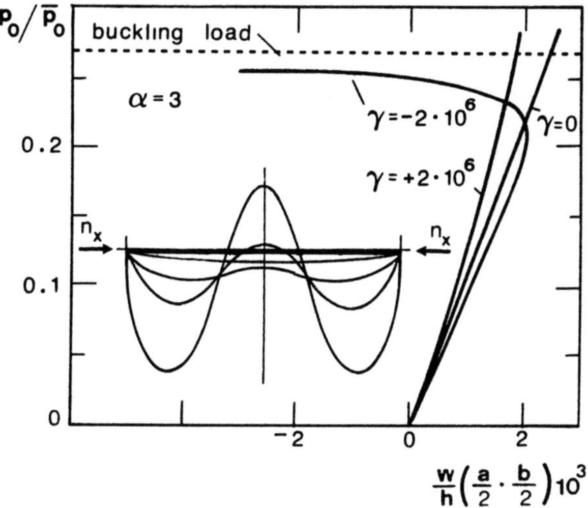

$$\frac{w}{h}\left(\frac{a}{2}\cdot\frac{b}{2}\right)10^3$$

Fig. 4.31. Deflection at the centre of plate with an aspect ratio of $\alpha = 3$.

Even if the deformations are unrealistic in the vicinity of the Euler buckling load, it is still very important to determine the buckling values themselves, because then it can be seen when instabilities are to be expected, independent of the fact that more extensive non-linear theories are based on a knowledge of the bifurcation point. It is therefore necessary to study when the denominator in (4.102) becomes zero. For this we study the different cases of n_x and n_y described above. In this connection case 1 has no meaning, as already pointed out. We set the denominator of (4.102) to zero, which gives the required singularity of the solution:

$$\left[\left(\frac{m}{a}\right)^2+\left(\frac{n}{b}\right)^2\right]^2+\frac{n_x}{\pi^2 D}\left(\frac{m}{a}\right)^2+\frac{n_y}{\pi^2 D}\left(\frac{n}{b}\right)^2 = 0. \tag{4.103}$$

We consider now the stresses given by (4.32) as well as the so-called Euler buckling stress

$$\sigma_e = \frac{\pi^2 D}{hb^2}.$$

This is the critical buckling stress of a strip of a plate with a large aspect ratio. If we substitute $k_x = \sigma_x/\sigma_e$ and $k_y = \sigma_y/\sigma_e$, (4.103) becomes

$$[m^2+(na)^2]^2-(m\alpha)^2 k_x-(n\alpha^2)^2 k_y = 0. \tag{4.104}$$

Figure 4.32 shows the plots of the above equation, for various values of the ratio σ_x/σ_y. Under the assumption that they are compressive stresses, we obtain

$$k_x = k = \frac{[m^2+(\alpha n)^2]^2}{(m\alpha)^2+(n\alpha^2)^2\dfrac{\sigma_y}{\sigma_x}}. \tag{4.105}$$

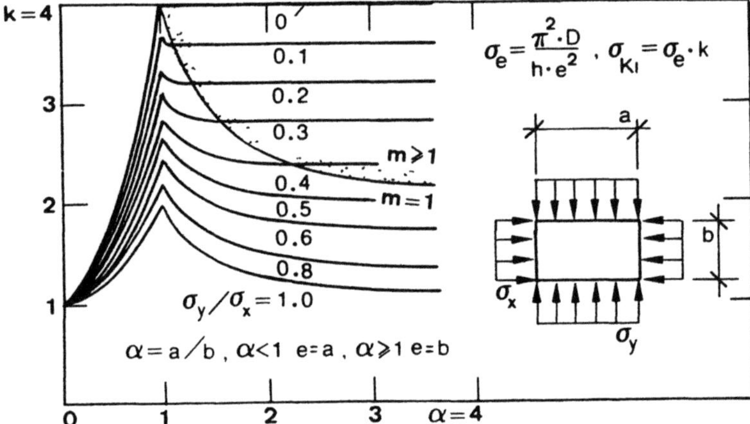

Fig. 4.32. Buckling results of simply supported rectangular plate subject to compressive load in both directions

The ideal buckling stress is then

$$\sigma_{Ki} = \sigma_e k.$$

Figure 4.33 shows another representation of the interaction diagram, where R_x and R_y are the ratios of the critical buckling stress to the uniaxial buckling stress in x and y directions, respectively:

$$R_x = \frac{\sigma_x}{\sigma_{xKi}} \quad \text{or} \quad R_y = \frac{\sigma_y}{\sigma_{yKi}}.$$

We read off from Fig. 4.32

$$\sigma_{xKi} = \sigma_e \cdot 4,$$

and for $k_x = 0, m = 1$ and $n = 1$ we get, from (4.104),

$$\sigma_{yKi} = \sigma_e \frac{(1 + \alpha^2)^2}{\alpha^4},$$

and subsequently

$$R_y + \left(\frac{2m\alpha}{1 + \alpha^2}\right)^2 R_x = \left(\frac{m^2 + \alpha^2}{1 + \alpha^2}\right)^2. \tag{4.106}$$

The safety margin for the buckling load can now be determined easily from the quotient $\overline{OB}/\overline{OA}$ and the distances $\overline{OB}$ and $\overline{OA}$ are as defined in Fig. 4.33. The critical buckling stresses have been obtained for a number of applications, and can be found in various sources (e.g. [5]). In most cases, approximation based on energy is a convenient solution method. Often, several load components have to be considered, making interaction between three or more such components necessary. For example, we have just seen that the interaction between inplane

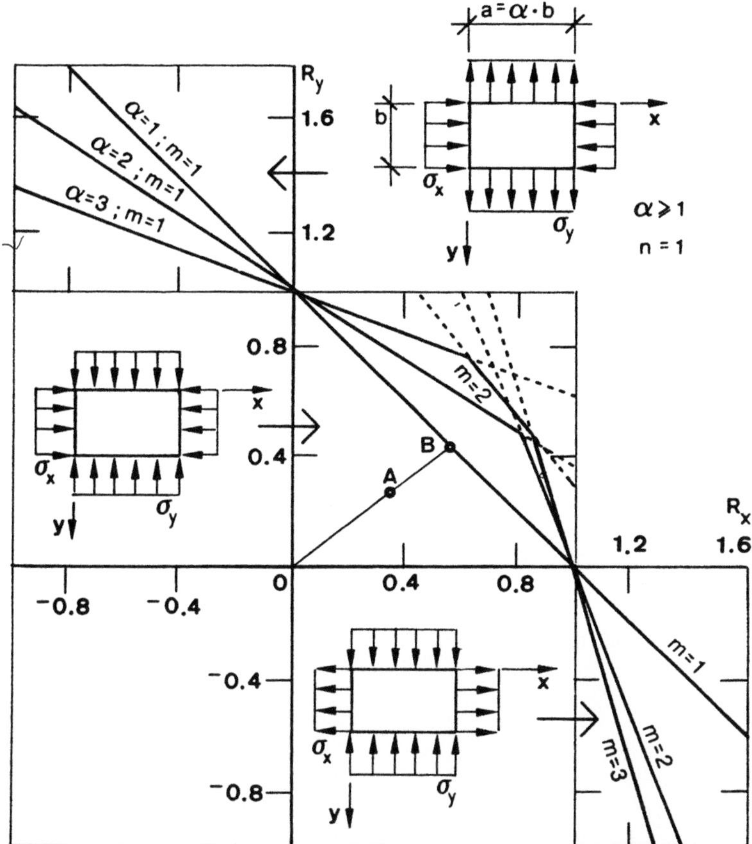

Fig. 4.33. Interaction diagram of buckling stresses for simply supported rectangular plate subject to compressive load in both directions.

stresses gives a straight line, whereas for the combination of shear and constant inplane stresses, we get a parabola. Figure 4.34 shows two typical interactions.

From (4.101) we can see that the inplane forces n_x, n_y, n_{xy} are introduced as constants. We now know, however, that these are not independent of one another due to the laws of membrane theory, and are coupled through the Airy stress function. Taking this into consideration, we obtain the final plate equation

$$D\Delta\Delta w = \left[p_z(x, y) + h\left(\frac{\partial^2 F}{\partial y^2}\frac{\partial^2 w}{\partial x^2} + \frac{\partial^2 F}{\partial x^2}\frac{\partial^2 w}{\partial y^2} - 2\frac{\partial^2 F}{\partial x\,\partial y}\frac{\partial^2 w}{\partial x\,\partial y}\right)\right], \qquad (4.107)$$

which is an equivalent statement of equilibrium. The deflection and the inplane forces are thus coupled. If the deflection is dependent upon the Airy stress function, then it is to be expected that the Airy differential equation will also be dependent on deflection while considering non-linear terms in the compatibility condition. The strain displacement relationships (4.29) will now have additional terms caused

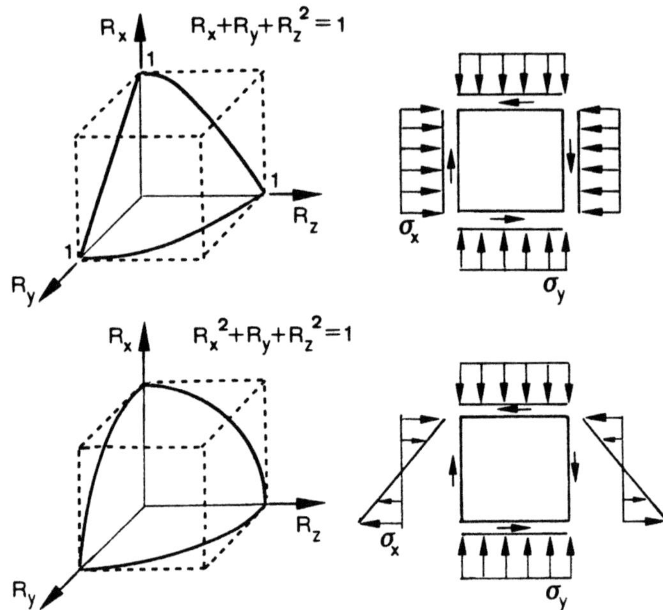

Fig. 4.34. Interaction curves for multi-parameter plate loading.

by deflection

$$\varepsilon_x = \frac{\partial u}{\partial x} + \frac{1}{2}\left(\frac{\partial w}{\partial x}\right)^2, \quad \varepsilon_y = \frac{\partial v}{\partial y} + \frac{1}{2}\left(\frac{\partial w}{\partial y}\right)^2,$$

$$\gamma_{xy} = \frac{\partial u}{\partial y} + \frac{\partial v}{\partial x} + \frac{\partial w}{\partial x}\frac{\partial w}{\partial y}. \tag{4.108}$$

Substitution of these relationships in the compatibility condition (4.30) gives

$$\frac{\partial^2 \varepsilon_x}{\partial y^2} - \frac{\partial^2 \gamma_{xy}}{\partial x \partial y} + \frac{\partial^2 \varepsilon_y}{\partial x^2} = \left(\frac{\partial^2 w}{\partial x \partial y}\right)^2 - \frac{\partial^2 w}{\partial x^2}\frac{\partial^2 w}{\partial y^2}. \tag{4.109}$$

The rest of the derivation follows the linear plate theory, so that finally we obtain the complete differential equation of the plate

$$\Delta\Delta F = \left[\left(\frac{\partial^2 w}{\partial x \partial y}\right)^2 - \frac{\partial^2 w}{\partial x^2}\frac{\partial^2 w}{\partial y^2}\right]\cdot E. \tag{4.110}$$

Equations (4.107) and (4.110) are called the Kármán differential equations, which describe the complete state of displacement, including that above the Euler buckling load. A general solution to these equations has not been found, but approximations have been obtained for a series of plate problems. There is no space here to develop solutions; the relevant literature must be consulted (e.g.

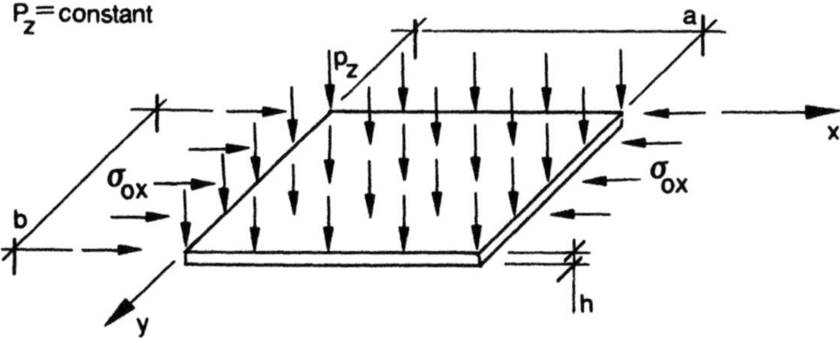

Fig. 4.35. Square plate subject to transverse loading and uniaxial compression ($p_z = p_0 =$ const.).

[17, 18]). To give an idea of the principles involved, a simple solution is derived below, which incorporates the essential features. Let us consider a freely supported square plate subjected to uniaxial compression, as in Fig. 4.35. We assume a sinusoidal function for the displacement field as follows:

$$w(x, y) = w_0 \sin \frac{\pi x}{a} \sin \frac{\pi y}{a}. \tag{4.111}$$

The inplane stress state due to the uniaxial compression can be expressed as a stress function of the form

$$F_0 = -\sigma_{0x} \frac{y^2}{2}.$$

This satisfies the homogeneous part of (4.110). The inhomogeneous differential equation is obtained using (4.111)

$$\Delta\Delta F = w_0^2 E \left(\frac{\pi}{a}\right)^4 \left[\left(\cos\frac{\pi x}{a}\cos\frac{\pi y}{a}\right)^2 - \left(\sin\frac{\pi x}{a}\sin\frac{\pi y}{a}\right)^2\right]. \tag{4.112}$$

With the use of the formula

$$(\cos\alpha\cos\beta)^2 - (\sin\alpha\sin\beta)^2 = \frac{1}{2}(\cos 2\alpha + \cos 2\beta)$$

we obtain

$$\Delta\Delta F = \frac{1}{2} w_0^2 \left(\frac{\pi}{a}\right)^4 E\left(\cos\frac{2\pi x}{a} + \cos\frac{2\pi y}{a}\right), \tag{4.113}$$

and a solution of this is given by

$$F_{\text{inh}} = \frac{1}{32} E w_0^2 \left(\cos\frac{2\pi x}{a} + \cos\frac{2\pi y}{a}\right).$$

We then obtain the complete inplane state of stress as

$$F = F_0 + F_{\text{inh}}, \tag{4.114}$$

or directly in terms of stresses

$$\sigma_x = \frac{\partial^2 F}{\partial y^2} = -\sigma_{0x} - \frac{E}{8} w_0^2 \left(\frac{\pi}{a}\right)^2 \cos\frac{2\pi y}{a},$$

$$\sigma_y = \frac{\partial^2 F}{\partial x^2} = -\frac{E}{8} w_0^2 \left(\frac{\pi}{a}\right)^2 \cos\frac{2\pi x}{a}. \tag{4.115}$$

These additional stress terms are thus dependent on w_0, and ensure that the edges remain straight if a deflection occurs. To solve (4.107) we use Galerkin's procedure

$$\int_0^a \int_0^a \left[D\Delta\Delta w - h\left(\frac{p_0}{h} + \frac{\partial^2 F}{\partial y^2}\frac{\partial^2 w}{\partial x^2} - 2\frac{\partial^2 F}{\partial x\,\partial y}\frac{\partial^2 w}{\partial x\,\partial y} + \frac{\partial^2 F}{\partial x^2}\frac{\partial^2 w}{\partial y^2}\right)\right] dw\,dx\,dy = 0. \tag{4.116}$$

In this expression we substitute (4.111) for the deflection and (4.114) for the stress state, using the same function for the virtual deflection as for the deflection $w(x, y)$ itself, giving

$$\delta w_0 \int_0^a \int_0^a \left[4D\left(\frac{\pi}{a}\right)^4 w_0 \sin^2\frac{\pi x}{a}\sin^2\frac{\pi y}{a} - p_0 \sin\frac{\pi x}{a}\sin\frac{\pi y}{a}\right] dx\,dy$$

$$- \delta w_0 \int_0^a \int_0^a h\left[\left(\sigma_{0x} + \frac{E}{4} w_0^2\left(\frac{\pi}{a}\right)^2 \cos\frac{2\pi x}{a}\right)\right.$$

$$\left. \cdot\left(\frac{\pi}{a}\right)^2 w_0 \sin^2\frac{\pi x}{a}\sin^2\frac{\pi y}{a}\right] dx\,dy = 0.$$

As the virtual displacement δw_0 cannot be equal to zero, the integrals must disappear. Using

$$\sigma_{xK_1} = 4\sigma_e \quad \text{and} \quad \bar{p}_0 = \pi^2 D \frac{h}{16}\left(\frac{\pi}{a}\right)^4,$$

and integrating, we obtain

$$\left(\frac{w_0}{h}\right)^3 - \frac{8}{3}\frac{1}{1-v^2}\frac{w_0}{h}\left(\frac{\sigma_{0x}}{\sigma_{xKi}} - 1\right) - \frac{2}{3}\frac{1}{1-v^2}\frac{p_0}{\bar{p}_0} = 0. \tag{4.117}$$

for the condition that the edges $y = 0$ and $y = b$ must remain straight. If we permit free displacement of these edges in the y-direction, this becomes

$$\frac{1}{2}\left(\frac{w_0}{h}\right)^3 - \frac{8}{3}\frac{1}{1-v^2}\frac{w_0}{h}\left(\frac{\sigma_{0x}}{\sigma_{xK_1}} - 1\right) - \frac{2}{3}\frac{1}{1-v^2}\frac{p_0}{\bar{p}_0} = 0. \tag{4.118}$$

If we also allow the two sides $x = 0$ and $x = a$ to deform freely, we obtain

$$\frac{w_0}{h} - \frac{p_0/\bar{p}_0}{4(\sigma_{0x}/\sigma_{xK_1} - 1)} = 0. \tag{4.119}$$

Figure 4.36 shows the functions (4.117), (4.118) and (4.119) for the case $p_0/\bar{p}_0 = 0$. They thus give information on the behaviour beyond the Euler buckling load σ_{xK_1}.

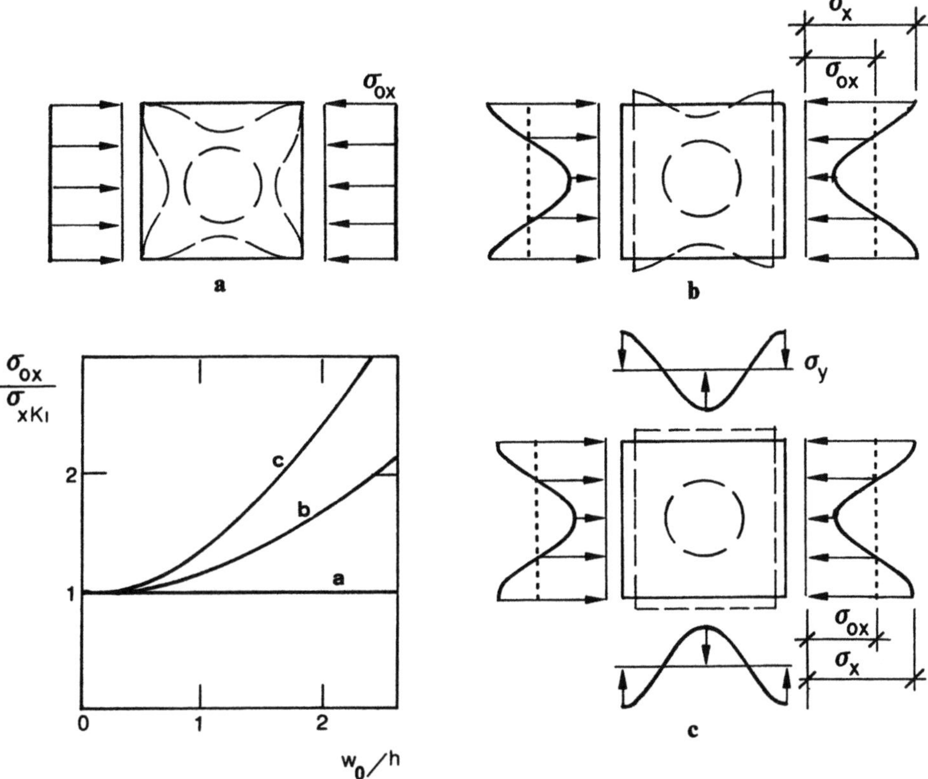

Fig. 4.36. Stresses beyond the Euler load $\sigma_{0x}/\sigma_{xK} \geqslant 1$ for various boundary conditions. **a** All edges free to move, **b** two opposite edges free to move, **c** all edges constrained in the plane of the plate.

We see that the behaviour in the post-buckling range is determined exclusively by the behaviour of the edges in the direction of the membrane. Here lies the difficulty of evaluating these peripheral displacements correctly while analysing a real design, especially when such a plate must be considered as part of a structure which is in general considerably more complex. If an analysis of such a structure is required, it can be done using non-linear finite element calculations. Nevertheless, solutions to the Kármán differential equation allow a deeper insight into the behaviour of the structure beyond the linear buckling load, and thereby enhance our understanding of the results obtained by numerical methods.

Figure 4.37 shows equation (4.117) for various levels of transverse load. It is clear that with a greater transverse loading, the deflection curve is constant, which means, for example, that load-bearing performance will not experience a bifurcation if the plate has an initial deflection. To complement the results of second-order stress theory, Fig. 4.38 compares its results with those obtained from the solution of the Kármán differential equation. For this purpose, (4.117)–(4.119) must be modified, and the parameters γ and β must be introduced. We thus obtain,

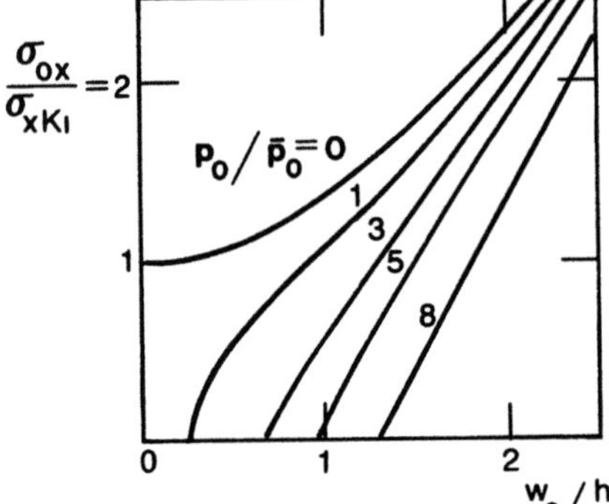

Fig. 4.37. Stress variation in square plate with large deformations under transverse and inplane loading.

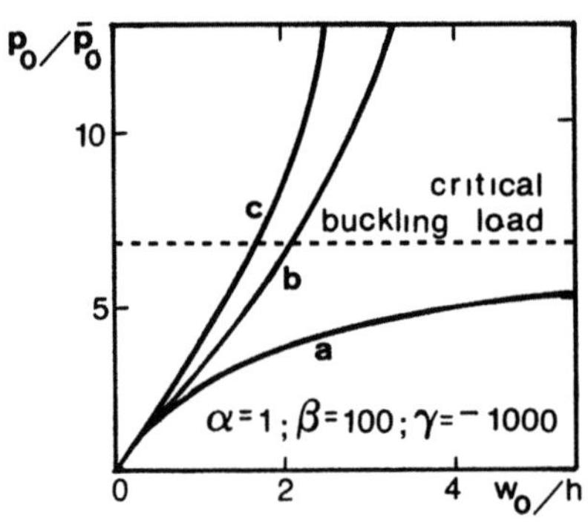

Fig. 4.38. Deflection of square plate for various boundary conditions. **a** All edges free to move, **b** two opposite edges free to move, **c** all edges constrained in the plane of the plate.

respectively,

$$\frac{p_0}{\bar{p}_0}=\frac{\left(\dfrac{w_0}{h}\right)^3+\dfrac{8}{3}\dfrac{1}{1-v^2}\dfrac{w_0}{h}}{\dfrac{2}{3}\dfrac{1}{1-v^2}\left[1-\dfrac{w_0}{h}\dfrac{\gamma}{\beta^2}\left(\dfrac{\pi}{2}\right)^4\right]},\tag{4.120}$$

$$\frac{p_0}{\bar{p}_0}=\frac{\dfrac{1}{2}\left(\dfrac{w_0}{h}\right)^3+\dfrac{8}{3}\dfrac{1}{1-v^2}\dfrac{w_0}{h}}{\dfrac{2}{3}\dfrac{1}{1-v^2}\left[1-\dfrac{w_0}{h}\dfrac{\gamma}{\beta^2}\left(\dfrac{\pi}{2}\right)^4\right]},\tag{4.121}$$

and

$$\frac{p_0}{\bar{p}_0} = \frac{4\dfrac{w_0}{h}}{1 - \dfrac{w_0}{h}\dfrac{\gamma}{\beta^2}\left(\dfrac{\pi}{2}\right)^4}. \tag{4.122}$$

We can see the different characters of the solutions clearly. The results of second-order stress theory are unusable and unrealistic in the region of the critical Euler load (linear buckling load), while the solution of the Kármán differential equation can give realistic results. In this representation we can also see the influence of the membrane boundary conditions on the load deflection relation of plates.

At this point, it is necessary to explain the concept of effective width of plates, which is very important in designing slender structures. It is very often required in the design of marine structures, e.g. in the relevant construction regulations of the Norwegian Classification Society Det norske Veritas. We consider the plane stress of state in the x-direction without transverse loading (Fig. 4.39) at the instant of buckling. The stress state, which has been constant up to this point, changes such that there occurs a reduction in stress at the plate centre associated with an increase in stress at the edges. With the notation in Fig. 4.39, substitution of the stress state (4.115) in (4.117) (without transverse load) gives

$$\sigma_x(y) = -\sigma_{0x} - (\sigma_{0x} - \sigma_{xK_1})\cos\frac{2\pi y}{b}.$$

The effective plate width is given by

$$\frac{b_m}{b} = \frac{\displaystyle\int_0^b \sigma_x(y)\,dy}{b\sigma_{x\,max}}. \tag{4.123}$$

Inserting the expression $\sigma_x(y)$, the stress term gives

$$\frac{b_m}{b} = \frac{\sigma_{0x}}{2\sigma_{0x} - \sigma_{xKi}},$$

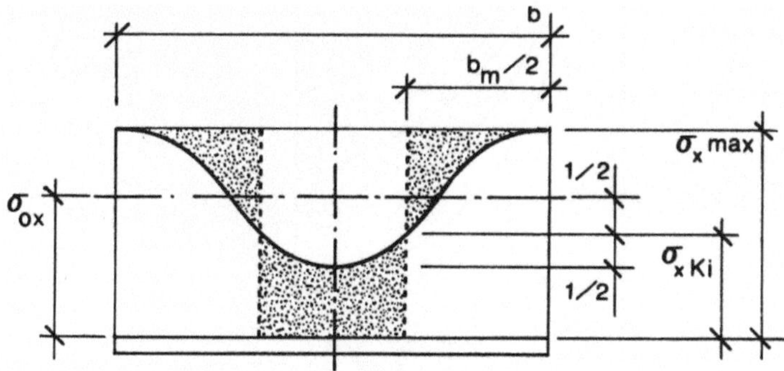

Fig. 4.39. Definition of effective width.

or, with $\Psi = \sigma_{0x}/\sigma_{xKi}$

$$\frac{b_m}{b} = \frac{1}{2 - 1/\Psi}.\tag{4.124}$$

Figure 4.40 presents result (4.124) in addition to the results calculated in [18] for various aspect ratios of plates.

We conclude these two sections on time-independent elastic problems with a simple example of the buckling of a cylindrical shell of length l subjected to axial pressure. We do not have space here for a comprehensive description of even the most important basic equations for shell buckling useful for problems in marine structures; relevant literature must be consulted (e.g. [12, 13, 19]).

We can, however, illustrate a simple special case using the material presented above. If we assume that a buckled shape which is axially symmetrical (i.e. uniform circular buckling), then we can deal with the problem in one dimension. We have established that in the case of axially symmetrical problems, the stress state can be described by the differential equation for beams on elastic foundation. With (4.76) and (4.86) we then obtain the homogeneous differential equation

$$D\frac{d^4w}{dx^4} + n_x\frac{d^2w}{dx^2} + B\frac{1-v^2}{r^2}w = 0.\tag{4.125}$$

A possible solution is

$$w(x) = C\sin\frac{m\pi}{l}x,$$

i.e. m half waves of buckling are assumed in the longitudinal direction of the tube.

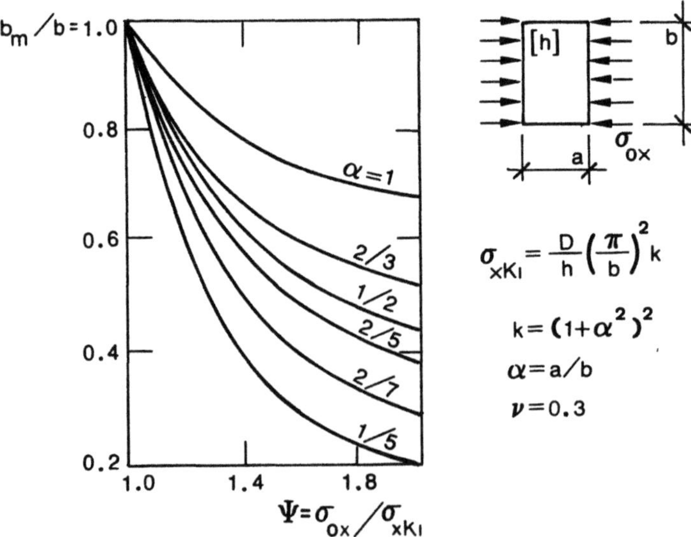

Fig. 4.40. Effective width for plates of various aspect ratios.

We then obtain

$$\sigma_{K_1} = \frac{n_x}{h} = D\left[\frac{1}{h}\left(\frac{m\pi}{l}\right)^2 + \left(\frac{l}{m\pi}\right)^2\frac{E}{Dr^2}\right].$$

(4.126)

Thus with

$$\left(\frac{m\pi}{l}\right)^2 = \sqrt{\frac{Eh}{Dr^2}}$$

the minimum buckling stress is given by

$$\sigma_{K_1} = \frac{E}{\sqrt{3(1-v^2)}}\frac{h}{r}$$

and the corresponding half wavelength is

$$\frac{l}{m} = \pi\sqrt[4]{\frac{r^2h^2}{12(1-v^2)}}.$$

The result (4.126) is important because it is also obtained if chequered pattern buckling is assumed, though this requires a considerably more sophisticated derivation. As already mentioned, it is not possible to give a full representation here of shell buckling theory for marine structures. In any case, it does not seem particularly urgent for another reason: as a rule, the results of linear and non-linear buckling theory cannot be verified experimentally. The reason for this is the extreme sensitivity of the cylindrical shell to imperfections, variable boundary conditions, etc., which make the actual critical buckling stresses considerably lower than theoretically expected. These analytical methods have not found practical use for shell-type structures. The results of buckling theory are indeed used, but they are corrected using empirical factors. We will return to such factors in Chap. 7, and shall present some important practical applications.

4.3 Time-Dependent Elastic Problems

In the preceding sections, we assumed that the loadings on the structural systems are quasi-static in nature. In reality, this applies to only a few cases. In many cases, it is necessary to determine the dynamic characteristics of the structured systems.

4.3.1 Natural Frequencies of Beams and Plates

Let us consider, for example, the freely supported beam (see Fig. 4.41) which undergoes time-dependent motions in the z-direction. We introduce the mass effect as the so-called 'd'Alembert inertia force', and establish the equilibrium of the forces as

$$\frac{\partial Q}{\partial x} = -\left(-\mu\frac{\partial^2 w}{\partial t^2}\right), \quad \frac{dM}{dx} = Q.$$

(4.127)

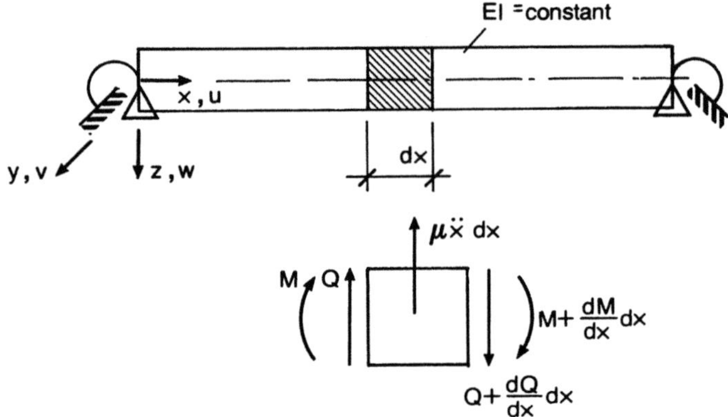

Fig. 4.41. Equilibrium of a vibrating beam.

The differential equation of the bending of the beam is now

$$EI\,\frac{\partial^4 w(x,t)}{\partial x^4} = -\mu\,\frac{\partial^2 w(x,t)}{\partial t^2}, \tag{4.128}$$

where $\mu = \rho A$, A is the beam cross-sectional area, and ρ is the mass density of the material of the beam. We can eliminate time by substituting

$$w(x,t) = w(x)e^{i\omega t},$$

leading to

$$EI\,\frac{d^4 w}{dx^4} - \mu\omega^2 w(x) = 0. \tag{4.129}$$

The angular frequency is defined by $\omega = 2\pi f = 2\pi/T$, f being the frequency in Hz and T the period of oscillation. We then substitute

$$\alpha^4 = \frac{\mu\omega^2}{EI}$$

and obtain

$$\frac{d^4 w}{dx^4} - \alpha^4 w(x) = 0. \tag{4.130}$$

The solution of this homogeneous differential equation is

$$w(x) = C_1 \sinh \alpha x + C_2 \cosh \alpha x + C_3 \sin \alpha x + C_4 \cos \alpha x.$$

The boundary conditions (4.3) produce a homogeneous system of equations for the unknown constants C_1 to C_4. This system has a non-trivial solution only if the determinant of the coefficient matrix vanishes. This leads to the characteristic equation

$$\bar{c}_\varphi \frac{\sinh\dfrac{\alpha l}{2}\cos\dfrac{\alpha l}{2}+\cosh\dfrac{\alpha l}{2}\sin\dfrac{\alpha l}{2}}{\alpha l}+\cosh\frac{\alpha l}{2}\cos\frac{\alpha l}{2}=0. \qquad (4.131)$$

The lowest zero of this equation gives the minimum natural frequency f_c depending on the spring constants. Figure 4.42 shows this graphically.

We can clearly see that the lowest natural frequency of the supported beam is reached only at large values of the spring constant c_φ (dashed line). When choosing the 'fixed' boundary condition, therefore, proper allowance must be made so as not to overestimate this frequency. Figure 4.43 shows a further example of a

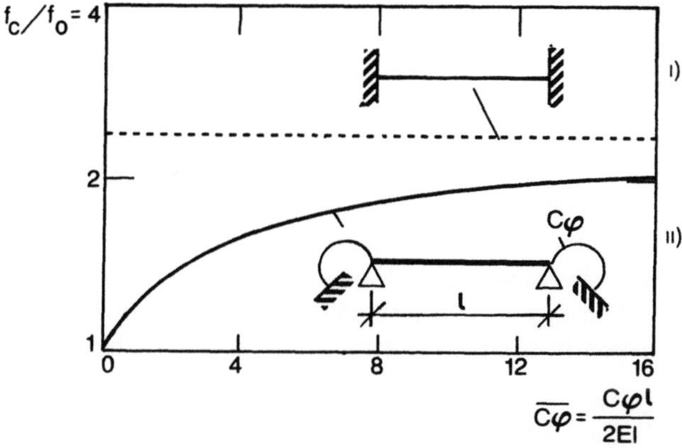

Fig. 4.42. Natural frequencies of ı) rotatıonally constrained and simply supported and iı) clamped–clamped beams. For reference frequency see Fıg. 4.43.

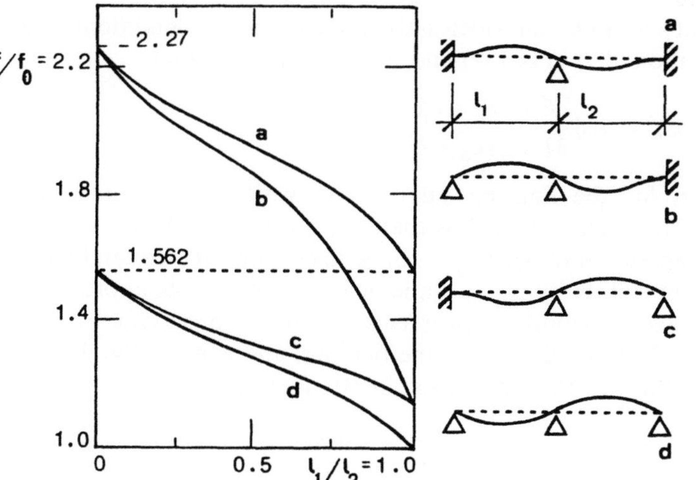

Fig. 4.43. Natural frequencies of two span gırders. Reference frequency $f_0=\dfrac{\pi}{2l^2}\sqrt{\dfrac{EI}{\mu}}.$

two-span girder with various supports and span lengths. The limiting case of a girder free at one end and fixed at the other is found at $l_1/l_2 = 0$ with $f/f_0 = 1.562$. Details for mode shapes for various simple cases can be found in Fig. 4.44. The effect of elastic support can be derived without further calculation from the differential equation for an elastically mounted beam. After expanding (4.129) with the spring constant k, this is

$$EI\frac{\mathrm{d}^4w}{\mathrm{d}x^4} + kw - \mu\omega^2 w = 0. \tag{4.132}$$

This equation can be referred back to (4.130) by taking

$$\alpha^4 = \frac{\mu\omega^2 - k}{EI}.$$

The natural frequency is now

$$f_{\mathrm{be}} = \sqrt{f^2 + \frac{k}{4\pi^2\mu}}, \tag{4.133}$$

f or $f = f_c$ being taken appropriately from Figs. 4.42, 4.43 or 4.44. This means that k increases the natural frequency of the system. The solution is valid only when $\mu\omega^2 \geqslant k$; for the other case, the solution of the beam which is elastically bedded is valid. The approximation methods of Ritz and Galerkin can also be applied, as described in the previous section; e.g. the Galerkin method gives

$$\int_0^{l}\left[EI\frac{\mathrm{d}^4w}{\mathrm{d}x^4} - (\mu\omega^2 - k)w\right]\delta w\,\mathrm{d}x = 0. \tag{4.134}$$

For the displacement field we must now choose a solution which fulfils the boundary conditions. We select the solution (4.9) which applies to the static deflection of the beam elastically supported at both ends. It does not exactly satisfy (4.132), but we can present a formulation which satisfies the boundary conditions. After substitution of the solution (4.9) in (4.134) and integrating we obtain

$$(\alpha l)^4 = 504\frac{6 + 7\bar{c}_\varphi + \bar{c}_\varphi^2}{31 + 11\bar{c}_\varphi + \bar{c}_\varphi^2}.$$

We obtain for the limiting case $\bar{c}_\varphi = 0$, which corresponds to free support, $(\alpha l)^2 = 9.877$, an error of far less than 1% compared with the exact solution of π^2. The same applies to the case $\bar{c}_\varphi = \infty$, i.e. fixed support. We can therefore appreciate the excellent suitability of the static deflection curves as approximation functions for solutions to vibration problems. Another important application is plate vibration. We proceed in the same manner, and expand the differential equation of plate (4.48) with the d'Alembert inertia force

$$\frac{\partial^4 w(x,y,t)}{\partial x^4} + 2\frac{\partial^4 w(x,y,t)}{\partial x^2\,\partial y^2} + \frac{\partial^4 w(x,y,t)}{\partial y^4} = -\frac{\mu}{D}\frac{\partial^2 w(x,y,t)}{\partial t^2}, \tag{4.135}$$

where μ is the mass distribution per unit surface area, $\mu = \rho h$. Time can be

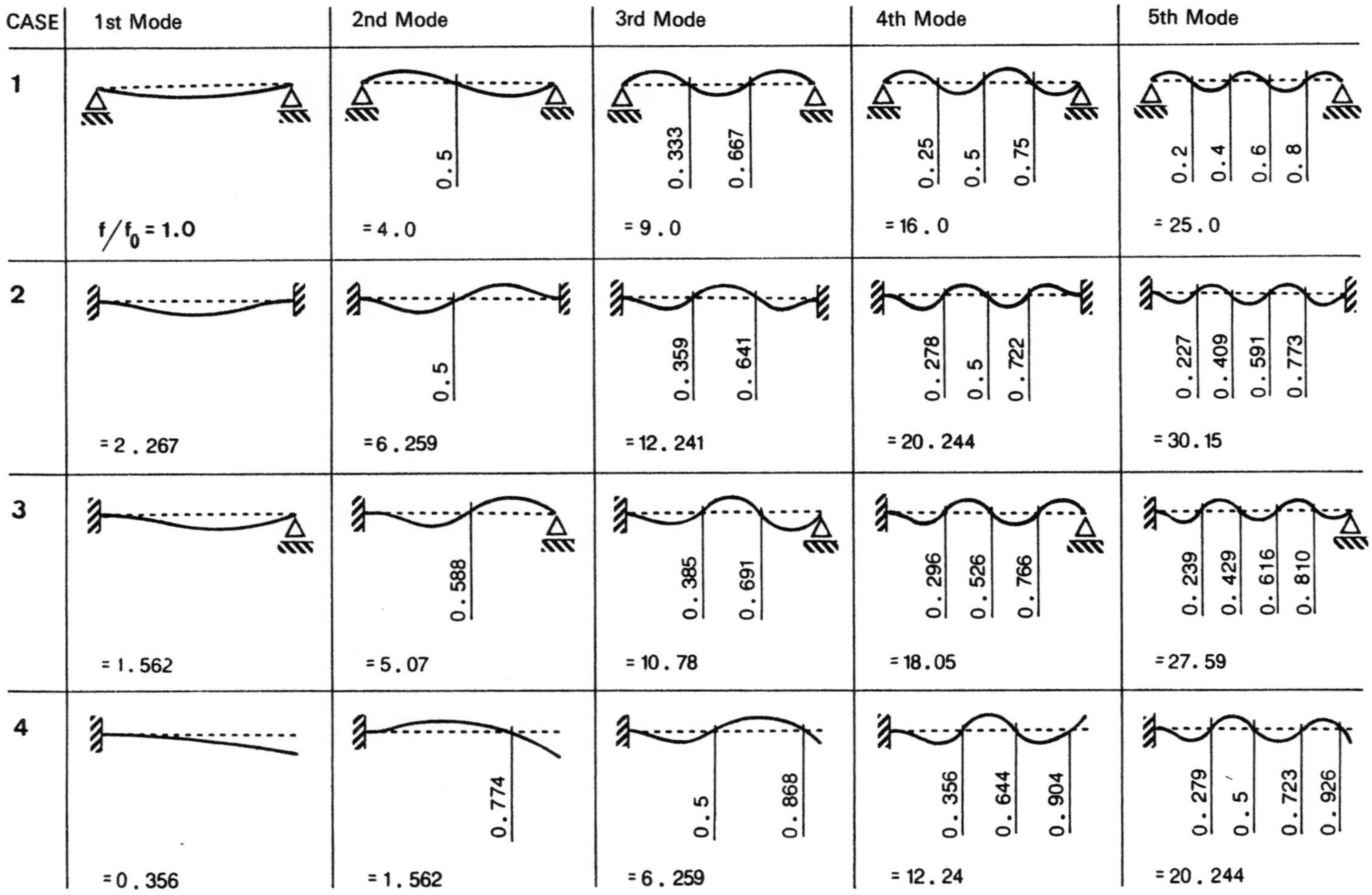

CASE	1st Mode	2nd Mode	3rd Mode	4th Mode	5th Mode
1	$f/f_0 = 1.0$	0.5 = 4.0	0.333 0.667 = 9.0	0.25 0.5 0.75 = 16.0	0.2 0.4 0.6 0.8 = 25.0
2	= 2.267	0.5 = 6.259	0.359 0.641 = 12.241	0.278 0.5 0.722 = 20.244	0.227 0.409 0.591 0.773 = 30.15
3	= 1.562	0.588 = 5.07	0.385 0.691 = 10.78	0.296 0.526 0.766 = 18.05	0.239 0.429 0.616 0.810 = 27.59
4	= 0.356	0.774 = 1.562	0.5 0.868 = 6.259	0.356 0.644 0.904 = 12.24	0.279 0.5 0.723 0.926 = 20.244

Fig. 4.44. Natural frequencies of beam for various boundary conditions. Reference frequency $f_0 = \dfrac{\pi}{2l^2}\sqrt{\dfrac{EI}{\mu}}$.

eliminated with

$$w(x, y, t) = w(x, y)e^{i\omega t}$$

and we obtain

$$\frac{\partial^4 w}{\partial x^4} + 2\frac{\partial^4 w}{\partial x^2 \partial y^2} + \frac{\partial^4 w}{\partial y^4} = \frac{\varrho h}{D}\omega^2 w(x, y). \tag{4.136}$$

For the freely supported plate of length a and width b, we can assume

$$w(x, y) = w_0 \sin\frac{m\pi x}{a}\sin\frac{n\pi y}{b}.$$

After substitution in the differential equation we obtain

$$f_0 = \frac{1}{2\pi}\left[\left(\frac{m\pi}{a}\right)^2 + \left(\frac{n\pi}{b}\right)^2\right]\sqrt{\frac{D}{\varrho h}}$$

and for the fundamental frequency of oscillation

$$f_0 = \frac{\pi}{2}\frac{1}{a^2}\left[1 + \left(\frac{a}{b}\right)^2\right]\sqrt{\frac{D}{\varrho h}}. \tag{4.137}$$

Figure 4.45 shows a number of natural frequencies of plates with different aspect ratios and boundary conditions. It is useful at this point to clarify the dimensions of the various quantities. Using Newton N for the unit of force and the centimetre as the unit of length, the plate flexural rigidity D has the dimension Ncm, and the density ρ is in kg/cm^3 ($7.85\cdot10^{-3}\,kg/cm^3$ for steel). As a force of 1 N corresponds to a mass of 1 kg at an acceleration of $1\,m/s^2$, we obtain the fundamental frequency as

$$f_0 = \frac{1}{a^2[cm^2]}\cdot\sqrt{\frac{D[Ncm]\cdot100[cm/s^2]}{\varrho[kg/cm^3]\cdot h[cm]}} = \left[\frac{1}{s}\right]. \tag{4.138}$$

It is interesting in this connection to study the effect of axial forces on the fundamental frequency of structures. Consider a freely supported beam subjected to an axial force for which (4.86) applies. Taking the mass forces into account, we obtain

$$EI\frac{\partial^4 w(x, t)}{\partial x^4} + N\frac{\partial^2 w(x, t)}{\partial x^2} = -\mu\frac{\partial^2 w(x, t)}{\partial t^2} \tag{4.139}$$

with $\mu = \rho A$. We introduce

$$w(x, t) = e^{i\omega t}w_0 \sin\frac{\pi x}{l}$$

as an approximate solution and obtain

$$EI\left(\frac{\pi}{l}\right)^4 - \mu\omega^2 - \left(\frac{\pi}{l}\right)^2 N = 0.$$

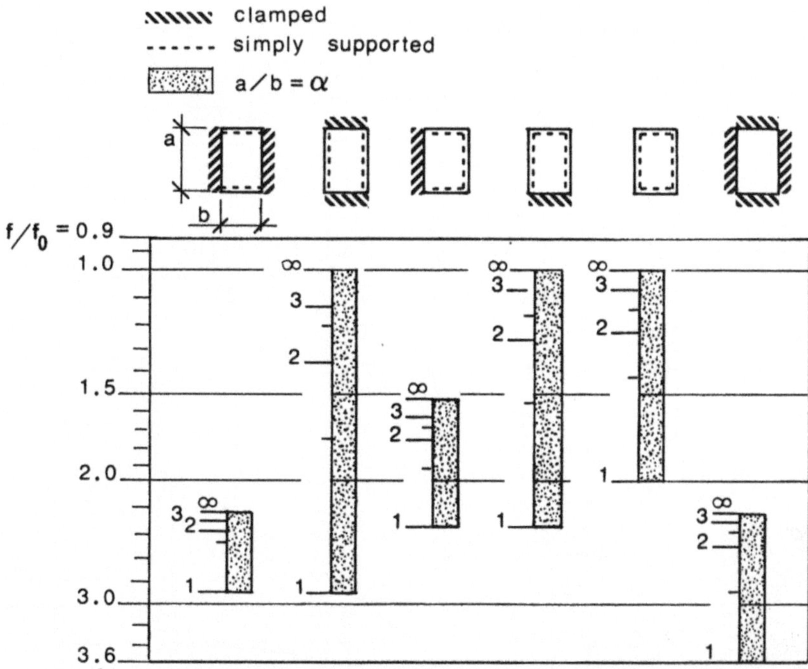

Fig. 4.45. Rectangular-plate natural frequencies f/f_0. Reference frequency $f_0 = \dfrac{h}{b^2}\pi\sqrt{\dfrac{E}{12(l-v^2)\rho}}$ in Hz, $f_0 = \dfrac{h}{b^2}$ in Hz, h in cm, b in cm, for steel.

With N_{Ki} from (4.91) and f_0 (or $\omega_0 = 2\pi f_0$) we obtain the interaction curve

$$\frac{N}{N_{\mathrm{Ki}}} + \left(\frac{\omega}{\omega_0}\right)^2 = 1. \tag{4.140}$$

This means that when the axial force N increases, the square of the fundamental frequency of the system is reduced by the same amount. We can also discuss here the 'Dunkerley straight line'. This relationship is sometimes used in measuring buckling loads in slender structures. The structural loading is changed, and the changes in the natural frequencies are measured. Using (4.140), it is possible to extrapolate to the buckling load. This is a method which is used particularly for large marine lattice-type structures. It is important, however, to check very carefully whether the characteristic shape which belongs to the lowest buckling load and that which belongs to the lowest natural frequency have the same character. The following example, which also demonstrates the application of the energy method, emphasizes this. We assume a freely supported beam with a spring in the centre (Fig. 4.46). The potential

$$\Pi = \frac{1}{2}\int_0^l \left[EI\left(\frac{d^2w}{dx^2}\right)^2 - N\left(\frac{dw}{dx}\right)^2 - \mu\omega^2 w(x)^2 \right] dx + kw^2(l/2) = \text{const.} \tag{4.141}$$

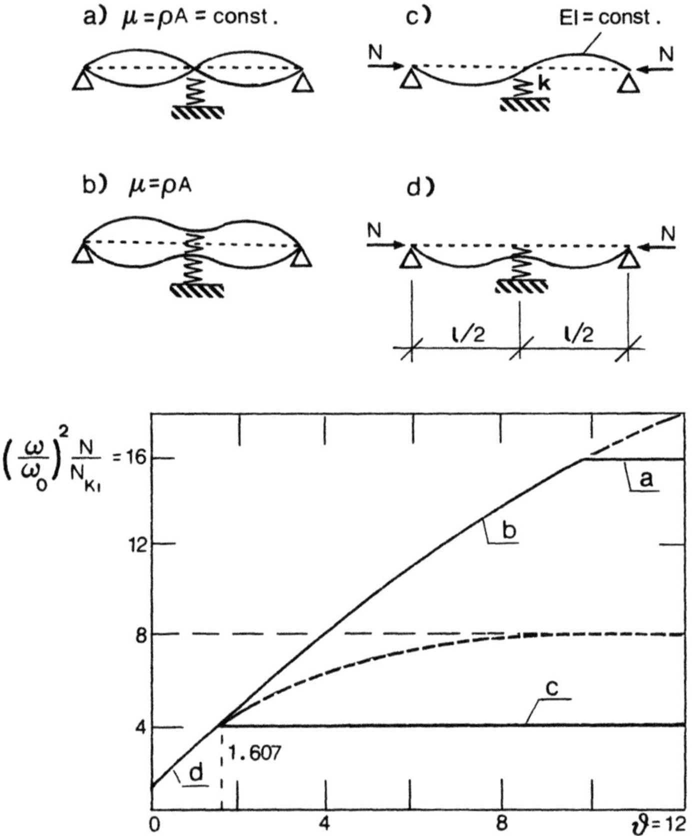

Fig. 4.46. Vibration of simply supported beam with a central spring under axial loading. Circular natural frequency and critical axial force as function of spring stiffness.

applies. We can check this expression with (4.16), but presently we do not want to use the closed form solution, so we substitute

$$w(x) = w_0 \sin\frac{\pi x}{l} + w_1 \sin\frac{3\pi x}{l}$$

in (4.141) and integrate to obtain

$$\Pi = w_0^2\left(\frac{\pi}{l}\right)^4 EI + w_1^2\left(\frac{3\pi}{l}\right)^4 EI + k(w_0 - w_1)^2\frac{2}{l}$$

$$- N\left[w_0^2\left(\frac{\pi}{l}\right)^2 + w_1^2\left(\frac{3\pi}{l}\right)^2\right] - \mu\omega^2(w_0^2 + w_1^2) = \text{const.} \qquad (4.142)$$

The coefficients w_0 and w_1 must be chosen to make the potential a minimum, i.e.

$\partial\pi/\partial w_0 = 0$, $\partial\pi/\partial w_1 = 0$. We then obtain

$$\left[\left(\frac{\pi}{l}\right)^4 EI + \frac{2k}{l} - N\left(\frac{\pi}{l}\right)^2 - \mu\omega^2\right]w_0 - \frac{2k}{l}w_1 = 0,$$

$$-w_0\frac{2k}{l} + \left[\left(\frac{3\pi}{l}\right)^4 EI + \frac{2k}{l} - N\left(\frac{3\pi}{l}\right)^2 - \mu\omega^2\right]w_1 = 0.$$

The determinant of the coefficient matrix must vanish for a non-trivial solution of the system of equations. With

$$\vartheta = \frac{kl^3}{\pi^4 EI}$$

we obtain the characteristic equation

$$81 + 164\vartheta - 90\frac{N}{N_{\text{K}_1}} - 82\left(\frac{\omega}{\omega_0}\right)^2 - 20\vartheta\frac{N}{N_{\text{K}_1}} - 4\vartheta\left(\frac{\omega}{\omega_0}\right)^2$$

$$+ 10\frac{N}{N_{\text{K}_1}}\left(\frac{\omega}{\omega_0}\right)^2 + 9\left(\frac{N}{N_{\text{K}i}}\right)^2 + \left(\frac{\omega}{\omega_0}\right)^4 = 0. \tag{4.143}$$

Next we consider the case $\omega/\omega_0 = 0$, and obtain for $N/N_{\text{K}i} \leqslant 4$ and $\vartheta \leqslant 1.607$

$$\frac{N}{N_{\text{K}_1}} = 5\left(1 + \frac{2}{9}\vartheta\right) - \sqrt{16 - \frac{64}{9}\vartheta + \frac{100}{81}\vartheta^2}. \tag{4.144}$$

The case $N/N_{\text{K}i} = 0$ gives

$$\left(\frac{\omega}{\omega_0}\right)^2 = 41 + 2\vartheta - 2\sqrt{400 + \vartheta^2}, \tag{4.145}$$

for the validity range $(\omega/\omega_0)^2 \leqslant 16$ and $\vartheta \leqslant 9.75$. The case $\vartheta = 0$ gives the familiar interaction curve (4.140). If the spring becomes very stiff, i.e. $\vartheta \geqslant 9.75$, then we have

$$\left(\frac{\omega}{\omega_0}\right)^2 + 4\frac{N}{N_{\text{K}_1}} = 16. \tag{4.146}$$

Figures 4.46 and 4.47 illustrate the situation. Up to a spring stiffness of $\vartheta = 1.607$ there is only one natural mode, irrespective of what combination of angular fundamental frequency and axial force exists. If the spring stiffness is $1.607 \leqslant \vartheta \leqslant 9.75$, then different natural modes can exist, depending on the magnitude of the axial force. Only when the spring stiffness increases does the mode shape become antisymmetric in every case.

As a result, we can state that the condition of comparable mode shapes for vibration and buckling is a prerequisite for the validity of a linear interaction curve.

4.3.2 Forced Oscillations

Having dealt with the problem of free undamped oscillations, the result of which was the search for the natural frequencies and natural modes, we now examine

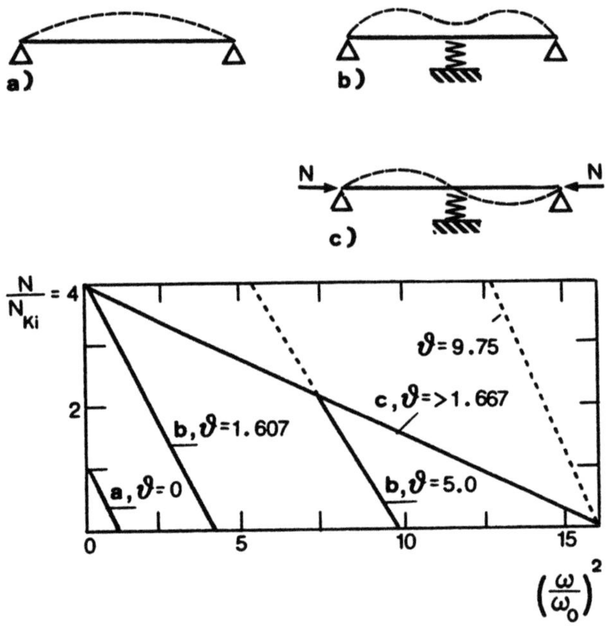

Fig. 4.47. Interaction curves of simply supported beam with a central spring.

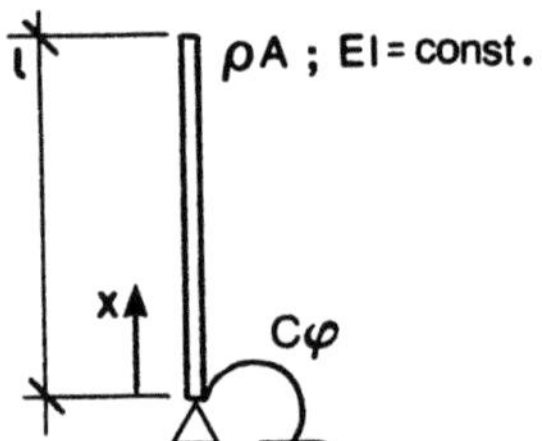

Fig. 4.48. Cantilever beam with rotational elastic restraint

the problem of forced oscillations without damping. Here we consider a cantilever girder supported either in a fixed or a torsionally elastic manner (Fig. 4.48).

The natural modes and frequencies are first determined in accordance with the earlier derivation, and the following frequency equation is obtained:

$$(1 + \cos\alpha l \cosh\alpha l) - \Psi l(\sin\alpha l \cosh\alpha l - \cos\alpha l \sinh\alpha l) = 0. \qquad (4.147)$$

$\Psi = (EI|c_\varphi|)\cdot\alpha l$ gives the natural frequencies which are shown against the spring restraint in Fig. 4.49. It is interesting to observe the effect of the torsional restraint on the natural frequency. While it is extremely large for the fundamental frequency, it has hardly any effect on higher ones. This becomes clear immediately, as in the case $c_\varphi = 0$ the system would undergo a rigid body rotation, which is associated with a frequency of zero.

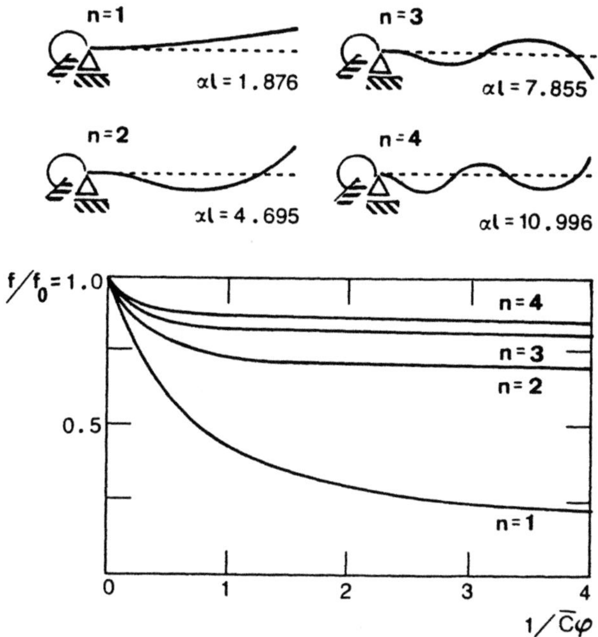

Fig. 4.49. Natural frequencies of elastically mounted cantilever beam. Reference resonant frequency
$$f_0 = \left(\frac{\alpha l}{2\pi}\right)^2 \frac{1}{l^2}\sqrt{\frac{EI}{\mu}}.$$

Examining the case of a harmonic excitation given by

$$w(x, t) = w(x)\sin\Omega t$$

we obtain, from (4.128),

$$EI\frac{\mathrm{d}^4 w}{\mathrm{d}x^4} - \varrho A\Omega^2 w(x) = 0. \tag{4.148}$$

We thus formally obtain the same differential equation as for free oscillation. The frequency of excitation Ω thus replaces the natural frequencies. With

$$\beta^4 = \frac{\varrho A\Omega^2}{EI}$$

we get

$$\frac{\mathrm{d}^4 w}{\mathrm{d}x^4} - \beta^4 w(x) = 0, \tag{4.149}$$

with the solution

$$w(x) = C_1 \sinh\beta x + C_2 \cosh\beta x + C_3 \sin\beta x + C_4 \cos\beta x.$$

We obtain constants C_1 to C_4 from the boundary conditions

$$w = w(x)\sin\Omega t|_{x=0}, \quad \frac{\mathrm{d}w}{\mathrm{d}x} = -\frac{EI}{c_\varphi}w''|_{x=0},$$

$$\frac{\mathrm{d}^2w}{\mathrm{d}x^2} = 0|_{x=l}, \quad \frac{\mathrm{d}^3w}{\mathrm{d}x^3} = 0|_{x=l}. \tag{4.150}$$

Substituting the coefficients C_1 into the solution of (4.149) gives the variation of

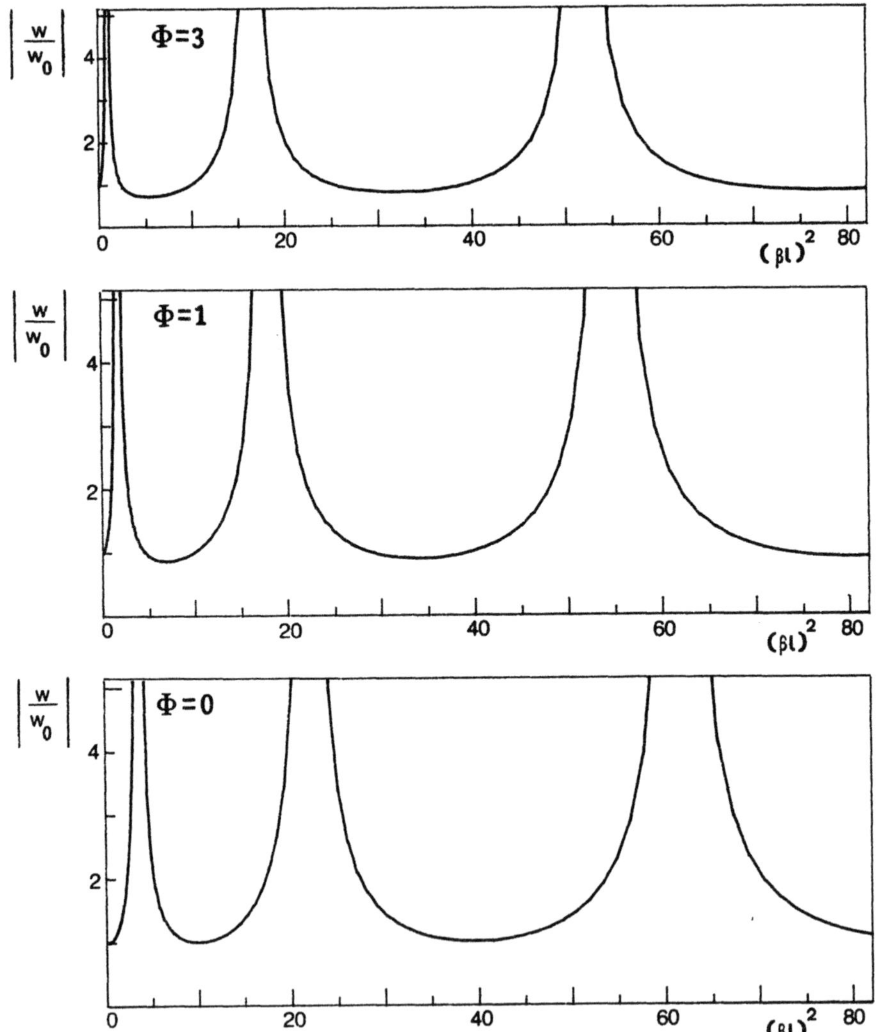

Fig. 4.50. Forced vibration of cantilever beam with elastic restraint $w =$ amplitude at point $x = 1$. Parameter $\Phi = \dfrac{EI}{c_\varphi l}\beta l$. $(\beta l)^2$ is the dimensionless excitation frequency

the amplitude at the free end of the cantilever beam, with the maximum deflection at the support $w_0 = w(0)$ against the dimensionless excitation frequency $(\beta l)^2$ (Fig. 4.50). The constant ϕ affects the first resonance point most severely, and the higher resonance points are nearly independent of ϕ. Apart from the fact that in the immediate neighbourhood of the resonance points, deflections are influenced by damping, the plots of Fig. 4.51 give a realistic picture of the phenomena. Of course, the damping of the system must be taken into consideration. To avoid complicating the problem further, let us model the system in Fig. 4.48 into a system with a single degree of freedom, i.e. a system consisting of a mass and a spring. The spring constant of the equivalent spring is then $k = 3EI/l^3$. If we demand that the lowest natural frequency of the system with a distributed mass be equal to that of a system consisting of a single mass, we obtain the value of the equivalent mass as

$$m = 0.242 \frac{\varrho A l}{1 + \dfrac{3}{\bar{c}_\varphi}}. \tag{4.151}$$

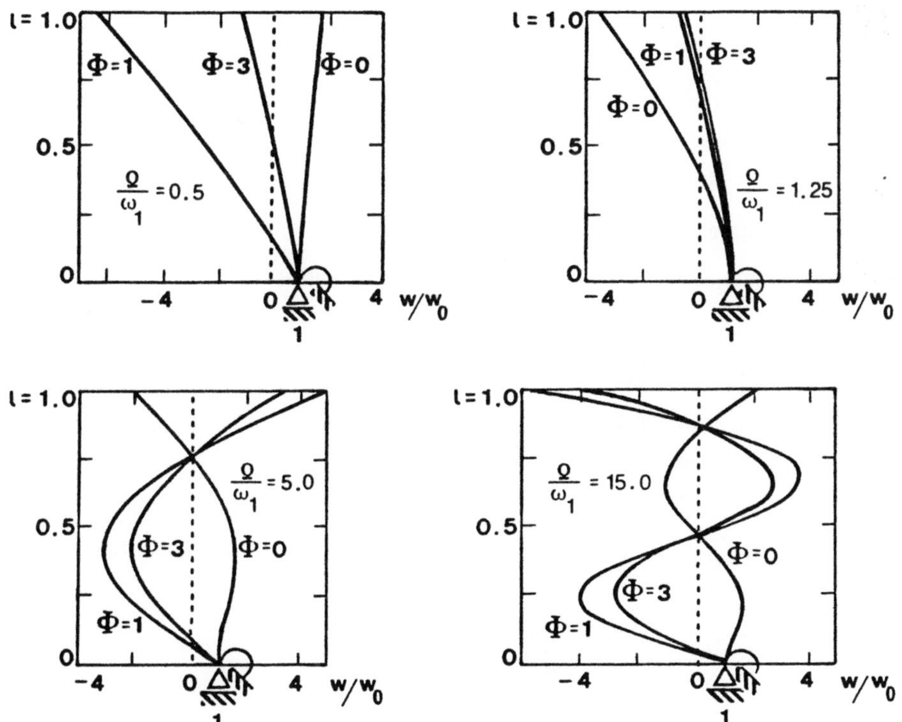

Fig. 4.51. Modes of vibration of undamped cantilever beam with elastic restraint for various excitations. Parameter $\Phi = \dfrac{EI}{c_\varphi l} \beta l$. Reference circular frequency $\omega_1 = (\alpha l) \dfrac{1}{l^2} \sqrt{\dfrac{EI}{\mu}}$; $\alpha l = 1.876$.

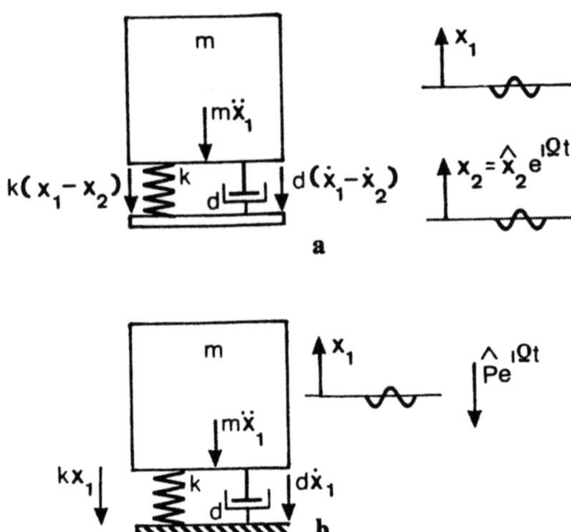

Fig. 4.52. Damped single mass oscillator, **a** with support excitation, **b** with mass excitation.

We can now study, in an elementary manner, the behaviour of the simplified system in the region of the lowest natural frequency, taking damping into account. Here we want to compare the behaviour of such a system with support excitation, as well as with excitation of the mass point. Figure 4.52 shows the two systems. For the system in Fig. 4.52a we obtain the differential equation

$$m\ddot{x}_1 + d(\dot{x}_1 - \dot{x}_2) + k(x_1 - x_2) = 0. \tag{4.152}$$

We rewrite this equation as

$$m(\ddot{x}_1 - \ddot{x}_2) + d(\dot{x}_1 + \dot{x}_2) + k(x_1 - x_2) = -m\ddot{x}_2 \tag{4.153}$$

or

$$m\ddot{x}_1 + d\dot{x}_1 + kx_1 = d\dot{x}_2 + kx_2. \tag{4.154}$$

We choose the solution form as

$$x_1 = \hat{x}_1 e^{(i\Omega t + \varphi)}, \tag{4.155}$$
$$x_2 = \hat{x}_2 e^{i\Omega t}.$$

After substitution in (4.154), the absolute value of the amplitude ratio is obtained as

$$\left| \frac{\hat{x}_1}{\hat{x}_2} \right| = \sqrt{\frac{1 + (2D\eta)^2}{(1 - \eta^2)^2 + (2D\eta)^2}}, \tag{4.156}$$

where

$$D = \frac{d}{2m\omega} = \frac{d}{2\sqrt{mk}}; \eta = \frac{\Omega}{\omega}.$$

The solution of (4.153) is interesting. If we substitute the complex solutions (4.155) into it, we obtain

$$\left|\frac{\hat{x}_1 - \hat{x}_2}{\hat{x}_2}\right| = \frac{\eta^2}{\sqrt{(1 - \eta^2)^2 + (2D\eta)^2}}.$$ (4.157)

With known oscillation parameters Ω/ω or D, this solution can be used to determine the movement x_2 by measuring relative movement $|\hat{x}_1 - \hat{x}_2|$, which is easily accomplished in practice, and thereby inferring the value of $\hat{x}_2$ (seismic principle). The equilibrium differential equation

$$m\ddot{x}_1 + d\dot{x}_1 + kx_1 = \hat{P}e^{i\Omega t}$$ (4.158)

applies to the system shown in Fig. 4.52b.

With (4.155) we obtain

$$\left|\frac{\hat{x}_1}{x_{St}}\right| = \frac{1}{\sqrt{(1 - \eta^2)^2 + (2D\eta)^2}},$$ (4.159)

with $x_{St} = \hat{P}/k$. Now it is not so much the deflections which are important, but the forces which are introduced into the foundation via the spring. We can find this force if we compare the right-hand sides of (4.154) and (4.158):

$$d\dot{x}_1 + kx_1 = Pe^{i\Omega t}$$

to give

$$|P| = \hat{x}_1 k\sqrt{1 + (2D\eta)^2}.$$

We use index 1 as we are referring to the system in Fig. 4.52b. From (4.159) we then obtain the value of the ratio of the force which is transmitted into the foundation to the force of the static loading as

$$\left|\frac{P}{\hat{P}}\right| = \sqrt{\frac{1 + (2D\eta)^2}{(1 - \eta^2)^2 + (2D\eta)^2}}.$$ (4.160)

This is the same transfer function as the ratio of the amplitudes with support excitation.

We will conclude this section with the 2-mass oscillator, which is to be used as a model for a number of applications. The system is shown in Fig. 4.53. The equations of motion are

$$m_1\ddot{x}_1 + d_2(\dot{x}_1 - \dot{x}_2) + k_2(x_1 - x_2) + k_1 x_1 = \hat{P}_1 e^{i\Omega t},$$
$$m_2\ddot{x}_2 - k_2 x_1 + k_2 x_2 = 0.$$ (4.161)

First we determine the natural frequencies of the system by considering the homogeneous part of the equations without taking the damping into account. Setting the determinant of the coefficient matrix to zero gives the natural frequencies of the system as

$$\omega^2_{1,2} = \frac{1}{2}\left[\left(\frac{k_1 + k_2}{m_1} + \frac{k_2}{m_2}\right) \pm \sqrt{\left(\frac{k_1 + k_2}{m_1} + \frac{k_2}{m_2}\right)^2 - 4\frac{k_1 k_2}{m_2 m_1}}\right].$$ (4.162)

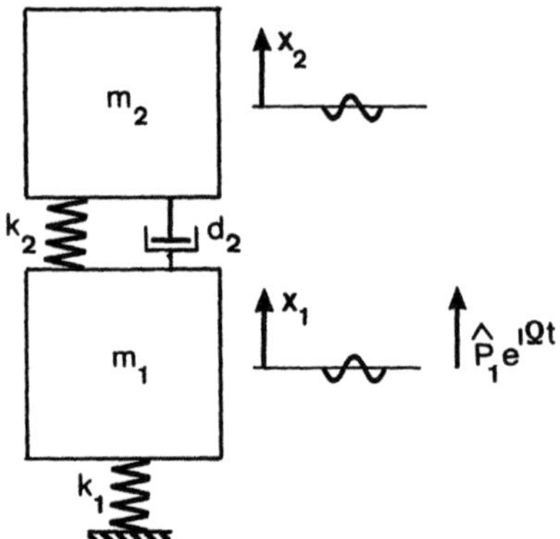

Fig. 4.53. Damped twin-mass oscillator.

With

$$\bar{\omega}_1^2 = \frac{k_1}{m_1}; \quad \bar{\omega}_2^2 = \frac{k_2}{m_2}; \quad \mu = \frac{m_2}{m_1}$$

we obtain

$$\omega_{1,2}^2 = \frac{1}{2}\{[\bar{\omega}_1^2 + (1+\mu)\bar{\omega}_2^2] \pm \sqrt{[\bar{\omega}_1^2 + (1+\mu)\bar{\omega}_2^2]^2 - 4\bar{\omega}_1^2\bar{\omega}_2^2}\}.$$

Taking the solution of (4.161) as given by (4.155) we obtain

$$\left|\frac{\hat{x}_1}{x_{St}}\right| = \sqrt{\frac{(v^2 - \eta_1^2)^2 + (2D_2\eta_1 v)^2}{[\eta_1^4 - \eta_1^2(1 + v^2 + \mu v^2) + v^2]^2 + [2D_2\eta_1 v(1 - \eta_1^2(1+\mu))]^2}}$$

$$(4.163)$$

and

$$\left|\frac{\hat{x}_2}{x_{St}}\right| = \sqrt{\frac{v^4 + (2D_2\eta_1 v)^2}{[\eta_1^4 - \eta_1^2(1 + v^2 + \mu v^2) + v^2]^2 + [2D_2\eta_1 v(1 - \eta_1^2(1+\mu))]^2}}$$

$$(4.164)$$

where

$$v^2 = \left(\frac{\bar{\omega}_2}{\bar{\omega}_1}\right)^2; \quad D_2 = \frac{d_2}{2\sqrt{m_2 k_2}}; \quad \eta_1 = \frac{\Omega}{\bar{\omega}_1}; \quad x_{St} = \frac{\hat{P}_1}{k_1}.$$

A detailed discussion of this system can be found in [20]. We delineate (4.163) and (4.164) with the parameters $\mu = 0.1$ and $v = 1$ (Fig. 4.54).

We recognize that with zero damping the deflection of the mass m_1 disappears at $\eta = 1$. This is the familiar cancellation effect. We can use this effect to design an oscillation damper. Flow around tubular structures leads, as we know, to eddy separation, the physics of which is described in Chapter 3. If the vortex frequency is near a natural frequency of the tubular steel structure, resonance conditions

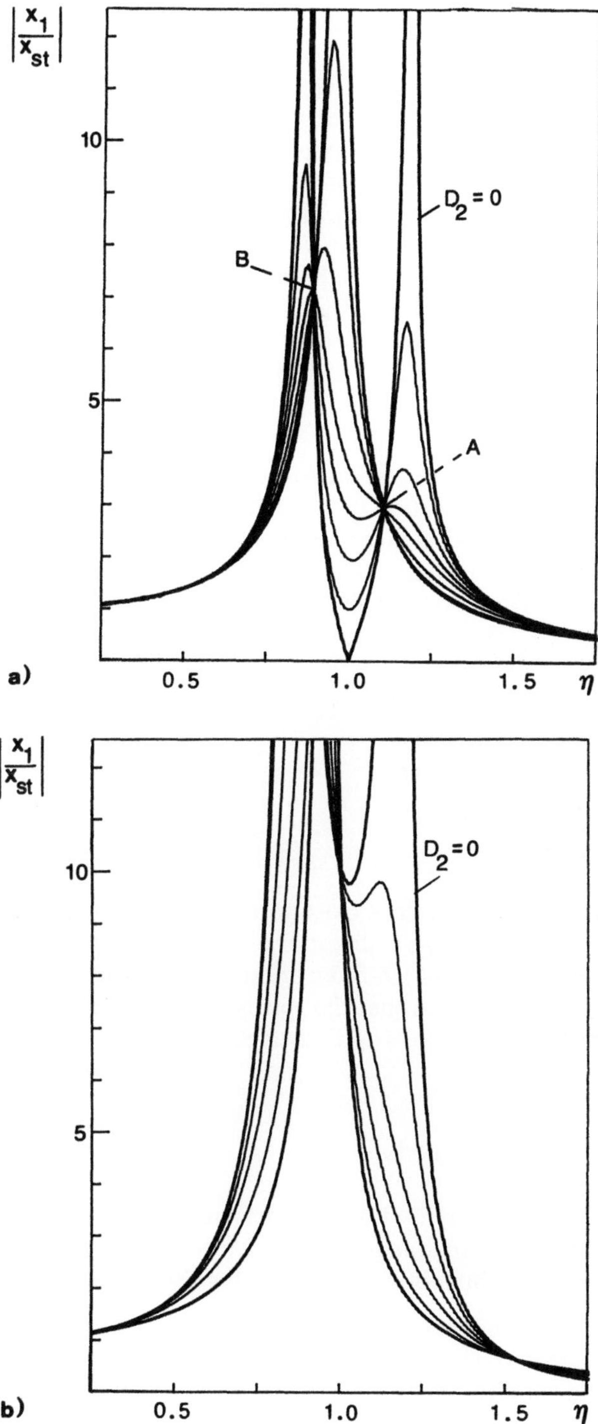

Fig. 4.54. Behaviour of twin-mass oscillator. **a** mass m_1, **b** mass m_2. Parameter, damping constant D_2.

occur. If an additional damped mass is applied in the region of maximum deflection, these oscillations can be damped to a negligible level by intelligent selection of parameters. Let us consider the steel structure of the semisubmersible structure shown in Fig. 4.1a. Vibration analysis can give us a relatively low resonant vibration with a distinct oscillation of the horizontal connecting tube between the two pontoons. When this semisubmersible structure is towed, a vortex-excited resonance occurs. To make the correct selection of parameters for the damper, the system first has to be reduced to a single-mass oscillator. The natural frequency of the oscillator is therefore made equal to the natural frequency of the tube which is to be damped, obtained from a global free vibration analysis of the semisubmersible structure:

$$m_{\text{Er}} = \frac{k}{\omega^2}. \tag{4.165}$$

Here we must use a static calculation to determine the spring stiffness k of the structure at the point of application of the damper in the direction of the deflections to be damped. From Fig. 4.43a we can see that the curves for four different damping values always go through points A and B. The ratio of the masses μ and the frequencies ν are therefore selected so that the intersections at A and B are at the same level (Fig. 4.55a). We achieve this by selecting

$$\nu = \frac{1}{1+\mu}. \tag{4.166}$$

The associated optimal damping is

$$D = \sqrt{\frac{\mu}{2}\left(1 - \frac{1}{2}\mu\right)}. \tag{4.167}$$

Tests have shown that even values close to optimal damping can give extremely good results. Figure 4.56 shows a diagram of the design of such a damper.

In general, damper masses of 10–50 kg are sufficient. When building these simple devices, it must be remembered that the springs are subject to fatigue, and careful consideration must be given to fatigue strength.

4.4 Ultimate Load Analysis

In addition to the stress distribution and deformation for a given load, it is also important to determine which loadings will cause failure of the structure or lead to its unserviceability. The following limit states should be considered:

— Stability failure in the elastic region.
— Plastic failure through reaching the limiting load without stability phenomena.
— Elastic/plastic stability failure.
— Fatigue due to crack formation.
— Occurrence of unserviceability (e.g. unacceptable levels of deformation).

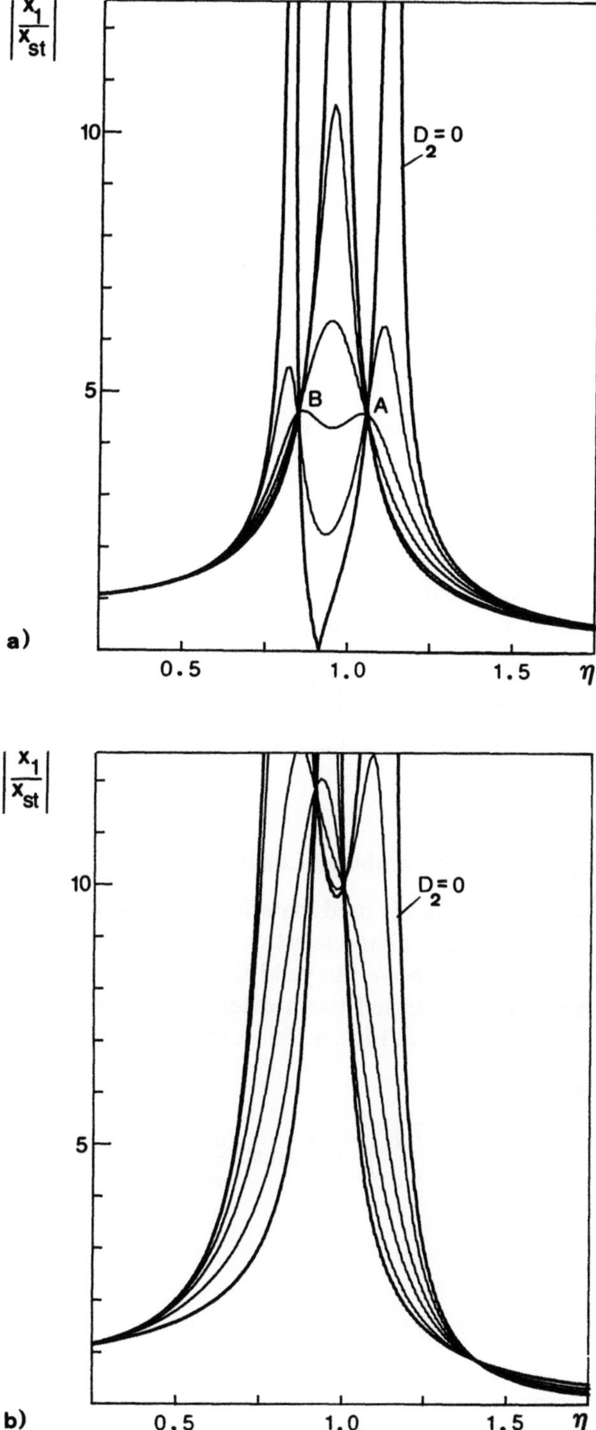

Fig. 4.55. Behaviour of optimized twin-mass oscillator. **a** mass m_1, **b** mass m_2. Parameter, damping constant D_2.

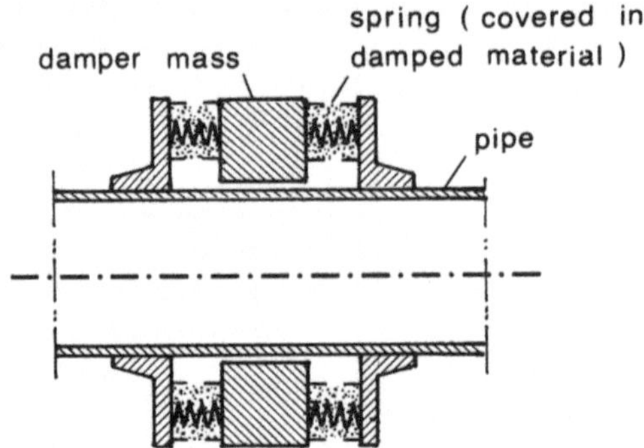

Fig. 4.56. Example of a vibration damper for preventing vortex-excited oscillations [DE Pat. 3, 433, 397 C2].

We have dealt with stability phenomena in Sect. 4.2, which are generally applicable for thin members. Such ideal stability failures do not normally occur in actual structures, because, in general, perturbations (e.g. due to constructional inaccuracies, eccentric load applications, or unexpected subsidiary stress conditions) cannot be avoided, giving rise to stress problems. When the point of maximum stress reaches the yield limit, the load-bearing limit has generally not been reached. It is therefore important to study the behaviour of load-bearing structures above the elastic load limit.

4.4.1 Plastic Capacity under Combined Loading

Steel, which is the most commonly used material in marine structures, has distinctive linear-elastic and ideal-plastic characteristics (Fig. 4.57), and therefore we do not need to consider hardening above the yield limit.

Let us consider a tube subjected to an external bending moment. Three stress states are important, as shown in Figs. 4.58a–c. Integrating the stress state (b) gives

$$M = W_{ep}\sigma_F,$$

$$W_{ep} = d^2 h \frac{1}{2}\left\{ \sqrt{1 - \left(\frac{d'}{d}\right)^2} + \frac{d}{d'}\arcsin\frac{d'}{d} \right\}. \tag{4.168}$$

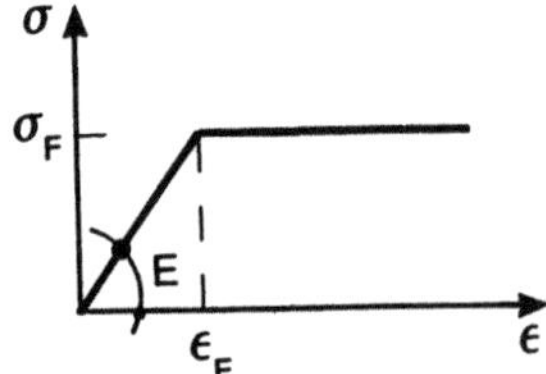

Fig. 4.57. Ideal elastic–plastic material law.

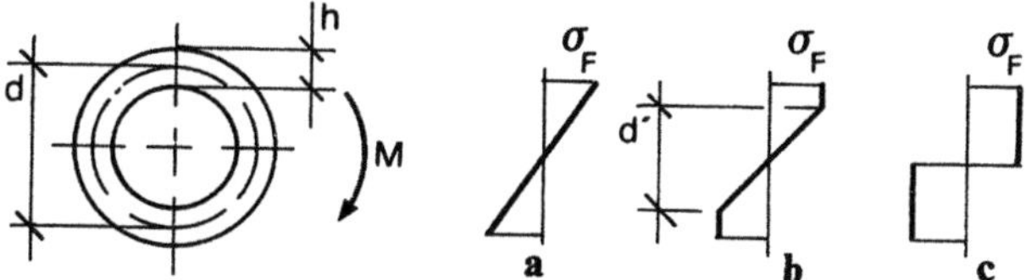

Fig. 4.58a–c. Stress distribution in tubular cross-section.

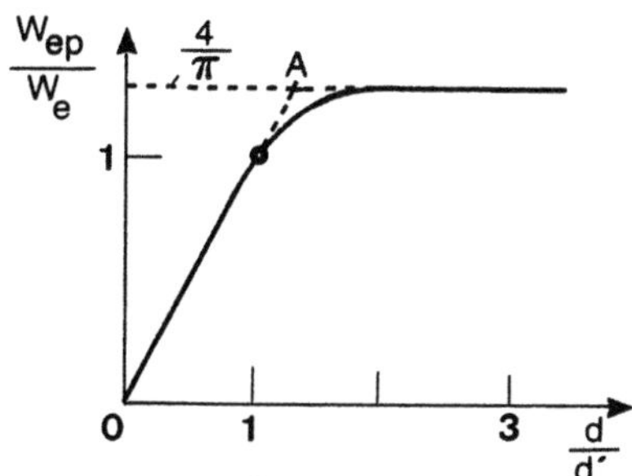

Fig. 4.59. Section modulus as a function of depth of the plastic zone.

We obtain the purely elastic limit, case (a), when $d' = d$

$$W_e = d^2 h \frac{\pi}{4} \tag{4.169}$$

and the fully plastic limits, case (c), when $d' = 0$

$$W_p = d^2 h. \tag{4.170}$$

The so-called 'plastic reserve' is therefore $\alpha = W_p/W_e = 4/\pi = 1.27$. Function (4.168) is shown non-dimensionally in Fig. 4.59 for the region $d/d' \geqslant 1$.

The fully-plastic moment is reached in practice, even when $d/d' \geqslant 2$. Calculations for complex structures can be simplified considerably if we assume linearity until point A, i.e. until the point where the fully-plastic moment is reached. Then, a deviation from linearity is interpreted in that no further increase in load can be resisted by the structure. Otherwise, an increase in load can only occur by its redistribution to areas which are yet to be loaded to their elastic limit. To fully evaluate the load-bearing limit of a tubular cross-section, axial as well as transverse forces must be considered in addition to moments. We therefore consider a fully plastic tubular cross-section under the influence of an axial force and a moment, as in Fig. 4.60. The axial force is obtained by

$$N = 2\sigma_F dh\varphi_0.$$

The fully plastic axial force N_p is equal to the elastic load limit, i.e. $N_F = N_p$. For the case $M = 0$ we find

$$N_p = \pi\sigma_F dh \tag{4.171}$$

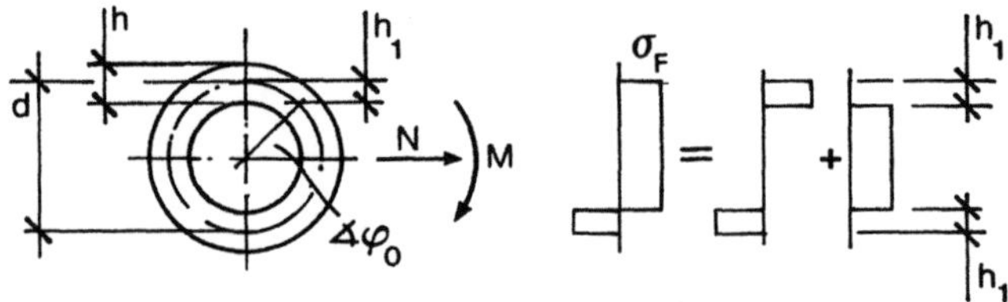

Fig. 4.60. Fully-plastic stress distribution in tube subject to axial load and bending moment.

and thus

$$\varphi_0 = \frac{\pi}{2} \frac{N}{N_p}.$$

We find the bending moment by integrating the stress state between $\varphi = \varphi_0$ and $\varphi = \pi/2$

$$M'_p = \sigma_F d^2 h \cos \varphi_0.$$

With the full plastic moment $M_p = W_p \sigma_F$ we find

$$\frac{M'_p}{M_p} = \cos\left(\frac{\pi}{2} \frac{N}{N_p}\right). \tag{4.172}$$

We can see that the interaction curve (Fig. 4.61),

$$\frac{M'_p}{M_p} + \left(\frac{N}{N_p}\right)^2 = 1, \tag{4.173}$$

which applies to a rectangular cross-section, differs only slightly from that for a tubular cross-section. As shown in Fig. 4.62, the elastic limit state is

$$M = \left(\sigma_F - \frac{N}{\pi dh}\right) W_e,$$

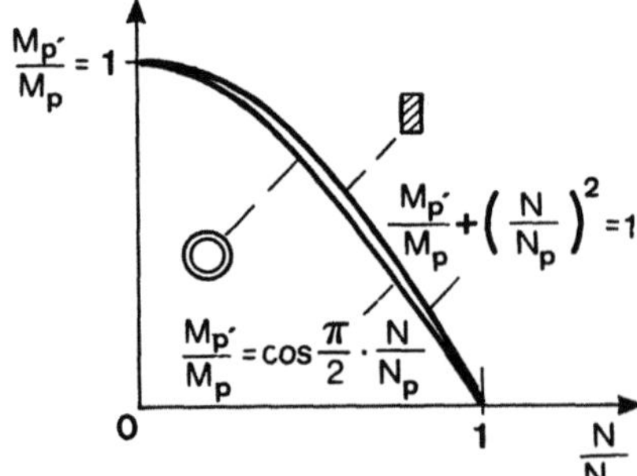

Fig. 4.61. Interaction curve between bending moment and axial force.

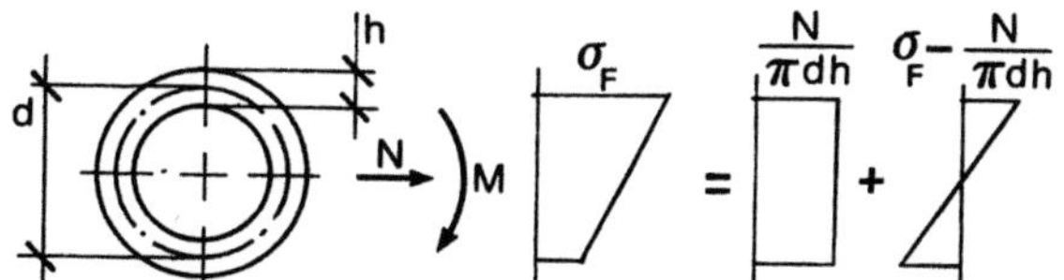

Fig. 4.62. Elastic stress distribution in tube with bending moments and axial loading.

and we obtain

$$\frac{M}{M_F}+\frac{N}{N_F}=1 \tag{4.174}$$

as the elastic interaction, where M_F is the elastic limit moment $W_e\sigma_F$ without axial force. A good approximation to the interaction between an axial force and a torque M_T can be obtained from von Mises' yield criterion

$$\sigma_F^2=\sigma^2+3\tau^2.$$

Here we can assume that $\alpha=1$ for torsional loading of thin-walled tubes, i.e. $M_{TF}=M_{Tp}$. Replacing stresses with forces and moments, we find

$$\left(\frac{N}{N_p}\right)^2+\left(\frac{M_T}{M_{Tp}}\right)^2=1, \tag{4.175}$$

where

$$M_{Tp}=\frac{\pi}{2}d^2h\frac{\sigma_F}{\sqrt{3}}.$$

We obtain the elastic limit load with

$$\frac{N}{N_p}+\frac{M_T}{M_{Tp}}=1, \tag{4.176}$$

as both the load components create constant stress states.

Without going into great detail, we also mention the additional relationship between torsion and bending, given by

$$\frac{M_p'}{M_p}+\left(\frac{M_T}{M_{Tp}}\right)^2=1. \tag{4.177}$$

Transverse forces are again taken into consideration using von Mises' yield criterion, so that we obtain

$$\left(\frac{N}{N_p}\right)^2+\left(\frac{Q}{Q_p}\right)^2=1 \tag{4.178}$$

where

$$Q_p=\pi dh\frac{\sigma_F}{\sqrt{3}}.$$

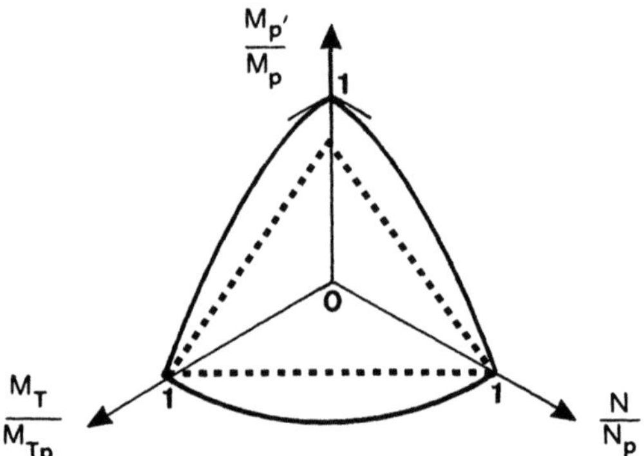

Fig. 4.63. Interaction between axial force, bending moments and torsional moments for elastic and fully-plastic load limit.

If all load components act, the interaction

$$\left(\frac{N}{N_p}\right)^2 + \left(\frac{M_T}{M_{Tp}}\right)^2 + \frac{M'_p}{M_p} + \left(\frac{Q}{Q_p}\right)^2 = 1 \tag{4.179}$$

can be regarded as a good approximation for the load limit. The elastic limit state is

$$\frac{N}{N_F} + \frac{M_T}{M_{Te}} + \frac{M}{M_F} + \frac{Q}{Q_F} = 1, \tag{4.180}$$

or, related to the fully-plastic sectional qualities, we obtain, for a tube,

$$\frac{N}{N_p} + \frac{M_T}{M_{Tp}} + \frac{M}{M_p}\frac{4}{\pi} + \frac{Q}{Q_p} = 1. \tag{4.181}$$

Figure 4.63 shows interactions between an axial force and the torque and bending moments for the elastic and fully-plastic limit states. We can now examine more closely the question of the load-bearing limit of simple structures.

4.4.2 Ultimate Loads of Simple Structures

We again consider the horizontal stiffening of the semisubmersible structure in Fig. 4.1a, which we have already used several times, and study its load-bearing limit (Fig. 4.64). First we consider the simply supported beam as an analytical model; for this the maximum bending moment at the middle of the beam is

$$M = \frac{ql^2}{8}.$$

The load at which the yield point is always reached at the outermost fibre is thus

$$q_e = \frac{8\sigma_F W_e}{l^2}.$$

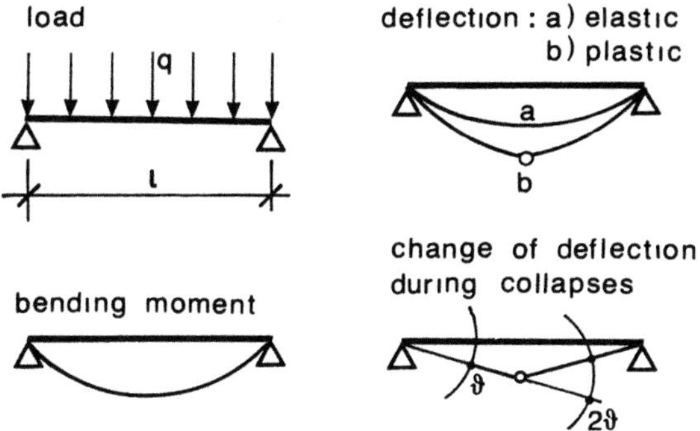

Fig. 4.64. Simply supported beam.

The load limit is reached when the cross-section is fully plastic, i.e.

$$q_p = \frac{8\sigma_F}{l^2} W_p = \frac{8}{l^2} M_p.$$

Then the structure has no further load-bearing reserve, and it behaves as a hinged mechanism, resulting in the collapse of the structure.

To determine the load limit, it is only necessary to determine the occurrence of the plastic hinge, i.e. the deformation state during collapse. The load limit can then in turn be determined using the principle of virtual displacement, or by using the elementary equilibrium laws. For this we will consider the beam to be fixed at both ends. The elastic load limit is

$$q_e = \frac{12}{l^2} \sigma_F W_e.$$

The location where yield first occurs is in the region of maximum moment, which in this case is at the supports. If the load is increased, so-called 'plastic hinges' occur, i.e. it becomes totally plastic at the supports

$$q_1 = \frac{12}{l^2} \sigma_F W_e \alpha.$$

A new static system now exists, namely a beam freely supported at both ends with two end moments of the same magnitude as the cross-sectional yield moments, where $M_p = \sigma_F W_p$ and $W_p = W_e \cdot \alpha$.

We imagine the new system as having slip couplings at the supports, the slip moment of which corresponds exactly to the fully-plastic moment. While, previously, the deflection halfway along the beam corresponded to the deflection of the fixed beam, i.e.

$$w = \frac{ql^4}{384EI},$$

now, with plastic hinges at the supports it is equal to that of the simply supported beam

$$w = \frac{5ql^4}{384EI},$$

i.e. five times greater. The external load can now be increased until the midspan yield moment is reached, so that the total load limit is

$$q_p = \frac{16}{l^2} M_p.$$

The increase in external load beyond the elastic load limit is therefore, with $a = 4/\pi$,

$$\frac{q_p}{q_e} = \frac{16}{3\pi}.$$

These results are shown in Fig. 4.65.

This load can also be determined using the 'kinematic chain' which occurs at the moment of collapse. We take for this purpose the virtual deformation state where the sum of the virtual external work and that of the internal work, which occurs only in the plastic hinges, disappears. We consider the case of the simply supported beam and find from Fig. 4.64

$$2M_p\vartheta - 2\frac{ql}{2}\vartheta\frac{l}{4} = 0, \tag{4.182}$$

from which the load can be immediately determined. The simplicity of this method is also illustrated by, for example, the case of a beam clamped at both ends. Here, all

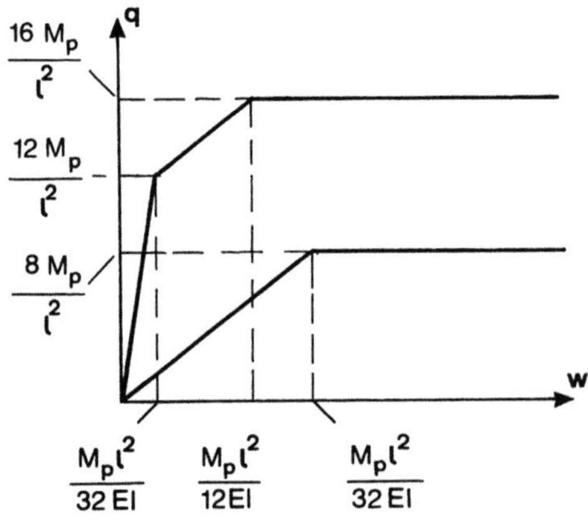

Fig. 4.65. Load–deflection curve of simply supported beam until failure.

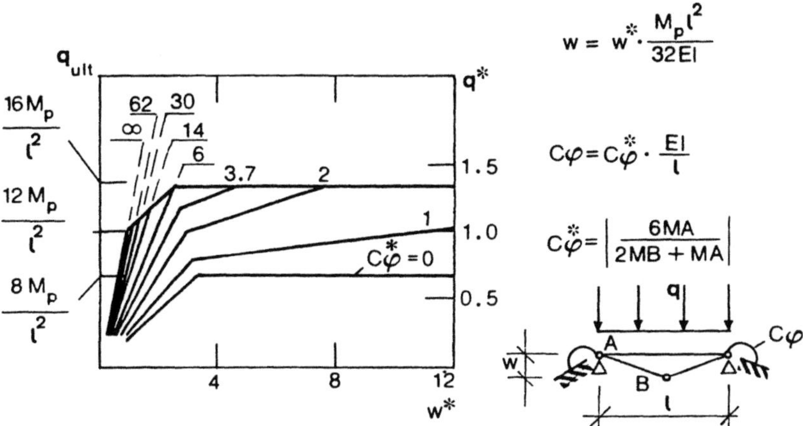

Fig. 4.66. Load–deflection behaviour of simply supported beam with rotational restraints at supports.

that is necessary is to add the work of the two yield moments at the supports

$$4M_{\mathrm{p}}\vartheta - 2\frac{q_{\mathrm{p}}l}{2}\vartheta\frac{l}{4} = 0, \tag{4.183}$$

and the load of the beam is immediately obtained.

We now study the effect of partial fixing. The calculation is as before, except for the location of the first plastic hinge. With a very weak edge-mounting, this occurs in the centre of the field, and with very stiff mounting it will be at the edge. Results are shown in Fig. 4.66, with the two limiting cases freely supported and fixed. We can see immediately that with partial support, the fixed girder load is always reached. If the partial support is very small, very large rotation is necessary at the edges, which is not compatible with the system. This is the so-called 'Stüssi Paradox', which for a long time delayed the introduction of the load method. If we consider the ultimate load of a partially supported girder, as spring stiffness disappears, the result is always a fixed-system load, which is twice as high as that of a simply supported girder, and thus contradictory. In other words, it is always necessary to check whether the edge support is sufficient to absorb a yield moment, even with rotation and very small amounts of torsion. This always happens if $c_\varphi \geqslant 2$. However, this also means that, in contrast to the elastic calculation, an increase in the degree of support does not mean an increase in the load capacity. We cannot go into the calculation of complex structures, and in particular the discussion of the 'loading laws', and must therefore refer the reader to the literature [21]. However, we still want to consider the important case of the behaviour of thin structures under axial loads. As already described in Sect. 4.2, stability failure is possible in such cases. Furthermore, we wish to know under what conditions the yield point is reached at the point of maximum loading if axial and transverse loadings occur, i.e. if it is a second order stress theory problem. We assume that when yield stress is reached, the load-bearing capacity is almost exhausted, but at least we obtain a lower limit for the estimate of this. We again consider the beam, simply supported at both ends, subjected to an axial load N and a transverse loading. We apply the interaction formula (4.174), using the moment while taking into consideration the axial forces and any transverse loading using (4.92),

$$\frac{M}{M_F}\,\frac{1+\Psi\dfrac{N}{N_{Ki}}}{1-\dfrac{N}{N_{Ki}}}+\frac{N}{N_F}=1. \tag{4.184}$$

We substitute $N_F/N_{Ki} = \delta$, and obtain

$$\frac{M_0}{M_F}\,\frac{1+\Psi\dfrac{N}{N_F}\delta}{1-\dfrac{N}{N_F}\delta}+\frac{N}{N_F}=1, \tag{4.185}$$

as in Fig. 4.67. We can also choose another representation

$$\frac{N}{N_F}=\frac{1+\delta\left(1+\dfrac{M_0}{M_F}\Psi\right)-\sqrt{\left[1+\delta\left(1+\dfrac{M_0}{M_F}\Psi\right)\right]^2-4\delta\left(1-\dfrac{M_0}{M_F}\right)}}{2\delta}, \tag{4.186}$$

which is known as the Perry–Robertson formula. Today, it is often used as the basis for standards, and is suitable for design purposes. Let us choose an example with $\delta^2 = 2$ and a tube of length $l = 20d$. The results are indicated with triangles in Fig. 4.21.

We conclude this section with a treatment of the initially deformed beam and its load-bearing capacity. This is of particular importance because constructional deformations, or those caused during operation (e.g. by mooring impacts) can always be present. Let us consider a tube subjected to compressive forces. For the

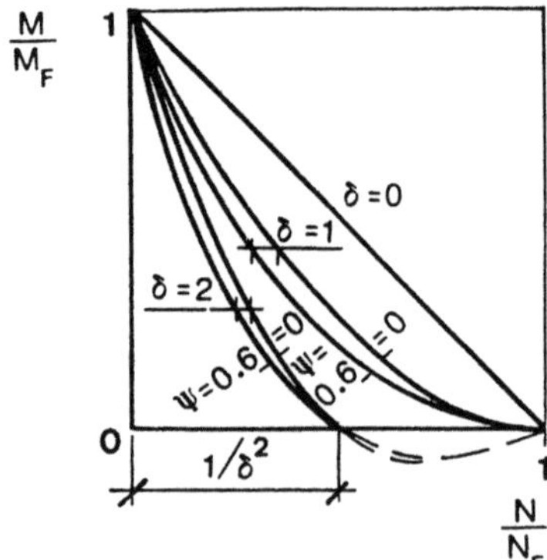

Fig. 4.67. Second-order stress theory interaction curves (between bending moments and axial forces).

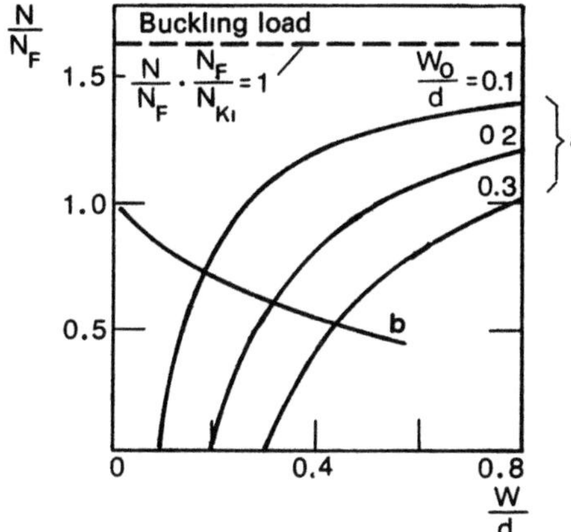

Fig. 4.68. Estimate of loads for simply supported beam with initial deflection subject to axial forces. **a** Load–deflection from (4.187), **b** from (4.189). The results of (4.188) are indicated by circles. Tube 180 × 10 mm, length 3400 mm, (term) = 450 N/mm².

sake of simplicity, it is taken to be simply supported at both ends. It is given a sinusoidal initial deformation with a maximum deformation of w_0. As a good approximation we find, taking Sect. 4.2 into account,

$$\frac{w}{d} = \frac{w_0}{d} \frac{1}{1 - \delta \dfrac{N}{N_F}}. \tag{4.187}$$

We show this result non-dimensionally in Fig. 4.68. As a lower estimate of the validity of this relationship, we use the Perry–Robertson formula in the representation (4.186), taking into account that $M_0 = Nw_0$, and obtain

$$\frac{N}{N_F} = \frac{1 + \delta + \eta - \sqrt{(1 + \delta + \eta)^2 - 4\delta}}{2\delta}, \tag{4.188}$$

where $\eta = A\omega_0/W$, where W is the section modulus and A the tube cross-sectional area. An upper estimate is obtained by assuming that the relationship stays constant from the onset of yield right up to the fully-plastic state. This is the same assumption as in the linear case at the beginning of the section. We can therefore use the interaction formula for the tube (4.172) and take

$$M'_p = Nw.$$

We thus obtain

$$\frac{w}{d} = \frac{1}{\pi} \frac{N_F}{N} \cos \frac{\pi}{2} \frac{N}{N_F}. \tag{4.189}$$

The intersection of (4.187) and (4.189) gives the required upper estimate. More rigorous calculations show that these estimates define the actual load-bearing capacity very precisely [16].

4.5 Numerical Methods

In the preceding sections we have surveyed some important structural analysis methods, mainly using the closed form solutions of differential equations of equilibrium. We have seen that closed form solutions are usable only for relatively simple applications. In general, therefore, we have to resort to numerical methods. The finite element method is the most important of these. This method is not associated with any specific mechanical principle. Finite element modelling can be applied methodically to force or displacement method, or to mixed methods, and it is to these we restrict ourselves. We develop closed form solutions where possible, and where this is not opportune, we present the finite approximation solutions using the energy method. We presented the method of Ritz and Galerkin while describing the energy method. These methods can be applied directly to derive finite element approximations. While the classical Ritz method starts with a solution formulation for the whole analysis domain, local solution formulations are used in the finite element method. Similarly, the classical Ritz method is improved by refining the solution formulation for the whole calculation domain, while with the finite element method this is achieved by refining the element mesh, the individual solution formulations within the elements remaining unchanged.

4.5.1 Finite Element Method

We begin with an introductory example of the finite element technique using the displacement approach. The unknowns to be determined are the displacements at specific points, or nodes. For this purpose, the structure is divided into finite areas or volumes (the finite elements), which are connected at the nodes at which the loads act. The relationship between the forces p and the displacements u is determined by a linear operator K, called the stiffness matrix

$$p_K = K \cdot u_K. \tag{4.190}$$

The vector p_K encompasses all nodal forces, forces being understood to include moments as well. The term u_K contains all nodal displacements, including rotations. The matrix K contains specific spring constants dependent upon the type of structure to be dealt with. By solving the system of equations (4.190) for u_K, the displacements for the whole structure are obtained, from which we can then obtain all constituent values such as forces, moments, or stresses, directly. We consider a structure composed of tubulars (Fig. 4.69). First we separate this structure into individual members (trusses) and establish the equilibrium conditions. We do this in element-specific, local coordinates, as this leads to relatively simple expressions. For an element i we obtain the relationship between nodal displacements and nodal forces (Fig. 4.70) as

$$\begin{Bmatrix} p_1 \\ p_2 \end{Bmatrix} = \frac{EA}{l} \begin{bmatrix} 1 & -1 \\ -1 & 1 \end{bmatrix} \cdot \begin{Bmatrix} u_1 \\ u_2 \end{Bmatrix}.$$

If several members are connected to form the structure, forces and displacements must be known not only in the direction of the member, but also in any direction,

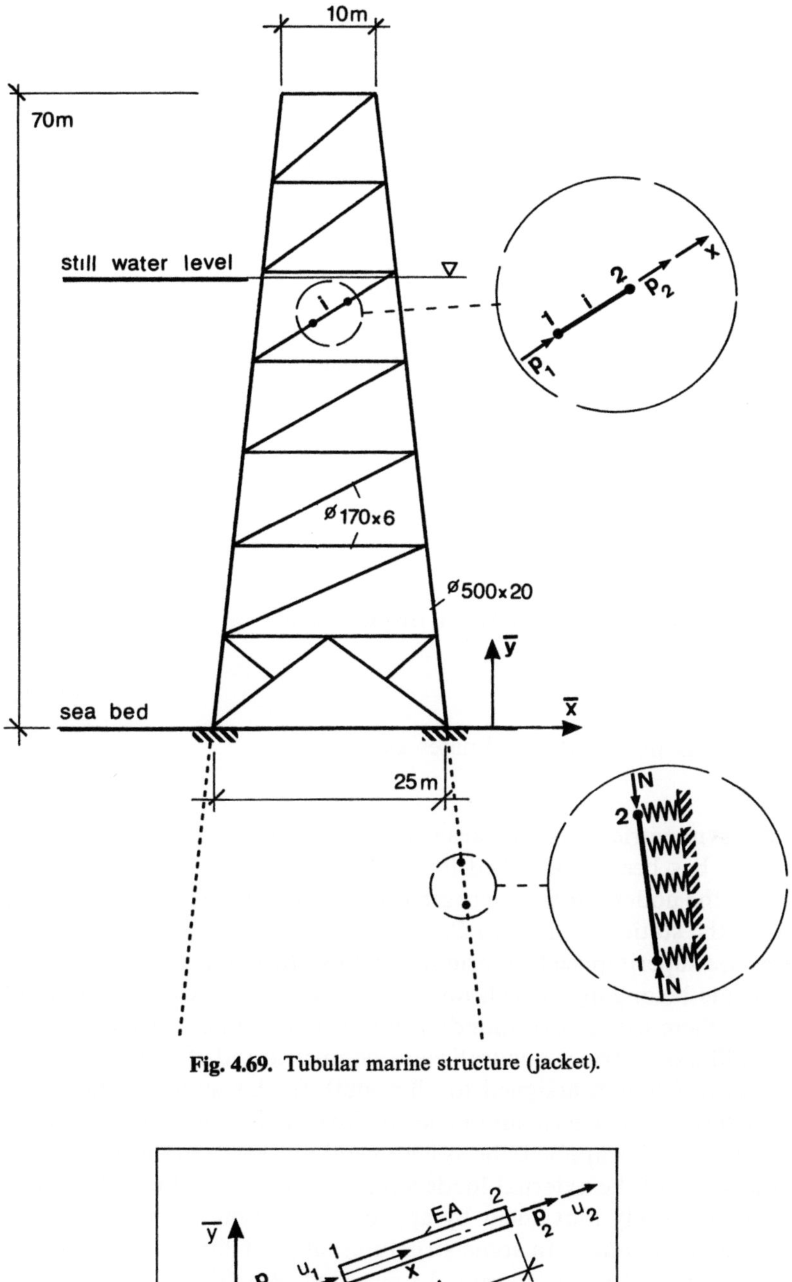

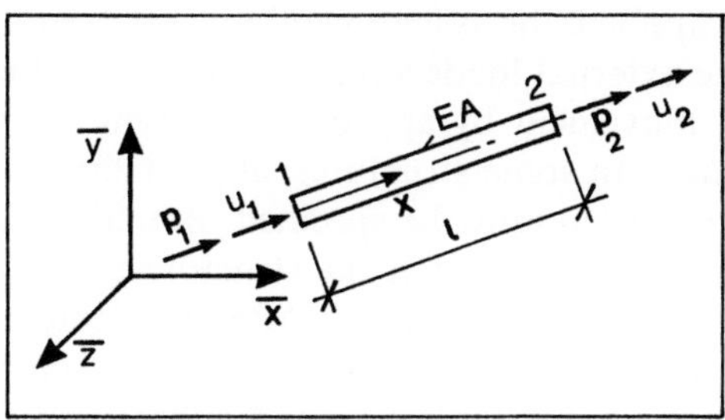

Fig. 4.69. Tubular marine structure (jacket).

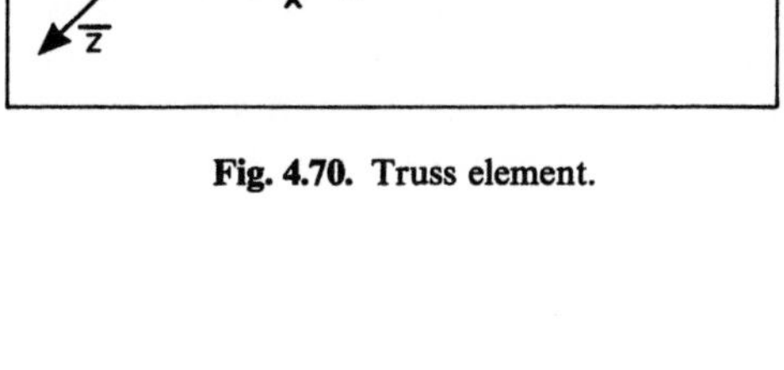

Fig. 4.70. Truss element.

because the static compatibility $\Sigma p = 0$ must be fulfilled at the nodes in addition to kinematic compatibility. The vector of nodal displacement in local coordinates can now be expressed in global coordinates as

$$u_K = \lambda \bar{u}_K. \tag{4.191}$$

The matrix λ contains the directional cosines of the truss axis from node 1 to node 2 in the global coordinate system, i.e.

$$\lambda = \left[\begin{array}{c|c} \cos\alpha\cos\beta\cos\vartheta & 0 \\ \hline 0 & \cos\alpha\cos\beta\cos\vartheta \end{array} \right], \tag{4.192}$$

$\bar{u}_K$ is the vector of the displacements in the global coordinate system, and α, β, ϑ are the angles between the local axis x and the global axes $\bar{x}, \bar{y}, \bar{z}$. The work done by a nodal force associated with nodal displacements is a scalar quantity, and is thus independent of the coordinate system, so we have

$$\bar{u}_K^T \bar{p}_K = u_K^T p_K.$$

With (4.191) we find

$$\bar{p}_K = \lambda^T p_K. \tag{4.193}$$

Thus, we obtain the relationship between displacements and forces in global coordinates as

$$\bar{p}_K = \lambda^T K \lambda \bar{u}_K, \tag{4.194}$$

and the stiffness matrix in global coordinates as

$$\bar{K} = \lambda^T K \lambda. \tag{4.195}$$

Both matrices are symmetric, which makes calculation considerably easier. When the stiffness matrix has been established for each tubular element, the elements must be coupled to the nodes (kinematic compatibility). This coupling is performed by the creation of the static equilibrium $\Sigma p = 0$, which means that the individual stiffness matrices are superimposed to form a system stiffness matrix, by summing the stiffness elements belonging to each node (where there are several degrees of freedom per node, there are a corresponding number of stiffness elements). The resultant system stiffness matrix is quadratic and symmetric and of size $n \cdot n$ ($n =$ the number of degrees of freedom assigned to all nodes). In this state, the matrix is singular, i.e. its determinant is zero, and hence its inverse does not exist. In other words, if a very small load is applied, the system would be set in motion as a rigid body, so that the work of the external loads would no longer be finite. We must therefore prevent rigid body movement by appropriate specification of displacement constraints at some nodes. In some structures, at least three constraints (in the x- and y-directions and rotation) must be specified. Boundary conditions for certain nodes are therefore introduced, so that the structure to be investigated is supported in a manner which is at least stable. After inserting the boundary conditions, we carry out the inversion and find the required displacement vector by multiplication with the force vector (inversion and subsequent multiplication are equivalent to the solution of the system of linear equations). In general, the displacements are of interest only in a spatially fixed coordinate system. If, however, we want to determine the forces and stresses within the finite elements, we require the displacem nts in element coordinates. The stresses within the tubulars are constant, and thus the strains are also constant. The expansion of the displacement in terms of the displacements of the end nodes is

$$u_x = u_1 + (u_2 - u_1)\frac{x}{l},$$

$$\frac{du_x}{dx} = \varepsilon_x = \frac{1}{l}(u_2 - u_1).$$

The stresses are

$$\sigma_x = E\varepsilon_x = \frac{E}{l}[-1 \quad 1]\begin{Bmatrix} u_1 \\ u_2 \end{Bmatrix},$$

$$\sigma_x = EBu_K. \tag{4.196}$$

Since we have the displacement vector in $\bar{x}, \bar{y}, \bar{z}$-coordinates, we must replace u_K with $\bar{u}_K$:

$$\sigma_x = EB\lambda\bar{u}_K. \tag{4.197}$$

Such a model consisting of tubulars is assumed to transmit forces only axially. But now we must consider, for example, wave forces, which tend to bend the individual tubulars. We therefore require an element formulation which include bending moments. We know that the potential for a beam is

$$\Pi = \frac{1}{2}\int_0^l \sigma\varepsilon dV. \tag{4.198}$$

We must now make an assumption for the displacement field w. Using $\xi = x/l$ we choose this as

$$w(\xi) = \alpha_1 + \alpha_2\xi + \alpha_3\xi^2 + \alpha_4\xi^3$$
$$w(\xi) = \xi\alpha \tag{4.199}$$

where ξ and α are vectors. The degrees of freedom of the nodes are

$$u_K^T = [w_1\varphi_1 w_2\varphi_2].$$

We then find the relationship between the physical degrees of freedom of the nodes and the constant α to be

$$u_K = A\alpha, \tag{4.200}$$

where

$$A = \begin{bmatrix} 1 & 0 & 0 & 0 \\ 0 & 1/l & 0 & 0 \\ 1 & 1 & 1 & 1 \\ 0 & 1/l & 2/l & 3/l \end{bmatrix}. \tag{4.201}$$

We can then write (4.199) as

$$w(\xi) = \xi A^{-1} u_K \tag{4.202}$$

$$w(\xi) = (1 - 3\xi^2 + 2\xi^3)w_1 + (\xi - 2\xi^2 + \xi^3)\varphi_1 \cdot l$$
$$+ (3\xi^2 - 2\xi^3)w_2 + (-\xi^2 + \xi^3)\varphi_2 \cdot l. \tag{4.203}$$

The elongation of the beam fibres is linked to the deflection via the curvature (4.4) and the stress is linked to the elongation via Hooke's Law $\sigma = \varepsilon E$. We thus obtain

$$\Pi = \frac{1}{2} \int_V \varepsilon E \varepsilon \, dV,$$

$$\Pi = \frac{1}{2} \int_V z \frac{d^2 w}{dx^2} Ez \frac{d^2 w}{dx^2} \, dV. \tag{4.204}$$

With

$$B = -z \frac{d^2 w}{dx^2}$$

we find the potential to be

$$\Pi = \frac{1}{2} u_K^T A^{-1T} \int_V B^{*T} E B^* \, dV A^{-1} u_K. \tag{4.205}$$

The matrix B^* is called the kernel strain matrix. We now apply the so-called 'first law of Castigliano', which states that the derivatives of the potential with respect to the displacements give the forces which correspond to the displacements, i.e. in our case, to the nodal displacements thus

$$\frac{\partial \Pi}{\partial u_K} = A^{-1T} \int_V B^{*T} E B^* \, dV A^{-1} u_K = p_K.$$

If we compare this expression with (4.190), we see immediately that

$$K = A^{-1T} K^* A^{-1} \tag{4.206}$$

where the kernel stiffness matrix is

$$K^* = \int_V B^{*T} E B^* \, dV.$$

We therefore obtain B^* and K^* as

$$B^{*T} = -\frac{z}{l^2} [0 \ 0 \ 2 \ 6\xi] \tag{4.207}$$

and

$$K^* = \frac{EI}{l^3} \begin{bmatrix} 0 & 0 & 0 & 0 \\ 0 & 0 & 0 & 0 \\ 0 & 0 & 4 & 6 \\ 0 & 0 & 6 & 12 \end{bmatrix}. \tag{4.208}$$

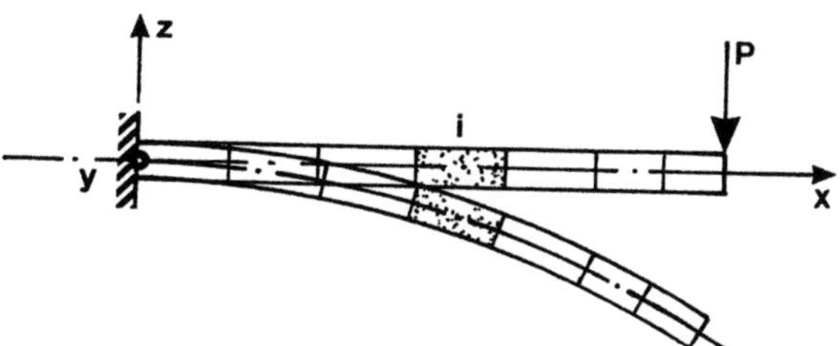

Fig. 4.71. Cantilever beam.

Integration over the cross-section of the beam gives the area moment of inertia I. By multiplication with A^{-1} in accordance with (4.201), we obtain finally

$$K = \frac{EI}{l^3} \begin{bmatrix} 12 & 6l & -12 & 6l \\ & 4l^2 & -6l & 2l^2 \\ & & 12 & -6l \\ \text{sym.} & & & 4l^2 \end{bmatrix} \tag{4.209}$$

which is a symmetric matrix. Furthermore,

$$p_K^T = [Q_1 M_1 Q_2 M_2]$$

gives the nodal forces. It is interesting that in (4.208) two columns and two rows are identically zero. These columns represent the rigid body movements, as a result of which no distortions occur within an element, and thus they make no contribution to the work due to deformation, or in this case, the potential. These rigid body movements can be clarified using Fig. 4.71. The cantilever beam shown is divided into a series of finite elements, of which we consider an element i. First, it must perform, in association with the adjacent elements, a translatory movement in the z-direction and a rotation about the y axis, then it can be additionally subjected to distortions. In other words, in the representation of the core stiffness matrix it can be seen immediately whether the selected process for the displacement field also contains the necessary rigid body displacements. We therefore calculate the jacket in Fig. 4.69 as a truss structure, and also as a frame structure with consideration of the flexural strength of the individual tubular elements (see Figs. 4.72–4.75). The displacement of the jacket head is practically the same in both cases. This means that we can determine the basic deformation behaviour with a truss structure, but since the flexural strength contributes greatly to the determination of the stress, it cannot be ignored.

It may also be noted that the solution (4.199) fulfils exactly the differential equation of equilibrium of the beam. The stiffness matrix derived from this therefore gives exact solutions. If Sects. 4.1 and 4.4 are examined, it will be seen that the majority of the differential equations are of the type

$$y'''' + \left(\frac{a}{l}\right)^2 y'' - \left(\frac{b}{l}\right)^4 y = \text{const.} \tag{4.210}$$

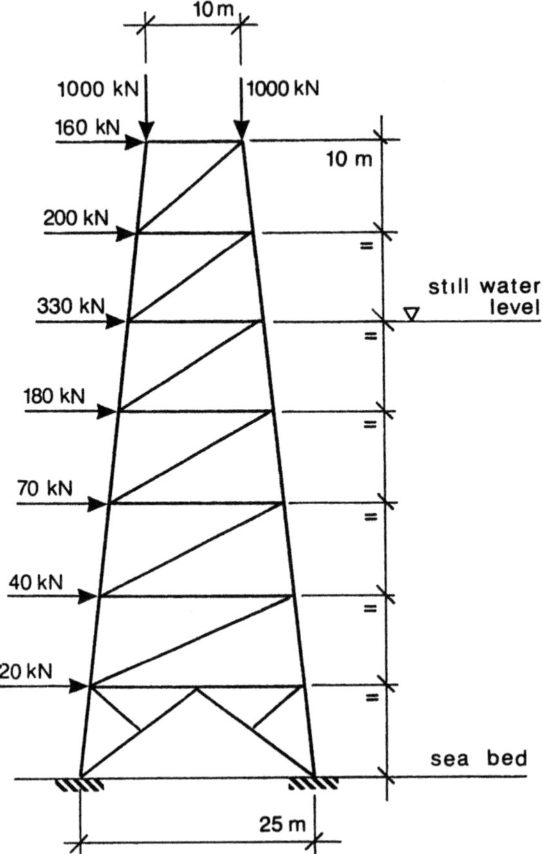

Fig. 4.72. Mathematical model of a jacket.

It thus seems reasonable to use this differential equation as the basis for a whole class of finite elements, with only the factors a and b and the constant on the right-hand side describing the physical factors. The potential associated with (4.210) is

$$\Pi = \frac{1}{2} \int_0^l \left\{ \left(\frac{d^2 y}{dx^2} \right)^2 - \left(\frac{a}{l} \right)^2 \left(\frac{dy}{dx} \right)^2 - \left(\frac{b}{l} \right)^4 y^2 - 2y \, \text{const} \right\} dx, \qquad (4.211)$$

which we can check immediately from Sect. 4.1.1. Again, the solution form (4.199) is selected, using y instead of w. As the degrees of freedom we introduce

$$y_k^T = [y_1 \, y_1' \, y_2 \, y_2'],$$

and find the solution as a function of the degrees of freedom y_K as

$$y = \xi A^{-1} y_K.$$

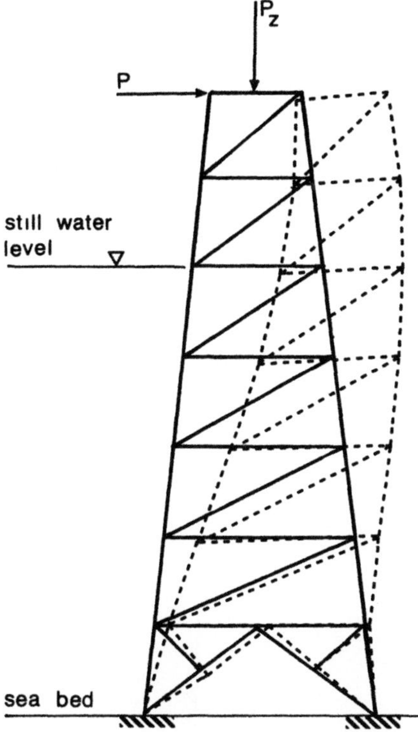

Fig. 4.73. Deformations.

We then find the potential as

$$\Pi = \frac{1}{2}y_K^T A^{-1T}\left\{\int_0^l \xi''^T\xi''\,dx - \left(\frac{a}{l}\right)^2\int_0^l \xi'^T\xi'\,dx - \left(\frac{b}{l}\right)^4\int_0^l \xi^T\xi\,dx\right\}$$

$$A^{-1}y_K - \int_0^l \xi\,\text{const}\,dx A^{-1}y_K. \tag{4.212}$$

The potential should be constant so that its variation with respect to y_K disappears

$$\frac{\partial\Pi}{\partial y_K} = A^{-1T}\left\{\int_0^l \xi''^T\xi''\,dx - \left(\frac{a}{l}\right)^2\int_0^l \xi'^T\xi'\,dx - \left(\frac{b}{l}\right)^4\int_0^l \xi^T\xi\,dx\right\}$$

$$A^{-1}y_K - \int_0^l \xi\,dx A^{-1}\,\text{const} = 0,$$

or

$$A^{-1T}\left[K_1^* - \left(\frac{a}{l}\right)^2 K_2^* - \left(\frac{b}{l}\right)^4 K_3^*\right]A^{-1}y_K = R^*A^{-1}\,\text{const}, \tag{4.213}$$

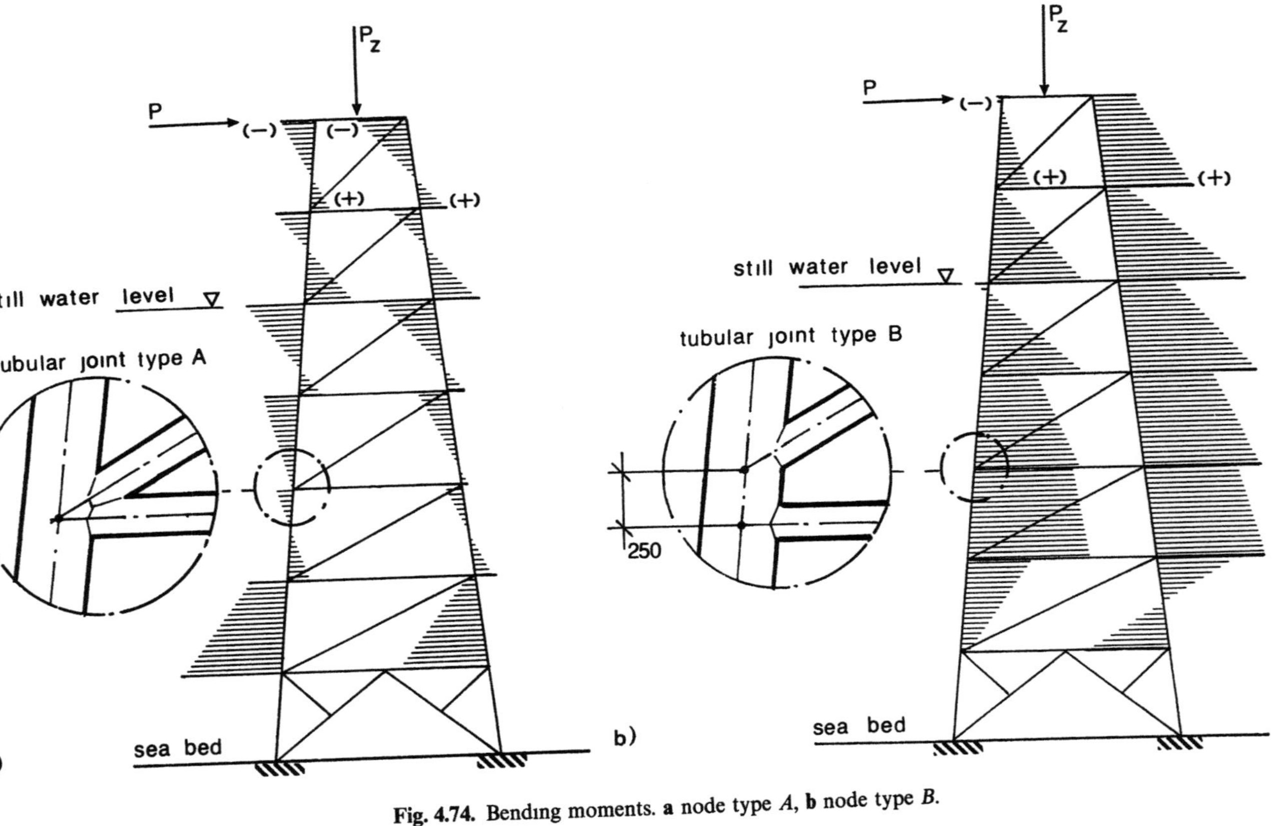

Fig. 4.74. Bending moments. **a** node type *A*, **b** node type *B*.

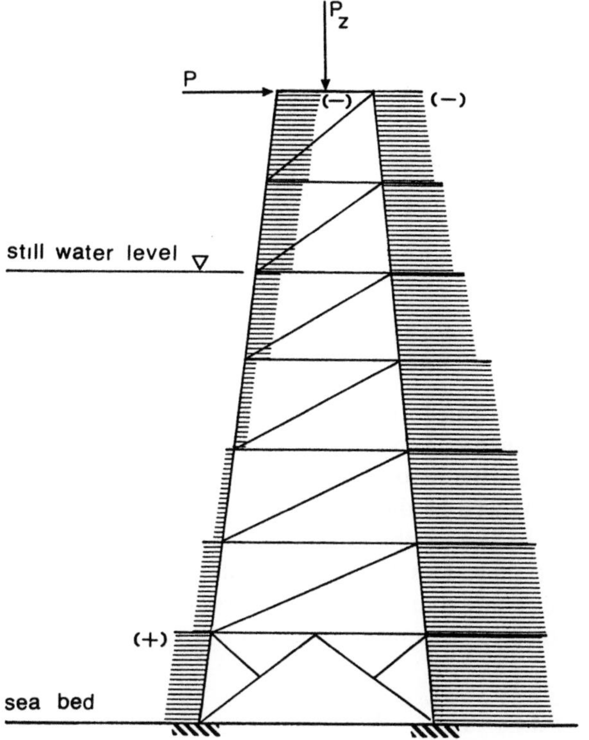

Fig. 4.75. Vertical forces.

or

$$\left[K_1 - \left(\frac{a}{l}\right)^2 K_2 - \left(\frac{b}{l}\right)^4 K_3 \right] y_K = R \, \text{const}, \tag{4.214}$$

where

$$K_1^* = \frac{1}{l^3} \begin{bmatrix} 0 & 0 & 0 & 0 \\ & 0 & 0 & 0 \\ & & 4 & 6 \\ \text{sym.} & & & 12 \end{bmatrix}; \qquad K_1 = \frac{1}{l^3} \begin{bmatrix} 12 & 6l & -12 & 6l \\ & 4l^2 & -6l & 2l^2 \\ & & 12 & -6l \\ \text{sym.} & & & 4l^2 \end{bmatrix};$$

$$\tag{4.215}$$

$$K_2^* = \frac{1}{30l} \begin{bmatrix} 0 & 0 & 0 & 0 \\ & 30 & 30 & 30 \\ & & 40 & 45 \\ \text{sym.} & & & 54 \end{bmatrix}; \qquad K_2 = \frac{1}{30l} \begin{bmatrix} 36 & 3l & -36 & 3l \\ & 4l^2 & -3l & -l^2 \\ & & 36 & -3l \\ \text{sym.} & & & 4l^2 \end{bmatrix};$$

$$\tag{4.216}$$

$$K_3^* = \frac{l}{420}\begin{bmatrix} 420 & 210 & 140 & 105 \\ & 140 & 105 & 84 \\ & & 84 & 70 \\ \text{sym.} & & & 60 \end{bmatrix};$$

$$K_3 = \frac{l}{420}\begin{bmatrix} 156 & 221 & 543 & -13l \\ & 4l^2 & 131 & -3l^2 \\ & & 156 & -22l \\ \text{sym.} & & & 4l^2 \end{bmatrix}; \tag{4.217}$$

$$R^* = \frac{l}{12}[12 \quad 6 \quad 4 \quad 3]\cdot\text{const}; \quad R = \frac{l}{12}[6 \quad l \quad 6 \quad l]\cdot\text{const}.$$

We now have a general transformation of the solution of the differential equation (4.210) into finite elements. From the differential equation of equilibrium, it is now possible to obtain the associated finite element. We need only to determine coefficients a and b as well as the RHS constant. We will now explain this using a simple example. The legs of the jacket shown in Fig. 4.69 are rammed into the sea-bed. To idealize these legs in terms of finite elements, we need the relevant differential equation. The legs are considered as beam elements, elastically supported by the soil and subjected to considerable axial force. The differential equation

$$\frac{d^4w}{dx^4} \pm \frac{N}{EI}\frac{d^2w}{dx^2} + \frac{k}{EI}w = \frac{q}{EI} \tag{4.218}$$

applies, which is a combination of (4.87) and (4.18), where q is a constant transverse load within an element. The constants belonging to this element are then

$$a^2 = \pm\frac{Nl^2}{EI}; b^4 = -\frac{kl^4}{EI}; \quad \frac{q}{EI} = \text{const}.$$

We can see that the coefficients of the matrix K_2 correspond to the geometric stiffness matrix, while K_3 corresponds to a mass matrix, because in the case of a vibrating beam, the constant b^4 becomes the mass expression

$$b^4 = \frac{\mu\omega^2 l^4}{EI}.$$

The most important elements are shown in Fig. 4.76, with their associated constants. We can therefore use a general beam element computer program so long as the stiffness matrix consists of the three parts (4.215)–(4.217). As both a and b can be eigenvalues, this computer program must be able to determine eigenvalues and eigenvectors. The question of the accuracy of the elements thus defined now arises. First, it should be noted that we can find an approximation only because we have chosen (4.199) as a solution form. We could also have used the exact solutions of the differential equation (4.210). That is possible, and was done, for example, in [22]. The stiffness matrices found with these solutions are exact-solutions, but have the disadvantage that they have very complicated

	Dgl.	a^2	b^4
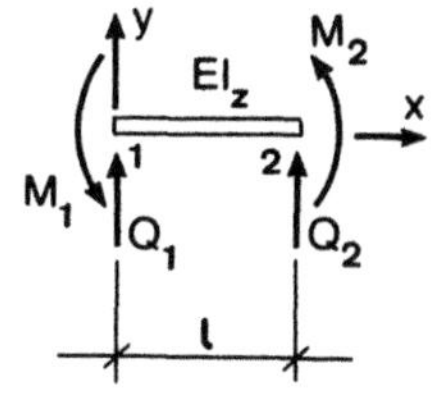 (diagram: beam with EI_z, nodes 1, 2, M_1, M_2, Q_1, Q_2, y, x, length l)	$EIw'''' = 0$ $EI_z = D$	0	0
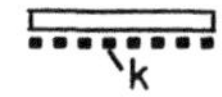 (diagram: beam on elastic foundation k)	$w'''' + \dfrac{k}{EI_z}\,w = 0$	0	$-\dfrac{kl^4}{EI_z}$
 (diagram: beam with compression H)	$w'''' + \dfrac{H}{EI_z}\,w'' = 0$	$\dfrac{Hl^2}{EI_z}$	0
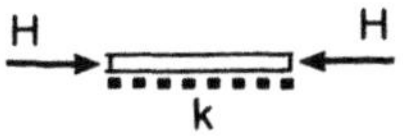 (diagram: beam with compression H on foundation k)	$w'''' + \dfrac{H}{EI_z}\,w'' + \dfrac{k}{EI_z}\,w = 0$	$\dfrac{Hl^2}{EI_z}$	$-\dfrac{kl^4}{EI_z}$
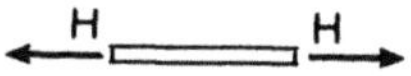 (diagram: beam with tension H)	$w'''' - \dfrac{H}{EI_z}\,w'' = 0$	$-\dfrac{Hl^2}{EI_z}$	0
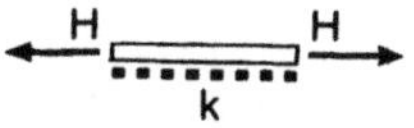 (diagram: beam with tension H on foundation k)	$w'''' - \dfrac{H}{EI_z}\,w'' + \dfrac{k}{EI_z}\,w = 0$	$-\dfrac{Hl^2}{EI_z}$	$-\dfrac{kl^4}{EI_z}$
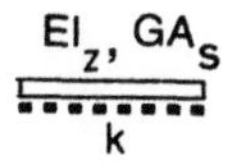 (diagram: beam EI_z, GA_s on foundation k)	$w'''' - \dfrac{k}{GA_s}\,w'' + \dfrac{k}{EI_z}\,w = 0$ $A_s = \text{shear area}$	$-\dfrac{k}{GA_s}\,l^2$	$-\dfrac{k}{EI_z}\,l^4$
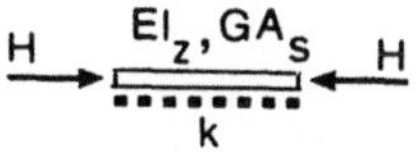 (diagram: beam EI_z, GA_s with compression H on foundation k)	$w'''' + \left(\dfrac{H}{EI_z} - \dfrac{k}{GA_s}\right)w'' + \dfrac{k}{EI_z}\,w = 0$	$\left(\dfrac{H}{EI_z} - \dfrac{k}{GA_s}\right)l^2$	$-\dfrac{k}{EI_z}\,l^4$
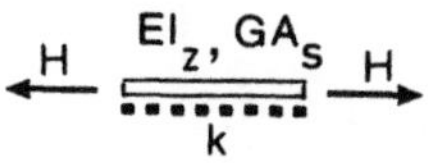 (diagram: beam EI_z, GA_s with tension H on foundation k)	$w'''' - \left(\dfrac{H}{EI_z} + \dfrac{k}{GA_s}\right)w'' + \dfrac{k}{EI_z}\,w = 0$	$-\left(\dfrac{H}{EI_z} + \dfrac{k}{GA_s}\right)l^2$	$-\dfrac{k}{EI_z}\,l^4$

Fig. 4.76. Beam elements.

expressions involving 8-fold differentiation. We can show, however, that the exact solution developed using a Taylor series is identical with the approximate solution given here, so long as higher terms of the series are neglected. The approximations given here are therefore consistent.

We now consider the elastically bedded beam, selecting simple support at both ends for the sake of simplicity. With the previously derived stiffness matrices, we obtain the eigenvalue problem

$$\left| \frac{1}{l}\begin{bmatrix} 4 & 2 \\ 2 & 4 \end{bmatrix} + \frac{Nl^2}{30lEI}\begin{bmatrix} 4 & -1 \\ -1 & 4 \end{bmatrix} + \frac{kl^3}{420EI}\begin{bmatrix} 4 & -3 \\ -3 & 4 \end{bmatrix} \right| = 0,$$

for which the characteristic equation is

$$\left(\frac{4EI}{l} + \frac{4Nl}{30} + \frac{4kl^3}{420} \right)^2 - \left(\frac{2EI}{l} - \frac{Nl}{30} - \frac{3kl^3}{420} \right)^2 = 0.$$

The smallest longitudinal force given by a root of the above is

$$N_{K_1} = 12\frac{EI}{l^2} + \frac{kl^2}{10}.$$

The exact solution is

$$N_{Ki} = (m\pi)^2 \frac{EI}{l^2} + \frac{kl^2}{(m\pi)^2}.$$

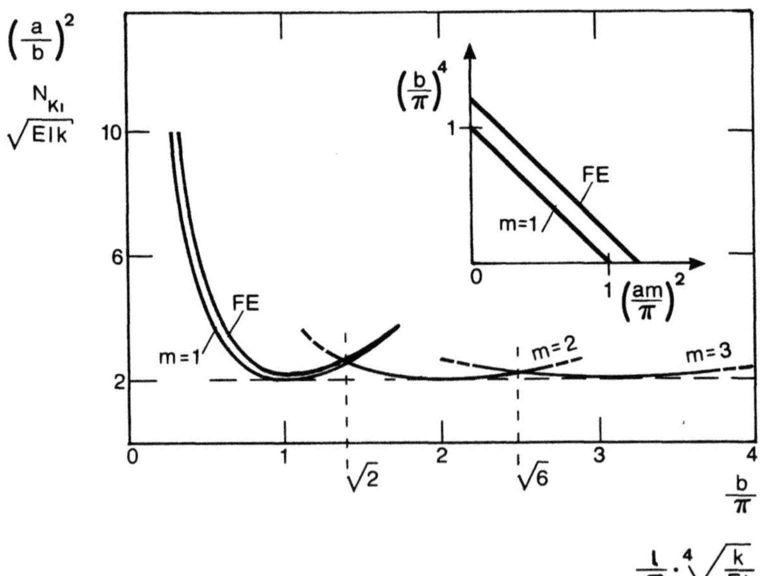

Fig. 4.77. Relationship between constants a and b.

The result is shown in Fig. 4.77. Dimensionless plotting was chosen, because with

$$\alpha = \frac{l}{\pi}\sqrt[4]{\frac{k}{EI}}$$

we obtain

$$\frac{N_{K_1}}{\sqrt{EIk}} = \frac{12}{\pi^2}\frac{1}{\alpha^2} + \frac{\pi^2}{10}\alpha^2 \tag{4.219}$$

for the approximation solution, and

$$\frac{N_{K_1}}{\sqrt{EIk}} = \left(\frac{m}{\alpha}\right)^2 + \left(\frac{\alpha}{m}\right)^2 \tag{4.220}$$

for the exact solution. This is also the so-called 'garland curve', familiar in plate buckling. The last equation is transposed, and the general values a and b are introduced to obtain

$$\left(\frac{a}{b}\right)^2 = \left(\frac{m\pi}{b}\right)^2 + \left(\frac{b}{m\pi}\right)^2. \tag{4.221}$$

The 'garland curve' is thus given a very general meaning. If a and b are eigenvalues, we can introduce

$$\left(\frac{b}{\pi}\right)^4 + \left(\frac{a}{\pi}m\right)^2 = m^4 \tag{4.222}$$

as interaction lines.

$$\left(\frac{a}{b}\right)^2 = \frac{12}{b^2} + \frac{b^2}{10}, \tag{4.223}$$

or

$$\frac{b^4}{120} + \frac{a^2}{12} = 1 \tag{4.224}$$

can be used as the approximate solution. When using the approximation with finite elements it should be noted that they can apply only to the region

$$\alpha = \frac{l}{\pi}\sqrt[4]{\frac{k}{EI}} \leqslant \sqrt{2},$$

i.e. an element can have only a certain length, which can be found from

$$l \leqslant \pi\sqrt[4]{\frac{4EI}{k}}.$$

This gives an important modelling rule, which is generally

$$b^4 \leqslant 4\pi^4.$$

For the case $a = 0$, one obtains the error

$$\Delta F = \left(\frac{120}{\pi^4} - 1 \right) \cdot 100 = 23.2\%,$$

or, if $b = 0$,

$$\Delta F = \left(\frac{12}{\pi^2} - 1 \right) \cdot 100 = 21.6\%,$$

These errors are drastically reduced by using several elements. When using two elements the error clearly drops below 1%. We wish to use the elements just described for the stability analysis of the jacket shown in Fig. 4.69. We apply the vertical forces p_z and carry out a vibration analysis for several load levels (cf. Fig. 4.78).

It can now be seen that the results obtained using the simplified model in Sect. 4.3.1 are confirmed for this considerably more complicated structure. The characteristic form of the lowest natural frequency differs considerably from that of the lowest buckling load. It is not therefore directly possible (e.g. empirically) to determine the buckling load of such a structure by measuring the vibrational behaviour for some vertical load of moderate size, and then making a linear extrapolation to obtain the buckling load. In our case, this would lead to extreme overestimation of the actual buckling load. This linear relationship between lowest natural frequency and the buckling load exists only at very high loads, which are

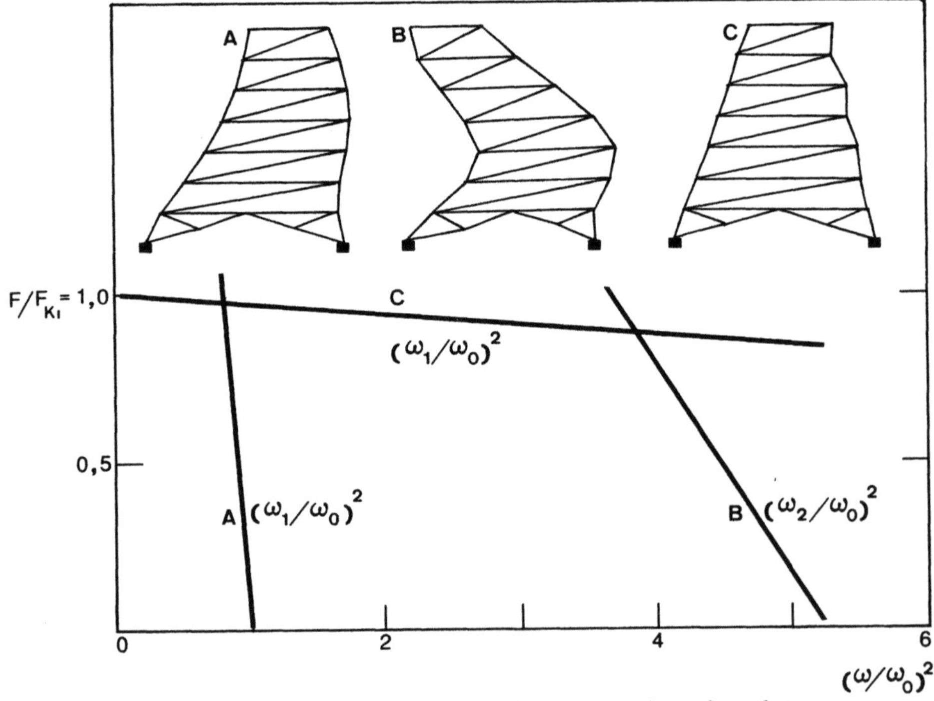

Fig. 4.78. Linear buckling load/natural frequency analysis of a jacket.

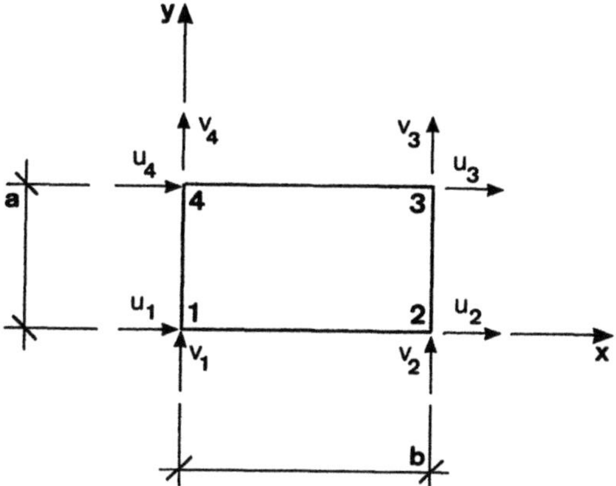

Fig. 4.79. Plane stress element.

of course impossible under test conditions, or at higher natural frequencies, the characteristic forms of which are similar to those of the buckling load. These are, however, difficult to measure. This property must be considered in real structures and, therefore, detailed calculations of the above type should accompany any measurement exercise.

In the early stages, the finite element method was used particularly in the analysis of plane load-bearing structures, and examples of applications to membranes, plates and shells are therefore particularly numerous. They can be classified according to shape (triangle or rectangle) or the type of displacement formulation, and there are very detailed reports available (e.g. [23–25]), to name only a few. Our derivation of the stiffness matrices will therefore only be explained using rectangular elements, without going into the numerous possibilities which permit, e.g. an isoparametric representation using numerical integration or special coordinates (e.g. area coordinates). We again begin from the elementary formula (4.190), and consider a rectangular plane stress element with four corner nodes (Fig. 4.79). Each node can move in the x and y directions. These movements lie exclusively in the plane of the plate. We therefore need a formulation for the displacement field with eight free coefficients

$$u = \alpha_1 + \alpha_2 \cdot x + \alpha_3 \cdot y + \alpha_4 xy,$$
$$v = \alpha_5 + (\alpha_6 - \alpha_3) \cdot x + \alpha_7 \cdot y + \alpha_8 xy. \tag{4.225}$$

Using the strain relations

$$\varepsilon = \begin{bmatrix} \partial u/\partial x & 0 \\ 0 & \partial v/\partial y \\ \partial v/\partial x & \partial u/\partial y \end{bmatrix}$$

we obtain the kernel strain matrix as

$$\boldsymbol{B^*} = \begin{bmatrix} 0 & 1 & 0 & y & 0 & 0 & 0 & 0 \\ 0 & 0 & 0 & 0 & 0 & 0 & 1 & x \\ 0 & 0 & 0 & x & 0 & 1 & 0 & y \end{bmatrix}. \tag{4.226}$$

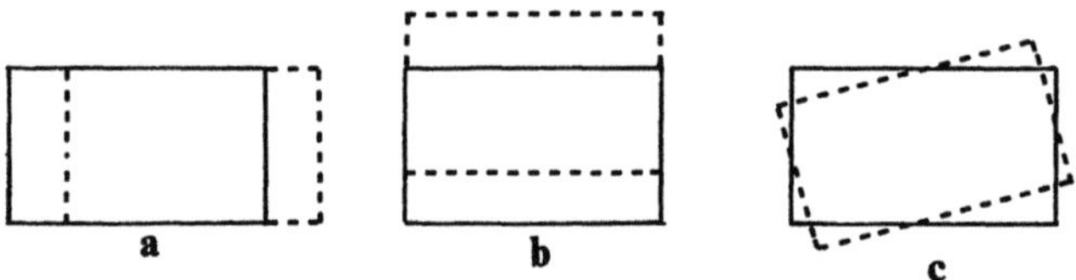

Fig. 4.80. Rigid body displacements of a plane stress element. **a** in x direction, **b** in y direction, **c** rotation.

The three columns of zeros indicate the three rigid body movements to which such an element is subject (Fig. 4.80). We obtain the correspondence of the coefficients to the nodal displacements with (4.200) as

$$u_K = A\alpha. \tag{4.227}$$

The coordinate system can also be placed at the centre of gravity of the element. The selection of coordinates affects only the matrix A

$$A = \begin{bmatrix} 1 & 0 & 0 & 0 & 0 & 0 & 0 & 0 \\ 0 & 0 & 0 & 0 & 1 & 0 & 0 & 0 \\ 1 & a & 0 & 0 & 0 & 0 & 0 & 0 \\ 0 & 0 & -a & 0 & 1 & a & 0 & 0 \\ 1 & a & b & ab & 0 & 0 & 0 & 0 \\ 0 & 0 & -a & 0 & 1 & a & b & ab \\ 1 & 0 & b & 0 & 0 & 0 & 0 & 0 \\ 0 & 0 & 0 & 0 & 1 & 0 & b & 0 \end{bmatrix}. \tag{4.228}$$

The shape function can also be indicated explicitly with

$$u = xy\alpha, \quad u = xyA^{-1}u_K, \quad u = Nu_K \tag{4.229}$$

or, writing $\xi = x/a$ and $\eta = y/b$,

$$N = \begin{bmatrix} (1-\xi)\cdot(1-\eta) & 0 & \xi(1-\eta) & 0 & \xi\eta & 0 & \eta(1-\xi) & 0 \\ 0 & (1-\xi)\cdot(1-\eta) & 0 & \xi(1-\eta) & 0 & \xi\eta & 0 & \eta(1-\xi) \end{bmatrix}.$$

It is now clear that the displacement at the boundaries of the element is compatible with that of the adjacent elements, but it is necessary to clarify the situation with regard to maintaining equilibrium of forces within an element. For this, the stress variation within the element must be determined:

$$\sigma = EB^*A^{-1}u_K, \tag{4.230}$$

Hookes' Law being defined with

$$E = \frac{1}{1-v^2} \begin{bmatrix} 1 & v & 0 \\ v & 1 & 0 \\ 0 & 0 & \dfrac{1-v}{2} \end{bmatrix}. \tag{4.231}$$

We obtain the stresses with (4.232).

$$\sigma = \frac{E}{1-v^2}
\begin{bmatrix}
\dfrac{-1}{a}(1-\xi) & -\dfrac{v}{a}(1-\xi) & \dfrac{1}{a}(1-\eta) & -\dfrac{v}{b}\xi & \dfrac{1}{a}\eta & \dfrac{v}{b}\xi & -\dfrac{\eta}{a} & \dfrac{v}{b}(1-\xi) \\[2mm]
-\dfrac{v}{a}(1-\eta) & -\dfrac{1}{b}(1-\xi) & \dfrac{v}{a}(1-\eta) & -\dfrac{\xi}{b} & \dfrac{v}{a}\eta & \dfrac{\xi}{b} & -\dfrac{v}{a}\eta & \dfrac{1}{b}(1-\xi) \\[2mm]
-\dfrac{1-v}{2b}(1-\xi) & -\dfrac{1-v}{2a}(1-\eta) & -\dfrac{1-v}{2b}\xi & \dfrac{1-v}{2a}(1-\eta) & \dfrac{1-v}{2b}\xi & \dfrac{1-v}{2a}\eta & \dfrac{1-v}{2b}(1-\xi) & -\dfrac{1-v}{2a}\eta
\end{bmatrix}
\cdot
\begin{Bmatrix}
u_1 \\ v_1 \\ u_2 \\ v_2 \\ u_3 \\ v_3 \\ u_4 \\ v_4
\end{Bmatrix},
\tag{4.232}$$

$$K = \frac{Et}{12(1-v^2)} \cdot
\begin{bmatrix}
\begin{matrix}4\beta \\ +2(1-v)\beta^{-1}\end{matrix} & \tfrac{3}{2}(1+v) & \begin{matrix}-4\beta \\ +(1-v)\beta^{-1}\end{matrix} & -\tfrac{3}{2}(1-3v) & \begin{matrix}-2\beta \\ -(1-v)\beta^{-1}\end{matrix} & -\tfrac{3}{2}(1+v) & \begin{matrix}2\beta \\ -2(1-v)\beta^{-1}\end{matrix} & \tfrac{3}{2}(1-3v) \\[3mm]
 & \begin{matrix}4\beta^{-1} \\ +2(1-v)\beta\end{matrix} & \tfrac{3}{2}(1-3v) & \begin{matrix}2\beta^{-1} \\ -2(1+v)\beta\end{matrix} & -\tfrac{3}{2}(1+v) & \begin{matrix}-2\beta^{-1} \\ -(1-v)\beta\end{matrix} & -\tfrac{3}{2}(1-3v) & \begin{matrix}-4\beta^{-1} \\ +(1+v)\beta\end{matrix} \\[3mm]
 & & \begin{matrix}4\beta \\ +2(1-v)\beta^{-1}\end{matrix} & -\tfrac{3}{2}(1+v) & \begin{matrix}2\beta \\ -2(1-v)\beta^{-1}\end{matrix} & \tfrac{3}{2}(1-3v) & \begin{matrix}-2\beta \\ -(1-v)\beta^{-1}\end{matrix} & \tfrac{3}{2}(1+v) \\[3mm]
 & & & \begin{matrix}4\beta^{-1} \\ +2(1-v)\beta\end{matrix} & -\tfrac{3}{2}(1-3v) & \begin{matrix}-4\beta^{-1} \\ +(1-v)\beta\end{matrix} & \tfrac{3}{2}(1+v) & \begin{matrix}-2\beta^{-1} \\ -(1-v)\beta\end{matrix} \\[3mm]
 & & & & \begin{matrix}4\beta \\ +2(1-v)\beta^{-1}\end{matrix} & \tfrac{3}{2}(1+v) & \begin{matrix}-4\beta \\ +(1-v)\beta^{-1}\end{matrix} & \tfrac{3}{2}(1-3v) \\[3mm]
 & & \text{sym.} & & & \begin{matrix}4\beta^{-1} \\ +2(1-v)\beta\end{matrix} & -\tfrac{3}{2}(1-3v) & \begin{matrix}2\beta^{-1} \\ -2(1-v)\beta\end{matrix} \\[3mm]
 & & & & & & \begin{matrix}4\beta \\ +2(1-v)\beta^{-1}\end{matrix} & -\tfrac{3}{2}(1+v) \\[3mm]
 & & & & & & & \begin{matrix}4\beta^{-1} \\ +2(1-v)\beta\end{matrix}
\end{bmatrix}
\tag{4.233}$$

If we apply the equilibrium conditions which have already been formulated in slightly modified form, to the stresses in the element, we obtain

$$\frac{\partial \sigma_x}{\partial x} + \frac{\partial \tau}{\partial y} = \frac{E}{1-v}\frac{1}{ab}(v_1 - v_2 + v_3 - v_4) = 0,$$

$$\frac{\partial \sigma_y}{\partial y} + \frac{\partial \tau}{\partial x} = \frac{E}{1-v}\frac{1}{ab}(u_1 - u_2 + u_3 - u_4) = 0.$$

This applies only when all displacements are the same, i.e. when there is a pure, stress-free rigid-body movement or a constant state of strain. In other words, in the special case of the constant state of strain, all conditions of elasticity theory, i.e. the compatibility of the displacements and the equilibrium of forces, are met. It is therefore to be expected that with increasing mesh refinement, the above element will give more accurate results for any stress state. The stiffness matrix is shown in (4.233).

We turn now to another important element, the plate element (Fig. 4.81). There is a very large variety of plate elements, because as a plate element can be subjected to a more complicated deformation, the number of possible formulations can be far more than for plane stress elements. The derivation is carried out again taking (4.190) into account and, as before, the analysis is restricted to a rectangular element. First the degrees of freedom must be defined. Then, the deflections and rotations about the x or y axis are suitable parameters for the degrees of freedom to describe a deflected plate. This gives us 12 degrees of freedom. It is remembered that a formulation with four degrees of freedom was selected for the beam, with a deflection and a rotation at each node. Since both beam and plate elements will be used later when modelling whole structures, it seems appropriate, for reasons of compatibility, to apply the formulation for the displacement field of the beam to the plate element. The subscript z or y indicates derivatives of w with respect to z or y, respectively. For clarity, the node number is shown as a superscript (Fig. 4.81).

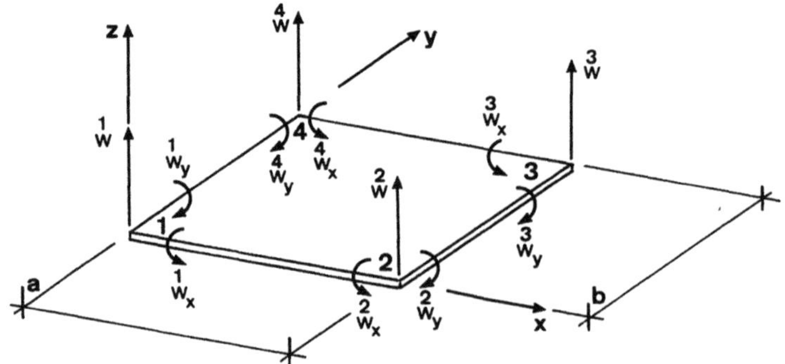

Fig. 4.81. Tubular joint modelling with shell elements.

The formulation with $\xi = x/a$ and $\eta = y/b$ is then:

$$
\begin{aligned}
w(\xi,\eta) = &(1 - 3\xi^2 + 2\xi^3)(1 - 3\eta^2 + 2\eta^3)\overset{1}{w} \\
&+ (1 - 3\xi^2 + 2\xi^3)(\eta - 2\eta^2 + \eta^3)b\,\overset{1}{w}_x \\
&- (\xi - 2\xi^2 + \xi^3)(1 - 3\eta^2 + 2\eta^3)a\,\overset{1}{w}_y \\
&+ (3\xi^2 - 2\xi^3)(1 - 3\eta^2 + 2\eta^3)\overset{2}{w} \\
&+ (3\xi^2 - 2\xi^3)(\eta - 2\eta^2 + \eta^3)b\,\overset{2}{w}_x \\
&+ (\xi^2 - \xi^3)(1 - 3\eta^2 + 2\eta^3)a\,\overset{2}{w}_y \\
&+ (3\xi^2 - 2\xi^3)(3\eta^2 - 2\eta^3)\overset{3}{w} \\
&- (3\xi^2 - 2\xi^3)(\eta^2 - \eta^3)b\,\overset{3}{w}_x \\
&+ (\xi^2 - \xi^3)(3\eta^2 - 2\eta^3)a\,\overset{3}{w}_y \\
&+ (1 - 3\xi^2 + 2\xi^3)(3\eta^2 - 2\eta^3)\overset{4}{w} \\
&- (1 - 3\xi^2 + 2\xi^3)(\eta^2 - \eta^3)b\,\overset{4}{w}_x \\
&- (\xi - 2\xi^2 + \xi^3)(3\eta^2 - 2\eta^3)a\,\overset{4}{w}_y.
\end{aligned}
\tag{4.234}
$$

Deformations of the element can be read off against the relevant degree of freedom of the node from Fig. 4.82. The expansion matrix is determined from (4.46). It is assumed that plate thickness is constant within an element, so that integration of the stiffness matrix can be carried out over the thickness

$$
K = \frac{h^2}{12} \int_0^a \int_0^b B^T EB\,\mathrm{d}x\,\mathrm{d}y.
\tag{4.235}
$$

For reasons of space, we cannot give the stiffness and strain matrix. They can be found in, for example, [26]. As a rule, not only the displacements, but also the moments of the plate are important for determining the stresses. These can be found directly from the expansion matrix

$$
m = EBu_K,
\tag{4.236}
$$

u_K being the vector of the 12 degrees of freedom. The mixed derivative of the deflection is necessary for describing the torsional moments of the plate. We determine these for node 1 using (4.234) as

$$
\frac{\partial^2 w}{\partial x\, \partial y} = \frac{36}{ab}(-\xi + \xi^2)(-\eta + \eta^2)\overset{1}{w}
$$

$$
+ \frac{6}{a}(-\xi + \xi^2)(1 - 4\eta + 3\eta^2)\overset{1}{w}_x
$$

$$
- \frac{6}{b}(1 - 4\xi + 3\xi^2)(-\eta + \eta^2)\overset{1}{w}_y.
\tag{4.237}
$$

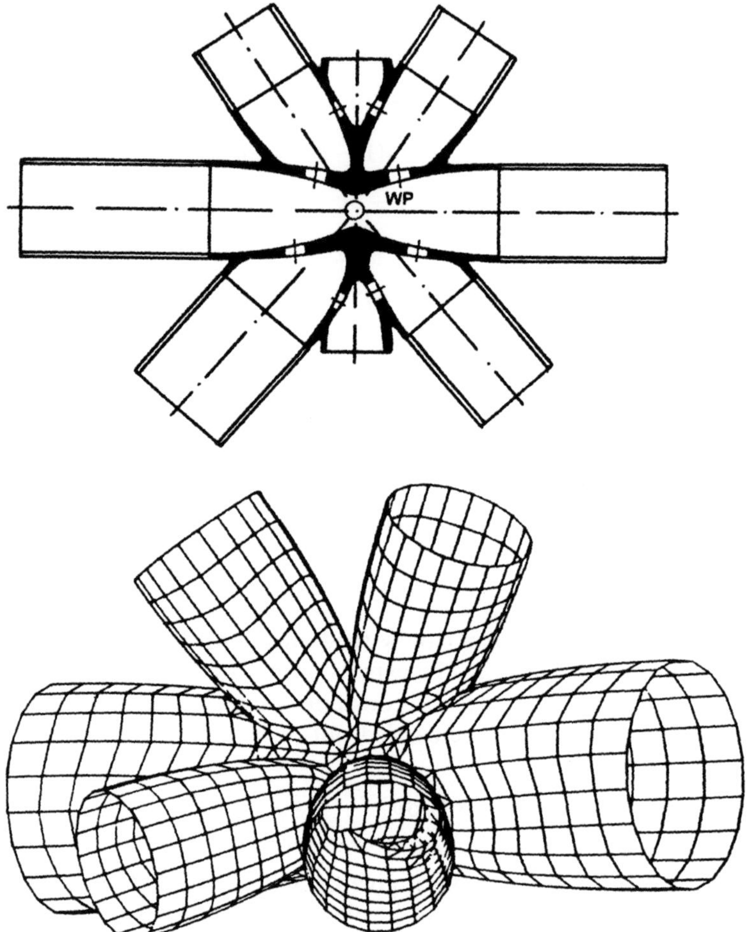

Fig. 4.82. Tubular joint modelling with shell elements.

We note that this expression vanishes at the nodes. The case of so-called 'pure torsion' of the plate cannot therefore be represented correctly with this element, which of course is disadvantageous in terms of general plate problems. This disadvantage can only be reduced by introducing the mixed derivative as an additional unknown at the nodes, so that this improved element now has 16 degrees of freedom. For the behaviour of these additional degrees of freedom we choose the shape function

$$w(\xi, \eta) = -(\xi - 2\xi^2 + \xi^3)(\eta - 2\eta^2 + \eta^3)ab \overset{1}{w}_{xy}$$
$$-(\xi^2 - \xi^3)(\eta - 2\eta^2 + \eta^3)ab \overset{2}{w}_{xy}$$
$$-(\xi^2 - \xi^3)(\eta^2 - \eta^3)ab \overset{3}{w}_{xy}$$
$$-(\xi - 2\xi^2 + \xi^3)(\eta^2 - \eta^3)ab \overset{4}{w}_{xy}.$$

(4.238)

In addition to (4.237) we obtain for node 1

$$\frac{\partial^2 w}{\partial x\,\partial y} = -(1 - 4\xi + 3\xi^2)(1 - 4\eta + 3\eta^2)\,\overset{1}{w}_{xy} \tag{4.239}$$

and corresponding expressions for the other nodes. We can easily show that the case of pure torsion is correctly represented by this element, the exact solution of which is

$$w(\xi, \eta) = (1 - 2\xi)(1 - 2\eta). \tag{4.240}$$

The conditions for the nodes are

$$\overset{1}{w} = -\overset{2}{w} = \overset{3}{w} = -\overset{4}{w} = 1, \quad \overset{1}{w}_x = \frac{\partial w}{\partial y} = -\overset{2}{w}_x = -\overset{3}{w}_x = \overset{4}{w}_x = -\frac{2}{b},$$

$$\overset{1}{w}_y = -\frac{\partial w}{\partial x} = \overset{2}{w}_y = -\overset{3}{w}_y = -\overset{4}{w}_y = \frac{2}{a},$$

$$\overset{1}{w}_{xy} = -\frac{\partial^2 w}{\partial x\,\partial y} = \overset{2}{w}_{xy} = \overset{3}{w}_{xy} = \overset{4}{w}_{xy} = -\frac{4}{ab}.$$

It is now clear that the boundary deformations of the elements with 12 and 16 degrees of freedom are identical and compatible. With 16 degrees of freedom, the twisting can now be modelled (Fig. 4.83).

The above examples show that when using available finite element programs, the important question of accuracy of the results can only be satisfactorily dealt with if the characteristics of the elements used are known precisely. Unfortunately, we have to say that many users of modern finite element software place little value on studying the characteristics of the elements they use, which often causes unreliable interpretation of the results.

4.5.2 Structural Modelling with Finite Elements

The quality of result obtained from finite element analysis depends decisively upon the accuracy with which the mathematical model describes actual conditions. The degree of refinement of the mathematical model must be tailored to the quality of the required result, but this often presents problems. On the one hand, the stress distribution is not usually known at the outset, and so the mathematical model has to be designed in terms of the fineness of the mesh; furthermore, the extent to which subsidiary stresses which are 'neglected' during measurement should be approximated is always questionable. We cannot describe this series of questions completely and restrict ourselves to some typical examples.

Once again, we consider the jacket in Fig. 4.69, which we have idealized with finite beam or truss elements. For various reasons, it is not always possible to achieve convergence of the lines of application of the members acting at a node. Much more commonly, the individual lines of application are misaligned. This misalignment, even if it is small relative to the member dimensions, can give rise to considerable additional stresses which must be taken into account in the finite

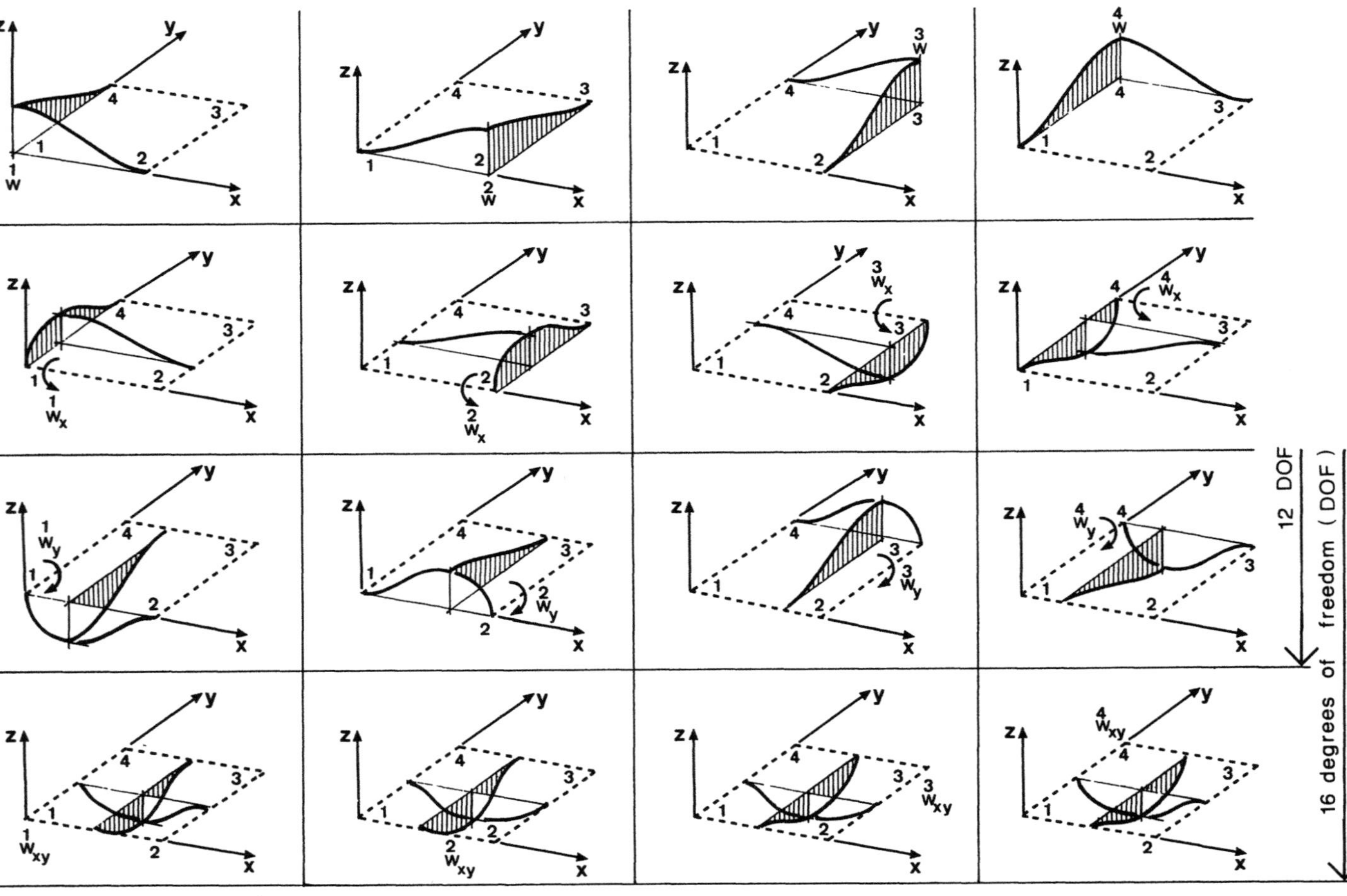

Fig. 4.83. Twist in a plate element with 12 and 16 degrees of freedom for the case of pure torsion.

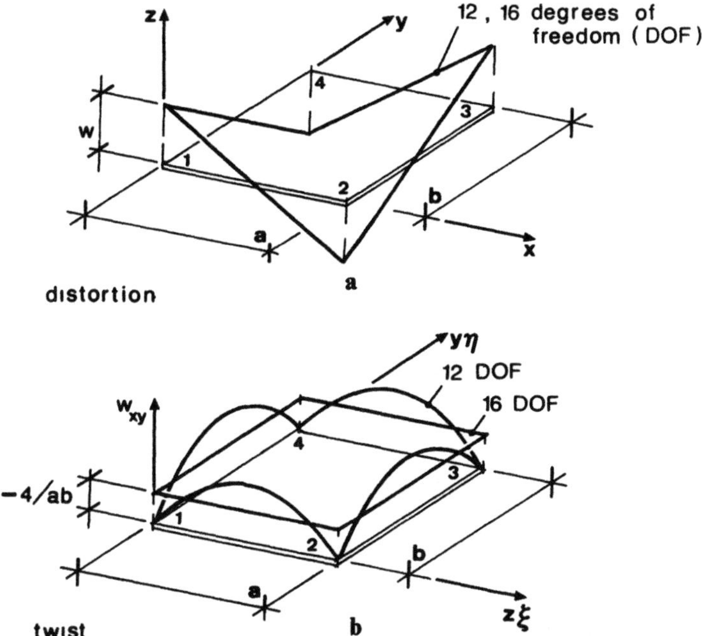

Fig. 4.84. Shape functions for a plate element.

element model. Figure 4.74b shows the results for the case of a misalignment of 50% of the member diameter for the jacket which has already been analysed.

Such a model does not allow the exact determination of the stress distribution in the region of the node. This requires a complex finite element model consisting of shell elements. The geometry of the intersections in particular is indeed very complex, but it can be described mathematically using the appropriate conic sections. Special mesh generators can therefore be used for mathematical models of nodes, for which only important parameters such as tube diameter, inclinations, misalignment, etc., need to be specified. The mesh itself is then created automatically.

Figure 4.84 shows a finite element model for a member node, which can give an accurate determination of the stress distribution. If an accurate stress distribution in the welded seams is required, finer models should be chosen with the aid of finite volume elements. This means, however, that an extremely high computational investment must be made, which is only warranted in certain exceptional cases. We can see already that a large complex structure requires various mathematical models, namely

— coarse modelling
— detailed modelling
— fine modelling.

Parts must be extracted from the detailed model or the coarse model for the last two, and the results from the coarse model must be used as boundary conditions.

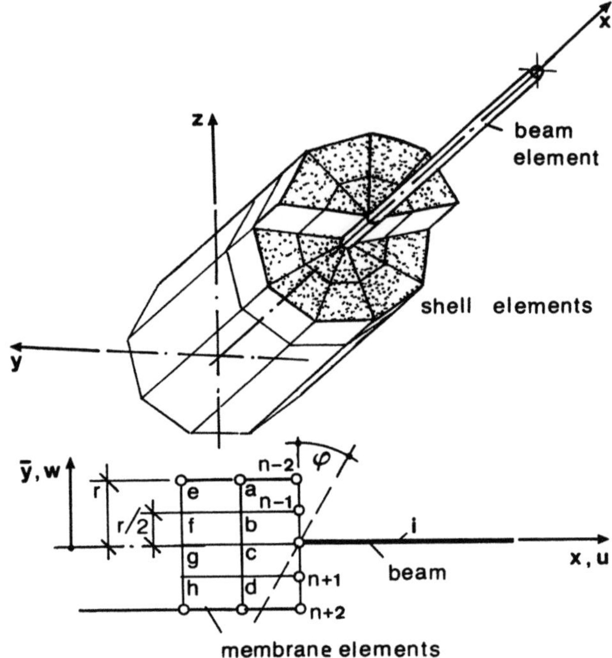

Fig. 4.85. Modelling of the zone between beam and membrane elements

It must also be remembered that this is an approximation process for all three forms, and that it therefore gives more accurate results as the mesh becomes finer. In other words, the procedure outlined above does not take these approximation characteristics into account, so errors often cannot be avoided. Sometimes it is a reliable procedure to progress directly from a coarse idealisation to a finer one within a mathematical model. This could be, for example, from a member node idealization by shell elements to a simple beam model. When transferring between quite different models, some special points should be considered. First, the interface between the two models must be chosen as an area which can be described accurately enough with the simpler model. Then the different models must be linked in such a way that the respective degrees of freedom can be connected, so that they are not subjected to unrealistic forced conditions. For example, the degrees of freedom of the beam element i at the node A (Fig. 4.85) are

$$u_{KB}^{T} = [u, w, \varphi]_{A}, \tag{4.241}$$

and those for the nodes $n-2, n-1, n, n+1, n+2$ of the plane stress elements a, b, c and d are

$$u_{KS}^{T} = [u_{n-2} v_{n-2} u_{n-1} v_{n-1} u_n v_n u_{n+1} v_{n+1} u_{n+2} v_{n+2}]. \tag{4.242}$$

We therefore require a kinematic link between these two vectors, such as

$$u_{KS} = A u_{KB}$$

where

$$A^{\mathrm{T}} = \begin{bmatrix} 1 & 0 & 1 & 0 & 1 & 0 & 1 & 0 & 1 & 0 \\ 0 & 1 & 0 & 1 & 0 & 1 & 0 & 1 & 0 & 1 \\ -r & 0 & -r/2 & 0 & 0 & 0 & r/2 & 0 & r & 0 \end{bmatrix}. \tag{4.243}$$

This means that the Bernoulli hypothesis of the continuous cross-sectional plane beam is forced in the transition between shell and plate model, and thus a kinematically perfect link is created between quite different models. Similar problems arise at the junction between the tubular structure and the platform deck. Very often, relatively thin tubes, which can generally be regarded as beam elements, have to be joined to large-diameter cylinders, which must be modelled by shell elements. The previously described principle of idealizing the actual transition between the parts of the structure with shell elements, and then carrying out the transition to beam elements in an undisturbed area, applies here. Occasionally, the situation occurs where very short, and hence very rigid, elements must be used, giving rise to the possibility of numerical ill-conditioning. Such a situation can arise in a tubular structure which has been reinforced in the region of the nodes, as in Fig. 4.86. The degrees of freedom u_K have to be transformed into degrees of freedom u_K^*. With the designations from Fig. 4.86 we obtain

$$u_K = I u_K^*, \tag{4.244}$$

where the factor I is

$$I = \begin{bmatrix} 1 & l_i & 0 & 0 \\ 0 & 1 & 0 & 0 \\ 0 & 0 & 1 & -l_j \\ 0 & 0 & 0 & 1 \end{bmatrix}. \tag{4.245}$$

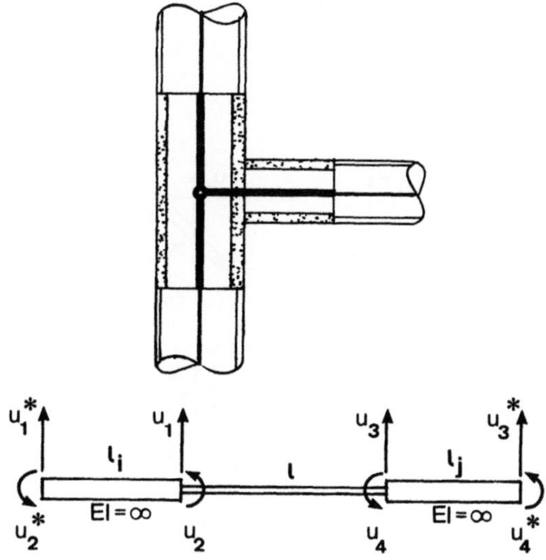

Fig. 4.86. Idealization of tubular joint using rigid elements (left).

Applied to the stiffness matrix of the beam, the transformation gives

$$\bar{K} = I^T K I.$$ (4.246)

After multiplying it out we obtain

$$\bar{K} = K + \overset{*}{K},$$

where K is the stiffness matrix and $\overset{*}{K}$ takes rigid ends into account

$$\overset{*}{K} = \frac{EI}{l^3}\begin{bmatrix} 0 & 12l_i & 0 & 12l_i \\ & 12l_i(l+l_i) & -12l_i & 12l_i l_j + 6ll_i + 6ll_j \\ & & 0 & -12l_j \\ \text{sym.} & & & 12l_j(l+l_j) \end{bmatrix}.$$ (4.247)

Such a transformation can also be carried out with the stiffness matrix from (4.214),

$$K = I^T\left[K_1 - \left(\frac{a}{l}\right)^2 K_2 - \left(\frac{b}{l}\right)^4 K_3 \right] I,$$ (4.248)

where, after multiplication, we obtain

$$K = (K_1 + \overset{*}{K_1}) - \left(\frac{a}{l}\right)^2 (K_2 + \overset{*}{K_2}) - \left(\frac{b}{l}\right)^4 (K_3 + \overset{*}{K_3}).$$

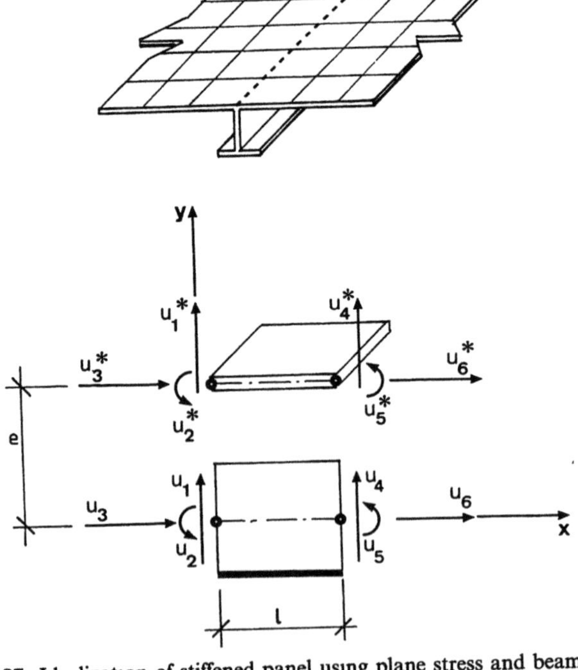

Fig. 4.87. Idealization of stiffened panel using plane stress and beam elements.

The elements of the matrix K_1 are already given in (4.247). For K_2 we obtain

$$K_2 = \frac{1}{30l} \begin{bmatrix} 0 & 36l_i & 0 & 36l_j \\ & 36l_i^2 + 6ll_i & -36l_i & 36l_il_j + 3ll_i + 3ll_j \\ & & 0 & -36l_j^2 \\ & \text{sym.} & & 36l_j^2 + 6ll_j \end{bmatrix} \qquad (4.249)$$

and for K_3

$$K_3 = \frac{l}{420} \begin{bmatrix} 0 & 156l_i & 0 & -54l_j \\ & 156l_i^2 + 44ll_i & 54l_i & 54l_il_j - 13ll_i - 13ll_j \\ & & 0 & -156l_j \\ & \text{sym.} & & 156l_j^2 + 44ll_j \end{bmatrix}. \qquad (4.250)$$

When using these transformations, the physical necessity must be justified for the application concerned. As mentioned several times, great care must be taken when using different finite elements in a single model. The problem can be studied using a frequently occurring detail of the modelling of reinforced decks. It is normal to idealize the individual girders using beam elements and the plating with plane stress elements (Fig. 4.87). The neutral axis of the beam is at a distance from the neutral plane of the plate, and it is usual to transform the bending axis of the beam to the neutral plane of the plate. This can be carried out with the transformation

$$u_K = I u_K^*,$$

where I is given by

$$I = \begin{bmatrix} 1 & 0 & 0 & & & \\ 0 & 1 & 0 & & 0 & \\ -e & 0 & 1 & & & \\ & & & 1 & 0 & 0 \\ & 0 & & 0 & 1 & 0 \\ & & & -e & 0 & 1 \end{bmatrix}. \qquad (4.251)$$

For the displacement of the plate element along its boundary we find

$$u = \beta_1 + \beta_2 x, \qquad (4.252)$$

i.e. a linear relationship. For the beam, the displacement at the upper boundary is

$$u_B = -e\frac{dw}{dx}. \qquad (4.253)$$

With the shape function (4.199) we obtain

$$u_B \simeq \alpha_2 + 2\alpha_3\xi + 3\alpha_4\xi^2, \qquad (4.254)$$

i.e. a quadratic curve. Apart from the choice of the constants β and α, which results from the nodal displacements, such a model is incompatible at the junction between beam and plate elements. Numerical calculations show that this effect disappears with a finer mesh, so in this case it is still generally possible to obtain useful results

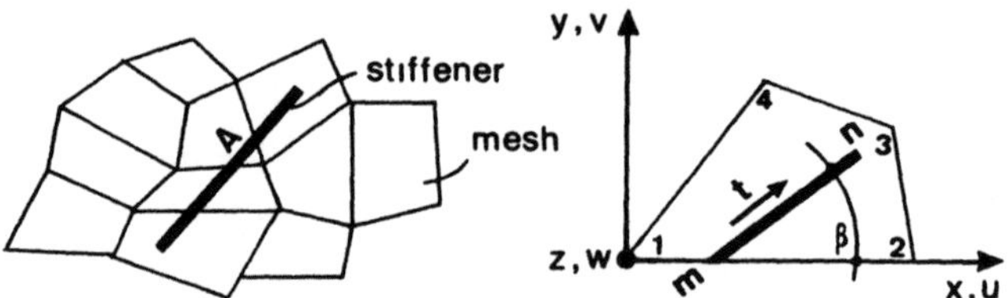

Fig. 4.88. Definition of Aranea element.

if we use, for example, the beam height for the plate element length. In many cases, it is too difficult to take each individual stiffness into account, because this means that each stiffness must be regarded as an element limit for the shell and plate elements. Here, what are known as Aranea elements can make the calculation much easier [27]. Discrete stiffnesses are inserted into any existing mesh, coinciding with previously defined nodes, as in Fig. 4.88. We first consider a mesh of rectangular plane stress elements. For the displacement field we use a rectangular element formulation. We obtain the expansions from

$$\varepsilon = B^*\alpha. \tag{4.255}$$

Only the expansion state on the line m–n can give a value, so we must perform a transformation between the coordinate system x, y and t

$$\varepsilon_t = \lambda\varepsilon. \tag{4.256}$$

We obtain the stiffness matrix as

$$K = \dot{A}^{-1T}EA\int_0^l B^{*T}\lambda^T\lambda B^*\,\mathrm{d}lA^{-1}. \tag{4.257}$$

The cross-sectional area A of the braces should always be constant, and can therefore be removed from the integral (4.257). A should not be confused with the correspondence matrix A, which permits the constants to be linked to the coordinates of the nodes. The transformation matrix λ is given by

$$\lambda = [\cos^2\beta \sin^2\beta \sin\beta\cos\beta]. \tag{4.258}$$

Of course, the same principle can be applied to plates, where, for example, a third-order polynomial with 12 constants should be chosen. One may ask why, in this and in the previous derivation, the shape function (i.e. the curve of the displacements with respect to the nodal degrees of freedom) was not used directly, rather than the more involved method of forming the core stiffness matrix with subsequent similarity transformation of the matrix A. Here, with the derivation of the Aranea elements, the answer becomes clear. As the integration takes place not within the whole surface described by any rectangle, but only along the line m–n, a complicated shape function for any rectangle is not necessary. A typical application is shown in Fig. 4.89, where a tube is introduced into a plate reinforced structure. While a mesh formulated in polar coordinates could be used to determine the stress state at the junction, a Cartesian system of coordinates would be more suitable for taking into account the reinforcements. The Aranea elements are particularly helpful here. The effect of the reinforcement on the stress distribution at the tube/deck junction can be seen clearly.

It must be remembered that the sometimes very complex modelling of the real structure with a finite element model produces a system of equations for discrete degrees of freedom. All continuous effects must be replaced by discrete values acting at the relevant nodes. This also applies to the loads. Concentrated individual loads and moments must be applied at the nodes according to their directions. It is more complicated to convert distributed loads into such nodal loads, so we illustrate this by way of example, considering again an individual element of the diagonal brace of the jacket shown in Fig. 4.69. The local coordinate system x, y does not correspond to the global coordinate system $\bar{x}, \bar{y}$, in which the load is defined.

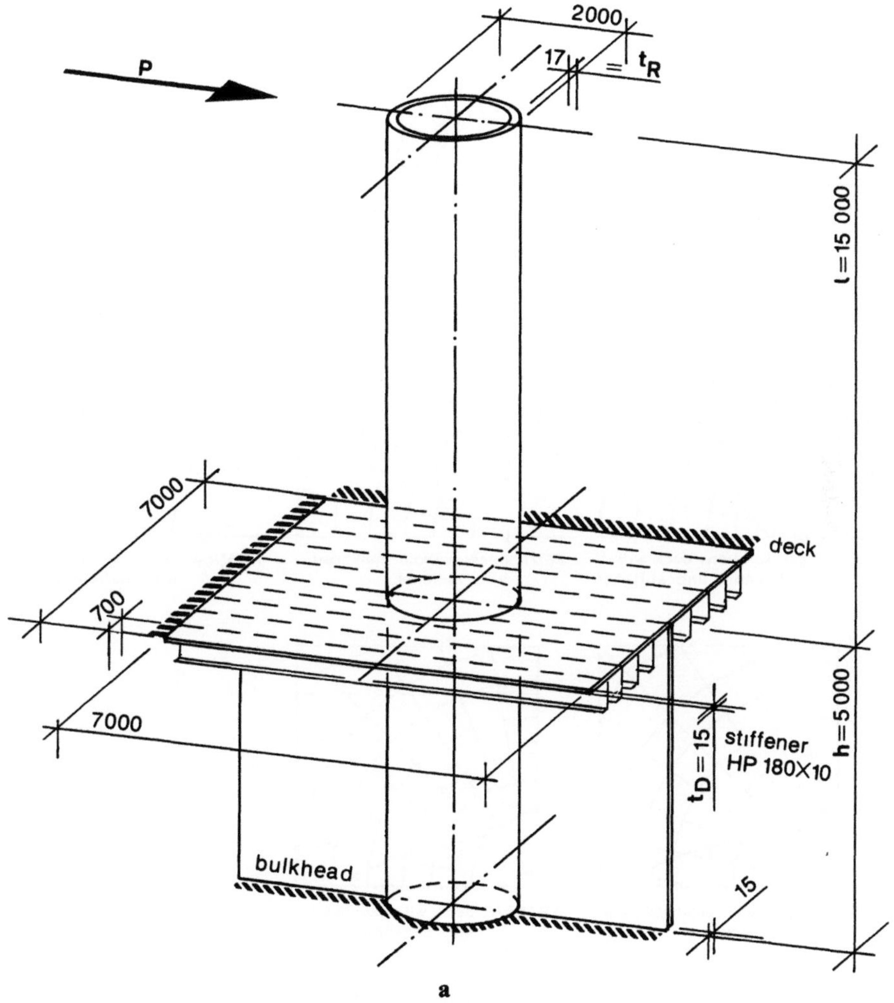

Fig. 4.89 a–c. Example application of Aranea elements.

(*continued on page 118*)

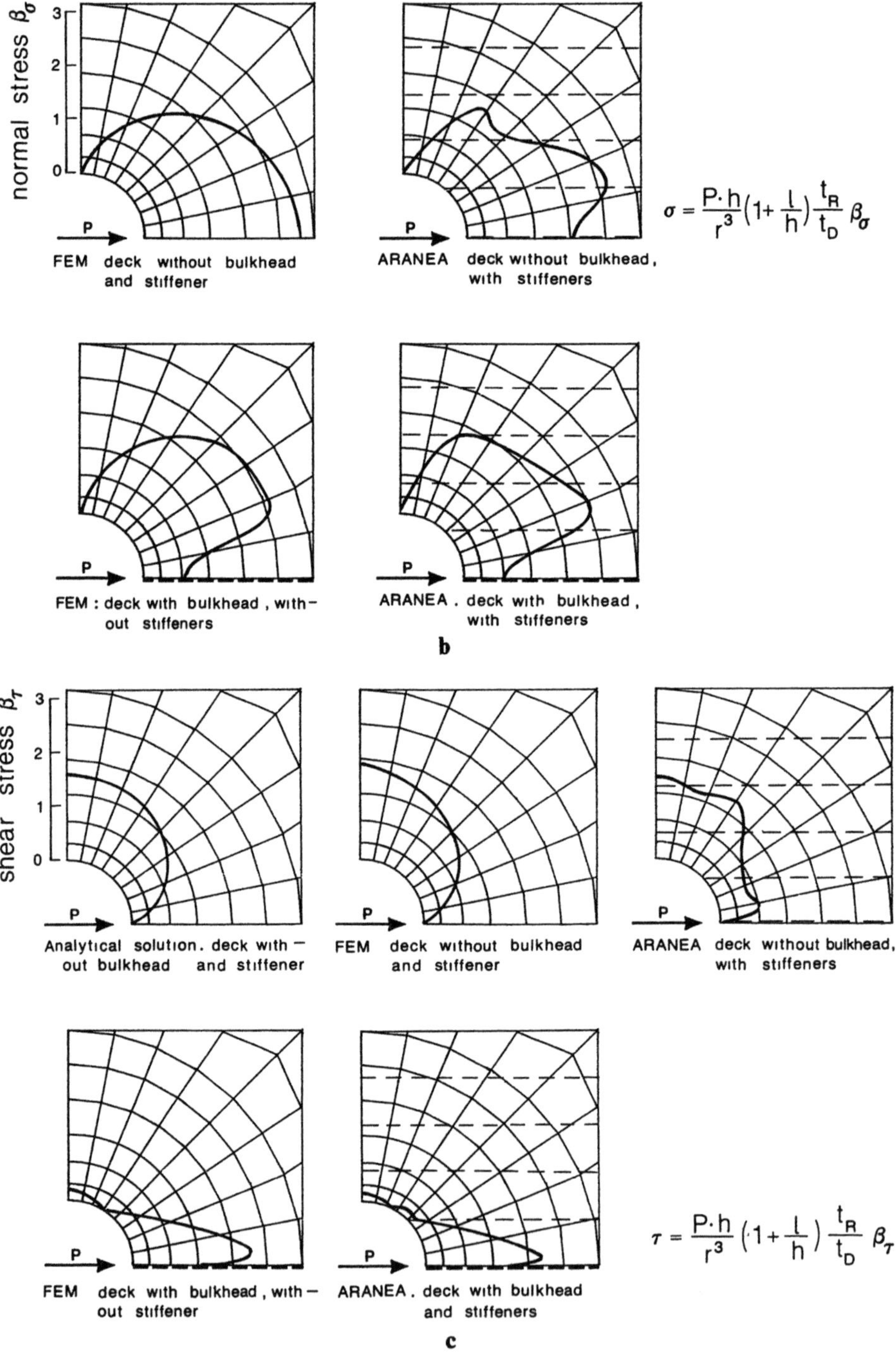

Fig. 4.89 (*continued*)

We assume that the work done by the nodal forces and moments on the relevant displacements and rotations is equal to the work done by the distributed loads on the displacements occurring in its region of application, i.e.

$$u_{\mathrm{K}}^{T} p_{\mathrm{K}} = \int_{\Gamma} w p\, \mathrm{d}x. \tag{4.259}$$

For the deformation, (4.199) applies for the beam, so that we obtain

$$p_{\mathrm{K}} = A^{-1T} \int_{\Gamma} \xi^{T} p\, \mathrm{d}x. \tag{4.260}$$

To analyse a concrete example, the load must be converted from global to local coordinates

$$p = \begin{Bmatrix} p_x \\ p_y \end{Bmatrix} = \lambda \begin{Bmatrix} \bar{p}_x \\ \bar{p}_y \end{Bmatrix}. \tag{4.261}$$

The variation of p_y over the length is given by

$$p_y = p_1(1 - \xi) + p_2\xi,$$

which in matrix notation can be written as

$$p_y = [1 - \xi \;\; \xi] \begin{Bmatrix} p_1 \\ p_2 \end{Bmatrix}. \tag{4.262}$$

With

$$w = \xi \cdot u_{\mathrm{K}},$$
$$\xi = [1 \;\; \xi \;\; \xi^2 \;\; \xi^3]$$

and

$$A^{-1T} = \begin{bmatrix} 1 & 0 & -3 & 2 \\ 0 & l & -2l & l \\ 0 & 0 & 3 & -2 \\ 0 & 0 & -l & l \end{bmatrix} \tag{4.263}$$

we obtain the equivalent nodal loads

$$p_{\mathrm{K}y} = \begin{Bmatrix} Q_1 \\ M_1 \\ Q_2 \\ M_2 \end{Bmatrix} = l \begin{bmatrix} 7/20 & 3/20 \\ l/20 & l/30 \\ 3/20 & 7/20 \\ -l/30 & -l/20 \end{bmatrix} \begin{Bmatrix} p_1 \\ p_2 \end{Bmatrix}. \tag{4.264}$$

These are, however, also the reaction values of the beam fixed at both ends, which is why they are called fixed-end forces, and can be found in the appropriate manuals for other load cases.

We conclude this chapter with a note on the modelling of jacket foundations. Jackets are usually anchored to the sea-bed with the aid of relatively deep piles. While the behaviour of the structural model is practically linear, this does not apply to the foundations. In the calculation we are dealing with a structure with linear behaviour, but which has boundary conditions which are partly non-linear. An economical approach is thus the structural behaviour of the jacket fixed to the ground and the behaviour of each individual pile to consider in separate models. The piles can then be replaced by a system of non-linear springs and the effect of these non-linear springs on the load-bearing behaviour of the whole jacket determined by iteration.

4.6 List of Symbols

A	cross-sectional area
A_1	coefficients
A	correspondence matrix
B	strain stiffness
B	strain matrix
B_1	constants
C_i	constants, integration function
D	flexural rigidity $\dfrac{Eh^3}{12(1-v^2)}$, damping
D_1	constants
E	modulus of elasticity
F	functional
F	Airy stress function
F	general solution formulation
G	shear modulus
H	horizontal force
I	moment of inertia
I	incidence matrix
K	stiffness matrix
L	length
M	bending moment
M_T	torsional moment
N	axial force
N_{K_1}	ideal critical force
P	force
Q	transverse force
R_x, R_y	stress relationships
T	period of oscillation
V	volume, vertical force
W	resistance moment
a	element length, factor
b	width, factor
b_m	effective width
c	spring constant

d	diameter
e	exponential function
f	function, frequency
g	acceleration due to gravity
h	plate thickness
k	bedding number, spring constant
k_x, k_y	buckling factors
l	length
m	index, half wave number
m_x, m_y	bending moments per unit length
n_x, n_y	inplane force per unit length
p	load vector
q	line load
q_x, q_y	transverse force per unit length
r	radius
r	load vector
t	time, coordinate
u	displacements
u	displacement vector
v	displacements
w	deflection
x, y, z	Cartesian coordinates
$\Delta, \Delta\Delta$	Laplace operators
Π	elastic potential
Φ	integration function
Φ	constant
Ψ	integration function
Ψ	factors
Ψ	stress relationship
Ω	circular frequency of excitation
α	constant, plastic form factor
α_m	aspect ratio
β	constant, slenderness ratio
γ	shear angle
γ	dimensionless longitudinal force
δ	variation, partial derivative
δ	ratio of yield load to column load
$\varepsilon_x, \varepsilon_y$	elongations in x and y directions
η	ratio of exciter frequency to resonant frequency
ϑ	virtual angle of rotation
λ	stiffness ratio
λ	transformation matrix
μ	mass density
μ	mass ratio
ν	transverse contraction number
ν	frequency ratio

ξ	dimensionless x coordinate
ρ	density
σ_x, σ_y	normal stress
σ_e	Euler buckling stress
τ	shear stress
φ	angle of rotation
ω	circular frequency

References

1   Timoshenko SP. Woinowsky-Krieger: Theory of plates and shells. New York: McGraw-Hill (1959)

2   Timoshenko SP, Goodier JN. Theory of Elasticity. New York: McGraw-Hill (1970)

3   Girkmann K. Flächentragwerke (Plane structures). 3. Aufl. Wien: Springer (1954)

4   Ritz W. Über eine neue Methode zur Lösung gewisser Variationsprobleme der mathematischen Physik (On a new method for the solution of certain variation problems of mathematical physics), J. reine u. angewandte Math. (1908) 135: 1–61

5   Petersen Ch. Statik und Stabilität der Baukonstruktionen (Statics and stability of structures). Braunschweig: Vieweg (1980)

6   Wolf E. Über eine Erweiterung des Begriffes der mittragenden Breite und Konsequenzen im Schiffbau (An extension of the concept of effective breadth and its consequences in shipbuilding). Schiffstechnik (1982) 29: 63–104

7   Schade H. The effective breadth concept in ship-structure design. Trans. SNAME (1953) 61: 410–430

8   Schmidt H, Peil U. Berechnung von Balken mit breiten Gurten (Calculation of beams with wide flanges). Berlin: Springer (1976)

9   Petershagen HJ. Beiträge zur Behandlung von Sonderproblemen bei schiffstechnischen Biegeträgern (Contributions to the treatment of special problems in girders subject to bending in ships) Jahrb Schiffsbautech. Ges., Bd. 59. Berlin: Springer (1965)

10 Hampe E. Statik der rotationssymmetrischen Flächentragwerke. Bd. 1 bis 4 (Statics of rotationally-symmetrical plane structures. Vols 1–4) Berlin: VEB Verlag für Bauwesen (1966)

11 Lehmann E. Festigkeit und Stabilität von dickwandigen Rohren, (Strength and stability of thick-walled tubes). Jahrbuch der STG, Bd. 77. Berlin: Springer (1983)

12 Axelrad E Schalentheorie (Shell theory). Stuttgart. Teubner (1983)

13 Flügge W. Statik und Dynamik der Schalen (Statics and dynamics of shells). Berlin: Springer (1962)

14 Pflügel A. Elementare Schalenstatik (Elementary statics of shells). Berlin: Springer (1960)

15 Kloppel K, Lie KN. Das hinreichende Kriterium für den Verzweigungspunkt des elastischen Gleichgewichtes (The adequate criterion for the bifurcation point of plastic equilibrium). Stahlbau (1943) 16: 17–32

16 Lehmann E. Berechnung schiffbaulicher und meerestechnischer Bauwerke nach der Spannungstheorie 2. Ordnung. (Calculation of ship and marine-structures according to second-order stress theory). Handb. der Werft, Vols XVI and XVII, Hamburg: Schiffahrtverlag Hansa (1982, 1984)

17 Schultz HG. Über die effektive Breite druckbeanspruchter Schiffsplatten (On the effective breadth of ship plates in compression). Schiff u Hafen (1964) 16. 730–739

18 Schultz HG. Neue Ergebnisse der Schiffsfestigkeitsforschung (Recent results of ship strength testing). Jahrbuch der STG, Vol. 58, Berlin: Springer (1964)

19 Kollar L, Dulacska E. Schalenbeulen (Buckling of shells). Dusseldorf: Werner (1975)

20 Den Hartog JP, Mesmer G. Mechanische Schwingungen (Mechanical vibrations). Berlin: Springer (1952)

21 Roik K. Vorlesungen über Stahlbau (Lectures on steel construction). Berlin: Ernst (1983)

22 Lehmann E. Analytische und halbanalytische Finite Elemente zur Konstruktionsberechnung schiffbaulicher Tragwerke. (Analytical and semi-analytical finite elements in structural calculations for ship structures). Jahrbuch der STG, Vol 70. Berlin: Springer (1976)

23 Bathe KJ. Finite-Elemente-Methode (Finite element methods). Berlin: Springer (1986)

24   Zienkiewıcz OC. Methode der finiten Elemente (Finite element method). Munchen: Hanser (1975)
25   Argyrıs J. Die Methode der finiten Elemente der elementaren Strukturmechanık (The method of finite elements ın elementary structural mechanics). Braunschweig: Vieweg (1986)
26   Przemienieckı JS. Theory of matrix structural analysis. New York: McGraw-Hıll (1968)
27   Lehmann E, Hung Ch. Das Aranea Element (The Aranea element). Schiffstechnik (1985) 32: 84
28. AISC Steel Construction Manual, Specification for the Desıgn, Fabrication and Erectıon of Structural Steel for Buildings. American Instıtute of Steel Constructıon (1970)

5. Environmental Conditions Affecting Marine Structures

As outlined in some detail in Chap. 2, in comparison with similar structures onshore, the characteristics peculiar to marine structures are closely limited by the hostile environmental conditions at sea, particularly the relatively high, horizontal seaway loadings, which are frequently ten times higher than wind or current loads. As far as seismic loads are concerned, there is actually a special physical feature of marine structures by which hydrodynamic masses and damping determine dynamic behaviour. However, the earthquake problem is not specific to marine technology, so it will not be included in this study. This section is a rational evaluation of the marine environment represented by waves, wind and currents.

Other environmental parameters must occasionally be considered, e.g. air and water temperatures, snow and ice formation, including sea ice and icebergs, or marine growths on parts of the structure close to the surface. These environmental parameters may be deterministically evaluated according to a design value concept.

Design values as specified in regulations for the most critical environmental parameters (e.g. the so-called '100-year design value' for wave height and the corresponding design wave period) are based on deterministic relationships developed in Chap. 3 between structures and elementary waves, so that, for example, design values for motions and forces affecting marine structures can be calculated. They provide the appropriate design values of stress and deformation when strength analysis methods developed in Chap. 4 are applied. Current regulations for marine structures, the characteristics of and latest developments of which will be mentioned briefly in Chap. 7, offer useful design formulae, frequently based on experience, in the context of the design value concept.

However, this simple concept is, in general, inadequate or ineffective for the evaluation of marine structure safety in a seaway. Hence, seaway effects in particular, and their stochastic (i.e. probabilistic and statistical) evaluation, will be dealt with in Chap. 6 in the context of a stochastic evaluation concept on the basis of considerations raised in this chapter.

The theory of random processes needed for an analytical description of the seaway is dealt with in Sect. 5.1 in the form of a brief survey of the most important definitions and relationships. This overview can only be very short and compact in the confines of this book, and we must provide the readers with some knowledge of the fundamentals of probability theory, which we give in Appendix 1, also in much abridged form.

Building on the theory of random processes, the special significance of the seaway is evaluated in Sect. 5.2 by a detailed explanation of the most important practical aspects of a stochastic description. Some practical design formulae for wind and currents which are of significance when evaluating marine structures are also being developed.

To indicate random numbers we use capital letters, and for realizations of random numbers lower-case letters (for details, see Appendix 1). Thus, X is a random number with realization x and, likewise, X is a random vector whose I components are the random numbers X_i, where $i = 1, 2, \ldots, I$. The realization of these random numbers are thus defined as values of x_i, which we interpret as components of vector x. We may thus symbolize an I-dimensional probability density function of X_i random variables, for example by $f_X(x)$, and the associated probability distribution by $F_X(x)$. Random vectors X in the text must not be confused with X matrices.

5.1 Evaluating Stochastic Processes

We will deal with concepts and relationships for stationary random processes in Sect. 5.1.1. From these, we go on to describe in Sect. 5.1.2 some of the characteristics of the stationary Gaussian random process as the most significant practical example, and in Sect. 5.1.3 we will look briefly at the well-known Poisson random process, as well as the Markov chain. Finally, in Sect. 5.1.4, we will develop a short summary of part of the stochastic dynamics of linear systems with one degree of freedom, which is important for further consideration. We need these relationships to understand the more recently developed state space model of the natural seaway in Sect. 5.2.1.2. Apart from that, however, the principles developed in Sect. 5.1.4 will have significance primarily for evaluating marine structures in a seaway, i.e. for Chap. 6.

The treatment of these themes will be as comprehensive as is necessary to understand stochastic approaches for the evaluation of marine structures in the marine environment. Of course, each theme can be developed further independent of the aims pursued here, so from the start the reader's attention is drawn to a selection from the wide-ranging specialist literature suitable for this purpose [1–6], with [2], [5] and [6] being particularly suitable as an introduction.

5.1.1 Stationary Random Processes

A random process $\{X(t)\}$ is the total of all random variables X that are observed at different times or in different places. We can define a random process, which is described by continuous random variables X at discrete points in time t_k, $k = 1, 2 \ldots, K$, by the joint probability density function of $f_X(x)$ of its random variables $X_k = X(t_k)$

$$f_X(x), \; x^T = (x(t_1), \ldots, x(t_K)) = (x_1, \ldots, x_K), \tag{5.1}$$

with the development of the process against time being characterized by an increase in K. Vectors are defined by analogy to the component notation in linear algebra basically as column vectors, so we indicate their space-saving representation in the form of row vectors with a superscript T indicating 'transposed'.

An observation $x(t)$ is termed a 'realization' or random sample of the random process, and the total of all random samples drawn is called the 'ensemble'. It is therefore the ensemble comprising an infinite number of random samples which is equal to the random process itself.

If we look at a certain point in time $t = t_1$, we can determine different characteristics of the random process $\{X(t)\}$ from the probability density function $f_{X1}(x_1)$ at that time, e.g. the expected value at time t_1 (see (A1.22)).

$$E[X_1] = m_{X1} = \int_{-\infty}^{\infty} x_1 \cdot f_{X1}(x_1)dx_1, \, x_1 = x(t_1) \tag{5.2}$$

or the mean square deviation (see (A1.22)).

$$E[X_1^2] = \int_{-\infty}^{\infty} x_1^2 \cdot f_{X1}(x_1)dx_1 \tag{5.3}$$

as well as the variance or square of the standard deviation σ_{X1} (see (A1.24)

$$\text{Var}[X_1] = \sigma_{X1}^2 = E[(X_1 - E[X_1])^2] = E[X_1^2] - E^2[X_1]. \tag{5.4}$$

If we look at the random process at two different times $t = t_1$ and $t = t_2$, then we define the covariance between the random variables $X(t_1) = X_1$ and $X(t_2) = X_2$ by means of

$$\begin{aligned}
\text{Cov}[X_1, X_2] &= E[(X_1 - E[X_1]) \cdot (X_2 - E[X_2])] \\
&= E[X_1 \cdot X_2] - E[X_1] \cdot E[X_2],
\end{aligned} \tag{5.5}$$

(see (A1.26)), with the autocorrelation function being defined as

$$R_{XX}(t_1, t_2) = E[X_1 \cdot X_2] = \int_{-\infty}^{\infty} \int_{-\infty}^{\infty} x_1 \cdot x_2 \cdot f_{X1 \, X2}(x_1, x_2)dx_1 \, dx_2. \tag{5.6}$$

In the case where the mean value of the random process at t_1 or t_2 is equal to zero, then the autocorrelation function is equal to the covariance of the process at these times (see (5.5)). If the autocorrelation function is zero, then the process is uncorrelated at times t_1 and t_2.

When considering two random processes $\{X(t)\}$ and $\{Y(t)\}$, a so-called cross-correlation function is defined as follows, analogous to (5.6)

$$R_{XY}(t_1, t_2) = E[X_1 \cdot Y_2] = \int_{-\infty}^{\infty} \int_{-\infty}^{\infty} x_1 \cdot y_2 \cdot f_{X1 \, Y2}(x_1, y_2)dx_1 \, dy_2 \tag{5.7}$$

$$= \langle x(t) \cdot y(t + \tau) \rangle = \lim_{T \to \infty} 1/(2T) \cdot \int_{-T}^{T} x(t) \cdot y(t + \tau)d\tau. \tag{5.8}$$

In the notation in (5.8) we have assumed that the time average of the product of the random samples from the two processes at any times $t = t_1$ and $t - \tau = t_2$ is equal to the expected value of the product of the random process levels at these times. Such conformity between the expected value of the process and the time average of a sample is generally termed 'ergodicity'. Ergodicity can be valid only if the random process simultaneously has the property of stationary behaviour, as explained in the following.

If we assume that the probability distributions of the random process are independent of time, then the random process is stationary (homogeneous if we interpret t as a location coordinate), and then

$$E[X_1] = E[X_2] = \cdots = E[X(t)] = \text{const.} \tag{5.9}$$

must also be true.

Further, according to (5.6) we obtain

$$E[X_1 \cdot X_2] = R_{XX}(\tau) = E[X(t) \cdot X(t + \tau)], \tag{5.10}$$

for $\tau = t_2 - t_1$, i.e. the autocorrelation function is only a function of τ. Conversely, a random process for which only (5.9) and (5.10) are valid, is termed weakly stationary (weakly homogenous). If the stationary random process has the expected value zero, then according to (5.4), (5.5) and (5.10), for all values of t

$$R_{XX}(0) = E[X^2] = \text{Var}[X] = \text{Cov}[X, X] = \sigma_X^2. \tag{5.11}$$

The autocorrelation function of a stationary random process is now used, applying a Fourier transform, to the definition of the spectral density or of the (auto) spectrum $S_{XX}(\omega)$ of the random process [7]

$$S_{XX}(\omega) = 1/(2\pi) \int_{-\infty}^{\infty} R_{XX}(\tau) \cdot \exp\{-i\omega\tau\} \, d\tau. \tag{5.12}$$

The inverse Fourier transform gives the autocorrelation function

$$R_{XX}(\tau) = \int_{-\infty}^{\infty} S_{XX}(\omega) \cdot \exp\{i\omega\tau\} \, d\omega. \tag{5.13}$$

According to (5.11), therefore, with (5.13)

$$\sigma_X^2 = \text{Var}[X] = R_{XX}(0) = \int_{-\infty}^{\infty} S_{XX}(\omega) \, d\omega. \tag{5.14}$$

is true for $\tau = 0$. By analogy to (5.12), a so-called cross-spectrum can of course also be defined with (5.7). To avoid confusion with the cross-spectrum, the term 'spectrum' in the following text should always be taken to mean 'auto spectrum'.

5.1.2 Stationary Gaussian Random Process

Our interest in Gaussian random processes is connected with the fact that linear functions of Gaussian random variables are themselves Gaussian random variables, so that the response characteristics of a linear system to excitation by a Gaussian random process are determined on the same principles as the characteristics of the Gaussian random process itself. In particular, when (5.9) and (5.10) are true,

Gaussian random processes are always stationary (homogeneous) in the stricter sense of the term.

Stationary Gaussian random processes have the special property of yielding all interesting characteristic numbers, if the spectral density (5.12) or the autocorrelation function (5.13) is given. The K-dimensional joint probability density function for (5.1) is

$$f_{\mathbf{X}}(x) = \frac{1}{\sqrt{(2\pi)^K |C_{\mathbf{X}}|}} \exp\left\{ -\frac{1}{2}(x - E[X])^T C_{\mathbf{X}}^{-1}(x - E[X]) \right\}. \tag{5.15}$$

for Gaussian random processes. Here

$$(x - E[X])^T = (x_1 - E[X_1], x_2 - E[X_2], \ldots, x_K - E[X_K]),$$
$$E[X]^T = (E[X_1], E[X_2], \ldots, E[X_K]), \tag{5.16}$$

$$C_{\mathbf{X}} = \begin{bmatrix} \mathrm{Var}[X_1] & \mathrm{Cov}[X_1, X_2] & \cdots & \mathrm{Cov}[X_1, X_K] \\ & \mathrm{Var}[X_2] & \cdots & \mathrm{Cov}[X_2, X_K] \\ \mathrm{sym.} & & \cdots & \vdots \\ & & & \mathrm{Var}[X_K] \end{bmatrix} \tag{5.17}$$

and $|C_{\mathbf{X}}| = \det C_{\mathbf{X}}$, with $C_{\mathbf{X}}$ being the covariance matrix, and det being the abbreviation for determinant. For a representation of (5.15) we illustrate in Fig. 5.1 the two-dimensional case

$$f_{\mathbf{XY}}(x,y) = \frac{1}{\sqrt{(2\pi\sigma_{\mathbf{X}}\sigma_{\mathbf{Y}})^2(1-\rho^2)}} \cdot \exp\left\{ -\frac{1}{2(1-\rho^2)}\left(\left[\frac{x - m_{\mathbf{X}}}{\sigma_{\mathbf{X}}}\right]^2 \right.\right.$$
$$\left.\left. - \frac{2\rho(x - m_{\mathbf{X}})(y - m_{\mathbf{Y}})}{\sigma_{\mathbf{X}}\sigma_{\mathbf{Y}}} + \left[\frac{y - m_{\mathbf{Y}}}{\sigma_{\mathbf{Y}}}\right]^2\right)\right\},$$

with $X_1 \to X$, $X_2 \to Y$, $\rho = 0$ (compare (A1.29)), $(m_{\mathbf{X}}, \sigma_{\mathbf{X}}) = (2.0.6)$, and $(m_{\mathbf{Y}}, \sigma_{\mathbf{Y}}) =$

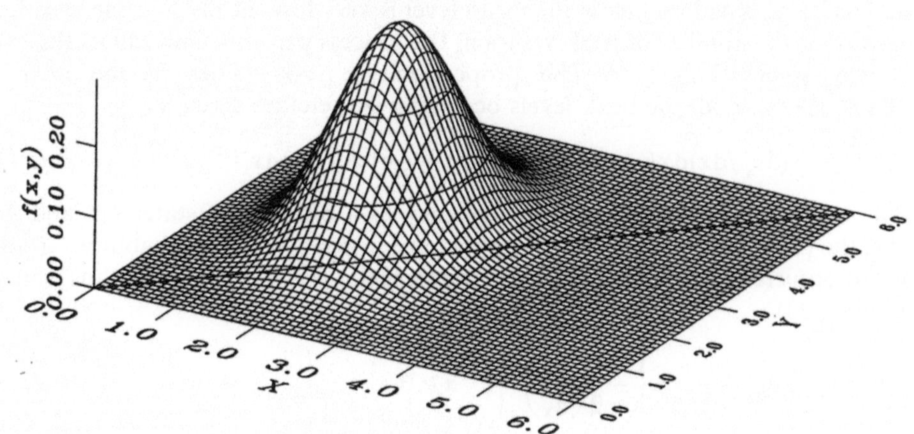

Fig. 5.1. Two-dimensional Gaussian density function exemplified with $m_{\mathbf{X}} = 2.0$ and $m_{\mathbf{Y}} = 3.0$.

(3.0.9). The well-known one-dimensional case follows as

$$f_X(x) = \frac{1}{\sqrt{2\pi}\sigma_X}\exp\left\{-\frac{1}{2}\left(\frac{x-m_X}{\sigma_X}\right)^2\right\}, \qquad (5.18)$$

and with the so-called standard random variable $U = (x - m_X)/\sigma_X$ we obtain the one-dimensional standard normal probability density function

$$\varphi(u) = \frac{1}{\sqrt{2\pi}}\cdot\exp\left\{-\frac{1}{2}u^2\right\}. \qquad (5.19)$$

A characteristic variable of the Gaussian random processes which is particularly important in practical applications is the up-crossing rate $v_x^+(a) = v_a^+$, i.e. a number which signifies how often a stationary Gaussian random process $\{X(t)\}$ exceeds a level $x = a$ per unit time. It can be determined relatively easily using a so-called Rice Formula [8]. From this, we present a particular result, i.e. the zero up-crossing rate (up-crossing rate of level $x = 0$) of the Gaussian random process, which is important for our purposes

$$v_0^+ = 1/(2\pi)\sqrt{\int_{-\infty}^{\infty}\omega^2\cdot S_{XX}(\omega)\,\mathrm{d}\omega/\int_{-\infty}^{\infty}S_{XX}(\omega)\,\mathrm{d}\omega}. \qquad (5.20)$$

This statement is true for stationary Gaussian random processes $\{X(t)\}$ with zero mean.

We now look at two levels, close to each other, $x = a$ and $x = a + \mathrm{d}x$. The difference

$$v_a^+ - v_{a+\mathrm{d}x}^+ = -(\mathrm{d}v_a^+/\mathrm{d}x)\mathrm{d}x \qquad (5.21)$$

then represents the number of relative maxima or peak values of the process occurring between the two levels per unit time. If we assume that the process is narrowband, then we mean with this concept that the probability of relative maxima or peak values below the mean level is very low. In the limiting case this means that the number of peak values in the process per unit time equals the rate of zero up-crossings (v_0^+). The proportion of peak values in the interval $a < x \leqslant a + \mathrm{d}x$ of all the peak levels occurring is therefore given by

$$-(\mathrm{d}v_a^+/\mathrm{d}x)\mathrm{d}x/v_0^+ = P[a < X \leqslant a + \mathrm{d}x] = f_X(a)\mathrm{d}x. \qquad (5.22)$$

Here, we have interpreted relative frequency as a probability statement and, by using the probability density function, we have introduced the probability $f_X(a)\mathrm{d}x$ for the occurrence of peak values in the interval $a < x \leqslant a + \mathrm{d}x$. We thus obtain (see, e.g. [2])

$$f_X(a) = \frac{a}{\sigma_X^2}\exp\left\{-\frac{1}{2}\left(\frac{a}{\sigma_X}\right)^2\right\}, \qquad a \geqslant 0, \qquad (5.23)$$

and for the probability of peak values in the Gaussian random process $\{X(t)\}$, which are greater than $x = a$, it follows that

$$P[X > a] = 1 - F_X(a) = \exp\left\{ -\frac{1}{2}\left(\frac{a}{\sigma_X}\right)^2 \right\}. \tag{5.24}$$

Equation (5.23) is the Rayleigh distribution density of peak values $X = a$ in the narrow-band random process $\{X(t)\}$, and (5.24) is the complementary Rayleigh distribution function $\bar{F}_X(a)$ of those peak values. In this context, the term amplitude is sometimes used instead of peak value, so (5.23) and (5.24) are also termed 'amplitude distributions' of the stationary Gaussian random process $\{X(t)\}$. The probability statement in (5.24) represents the ratio of up crossings of level $x = a$ to the up crossings of level $x = 0$, from (5.21), (5.22) and (A1.12) to (A1.14), therefore

$$P[X > a] = v_a^+/v_0^+. \tag{5.25}$$

Eliminating $P[X > a]$ from equations (5.24) and (5.25) would give a more general form of the Rice Formula than (5.20), which in turn may be verified with $a = 0$.

Assuming ergodicity, we may also interpret (5.24) as the distribution of populations of all amplitudes greater than a when recording an infinitely long random sample from the narrowband stationary Gaussian random process $\{X(t)\}$.

For non-narrowband processes, which we must occasionally deal with when looking at dynamic structural behaviour (see particularly Sect. 6.2.3.3), crossing rates are not so easy to determine. In such cases, it is shown that [9] the so-called regularity factor α of the spectrum, i.e.

$$\alpha = \sqrt{1 - m_2^2/(m_0 m_4)}, \quad 0 \leqslant \alpha \leqslant 1. \tag{5.26}$$

with the moments of the spectrum

$$m_i = \int_{-\infty}^{\infty} \omega^i \cdot S_{XX}(\omega)\,d\omega, \tag{5.27}$$

determines the distributions of the relative maxima, giving the Rayleigh distribution in the case where $\alpha = 0$ and the Gaussian distribution when $\alpha = 1$, as in Fig. 5.2.

We recognize from Fig. 5.2 that as narrowband patterns decrease (or wideband patterns increase) in a random process, we can also expect an increasing number

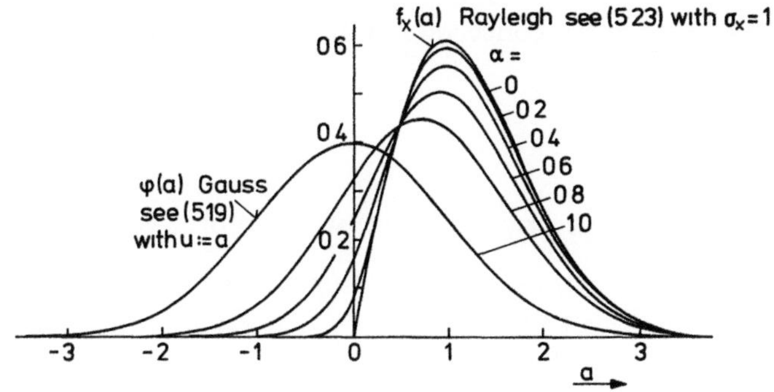

Fig. 5.2. Probability density functions of relative maxima of random processes having different bandwidths.

of relative maxima below the mean level of the process: the spectrum is spread over a wide frequency band. For our purposes, this statement can be taken as a clear definition of the concept of random process wideband patterns, but in this connection it can also be used to define the concept of 'white noise': by 'white noise' we mean a random process with uniform spectral density in the frequency range considered. From this it follows that, according to (5.26) $\alpha = 2/3$ for white noise in the frequency band $0 \leqslant \omega < \omega_{max}$.

5.1.3 Stationary Poisson Random Processes and Markov Chains

This section describes random processes of discrete (or non-continuous) random variables.

5.1.3.1 Poisson Random Processes

This is a process in which certain events occur at random, and we are interested in the frequency N with which these events may (probably) be observed. In this sense we talk about a (random) counting process $\{N(t)\}$, for which we now wish to postulate certain characteristics:

1. The numbers of events occurring at different, exclusive time intervals are independent of each other, i.e. the counting process $\{N(t)\}$ has independent increments. The stochastic distribution of the event number N in any of the time intervals observed is only dependent upon the length τ of the time interval under consideration, i.e. the counting process $\{N(t)\}$ has stationary increments.
2. At time $t = 0$, $N(0) = 0$, i.e. at the start of the counting process, none of the events observed has occurred.
3. The probability that exactly one event occurs in a relatively short time interval is proportional to

$$P[N(\tau) = 1] = P_1(\tau) = v\tau, \qquad v > 0, \tau \to 0.$$

4. The probability that exactly two or more events occur in a relatively short time interval, is $o(\tau)$, i.e. 'low of the order of τ'

$$P[N(\tau) \geqslant 2] = o(\tau).$$

Here, $o(\tau)$ represents every conceivable function $g(\tau)$ which has the following characteristic:

$$\lim_{\tau \to 0} g(\tau)/\tau = 0, \tau > 0.$$

Intuitively formulated, the probability of two or more events in τ is very low.

By means of postulates 1 to 4, a special counting process has been defined of which the characteristics are drawn directly from these postulates (see, for example, the detailed representation in [10], and also [1], [5] and [6]. We summarize here the results which interest us: thus, for every $n = 1, 2, \ldots,$

$$P_n(t) = P[N(t) = n] = (vt)^n \cdot \exp\{-vt\}/n!$$

$P_n(t)$ is termed a Poisson probability mass distribution of the discrete random number N. The stationary counting process $\{N(t)\}$ therefore has the name Poisson counting process, or simply Poisson process.

We now see in the mass distribution that with $m = n - 1$ as in (A1.22) we obtain

$$E[N(t)] = \sum_{n=0}^{\infty} n\cdot(vt)^n\cdot\exp\{-vt\}/n!$$

$$= (vt)\cdot \sum_{n=0}^{\infty} (vt)^{n-1}\cdot\exp\{-vt\}/(n-1)!$$

$$= (vt)\cdot \sum_{m=0}^{\infty} (vt)^m\cdot\exp\{-vt\}/m! = vt, \tag{5.28}$$

as the sum is equal to 1 (certain event) in every probability distribution over the whole range of definition of n. From this the expression mean event rate of Poisson process for the constant v follows, since $v = E[N(t)]/t$.

In practical applications of Poisson processes, we are frequently interested in the (random) waiting time T until the first event. With $n = 0$, the waiting time T with the Poisson mass distribution is

$$1 - F_T(t) = P[T > t] = P[N(t) = 0] = P_0(t) = \exp\{-vt\}, \tag{5.29}$$

as T may only be larger than t if there has been no event before t. This is the so-called 'exponential distribution' which, because of its simple, single-parameter structure, is used in practice whenever the basic counting process fulfils postulates 1 to 4. The waiting time T until the first event therefore has the distribution density as in (A1.15)

$$f_T(t) = dF_T(t)/dt = v\cdot\exp\{-vt\}, \quad t \geqslant 0. \tag{5.30}$$

Because of (A1.22), i.e.

$$E[T] = m_T = \int_0^{\infty} t\cdot v\cdot\exp\{-vt\}\,dt = 1/v, \tag{5.31}$$

we may also interpret v as the reciprocal of the average waiting time until the first event.

5.1.3.2 Markov Chains

The so-called Markov chain, which we wish to present here mainly in preparation for a modern fatigue model to be discussed in Sect. 6.2.2.3, is gaining in importance for more recent developments in marine technology.

First, we envisage a random process $\{D(t)\}$ which may only take on two discrete values, (states) 0 and 1, at discrete times $t = 1, 2, \ldots,$ (e.g. $D(t) = 0$ as an indicator of an intact structural element at time t, and $D(t) = 1$ as an indicator of an unacceptable crack length in the structural element). At time $t = 0$, the process is in the state $D = 0$ with a probability $\pi_{t=0}(D = 0) = \pi_0(0)$ and in the state $D = 1$

with a probability $\pi_{t=0}(D=1) = \pi_0(1)$. The probabilities that the process is in the state $D = 0$ or $D = 1$ at time $t = 1$ are correspondingly $\pi_1(0)$ and $\pi_1(1)$.

The probability that the process changes from state $D = 0$ into the $D = 1$ is p_{01}, and *vice versa* p_{10}, i.e. the probabilities that the process continues to stay in one or the other state are $p_{00} = 1 - p_{01}$ and $p_{11} = 1 - p_{10}$. Then at time $t = 1$ it is true that

$$\pi_1(0) = \pi_0(0) \cdot p_{00} + \pi_0(1) \cdot p_{10},$$
$$\pi_1(1) = \pi_0(0) \cdot p_{01} + \pi_0(1) \cdot p_{11}.$$

In matrix notation these equations may be written as

$$\pi_1^T = \pi_0^T P,$$

where

$$\pi_0^T = (\pi_0(0), \pi_0(1)), \; \pi_1^T = (\pi_1(0), \pi_1(1))$$

and

$$P = \begin{bmatrix} p_{00} & p_{01} \\ p_{10} & p_{11} \end{bmatrix}.$$

We term P a transition matrix which does not change in a stationary process $\{D(t)\}$. If we now define the Markov chain as $\{D(t)\}$, then the development of the process from $t = x$ to $t = x + 1$ is independent of all states that the process had at times $t < x$. Intuitively formulated, the Markov chain develops 'with no memory'. Therefore

$$\pi_2^T = \pi_1^T P,$$
$$\pi_3^T = \pi_2^T P,$$
$$\vdots \qquad \vdots$$

apply. In this way we may develop the stationary Markov chain to any time $t = x$. We thereby establish that there is a relation between the probability distribution at $t = x$ and $t = 0$ (see, for example, [5, 6])

$$\pi_x^T = \pi_0^T \cdot P^x. \tag{5.32}$$

Now we may take the basic statement made initially to simplify matters (that there are only two states $D = 0$ and $D = 1$), and regard (5.32) as a relation generally valid for any number of states $D = d_i$, where $i = 1, 2, \ldots, I$, with the vector π_x now including I components $\pi_x(i)$. The term P therefore represents a quadratic matrix of size $I \cdot I$ with the elements p_{ij}, which are to be interpreted as probabilities of the transition from state d_i into state d_j. Of course, for every i

$$\sum_{j=1}^{I} p_{ij} = 1, \tag{5.33}$$

must be true, as the process is certain, i.e. with 100% probability, to take place in any state d_j. The information contained in π_0 and P is necessary and adequate to describe the stationary Markov chain at any time $t = x$.

5.1.4 Linear Systems with One Degree of Freedom

We now consider the linear differential equation

$$m \cdot \ddot{s}(t) + d \cdot \dot{s}(t) + k \cdot s(t) = f(t),$$

which, in general, describes the equilibrium of the forces (inertial, damping, restoring, excitation) acting on a linear system at any point of time t. Divided by the mass m, this differential equation is given in another form

$$\ddot{s}(t) + 2\delta\omega_0 \cdot \dot{s}(t) + \omega_0^2 \cdot s(t) = f(t)/m, \tag{5.34}$$

with the undamped natural frequency $\omega_0^2 = k/m$, the mass m, the stiffness coefficient k and the damping measure (damping ratio) $\delta = d/(2\omega_0 \cdot m) = d/(2\sqrt{k \cdot m})$. We wish to deal first with harmonic excitation which is particularly important for practical problems, and to examine the system as steady-state. Then the linear system has the same oscillation frequency ω as the excitation force and the motion variable $s(t)$, as response to a harmonic excitation force

$$f(t) = \varphi(\omega) \cdot \exp\{i\omega t\}, \varphi(\omega) = \varphi_R(\omega) + i\varphi_I(\omega), \tag{5.35}$$

is obtained in the form,

$$s(t) = \sigma(\omega) \cdot \exp\{i\omega t\}, \sigma(\omega) = \sigma_R(\omega) + i\sigma_I(\omega), \tag{5.36}$$

and the generally complex amplitudes $\sigma(\omega)$ and $\varphi(\omega)$ are related by $H(\omega)$, i.e.

$$\sigma(\omega) = H(\omega) \cdot \varphi(\omega), \tag{5.37}$$

with the complex transfer or amplification function defined as

$$H(\omega) = 1/[m \cdot (\omega_0^2 - \omega^2 + 2i \cdot \delta \cdot \omega_0 \cdot \omega)]. \tag{5.38}$$

(We may convince ourselves that s is a particular solution of (5.34), from (5.36 to 5.38) by inserting s, $\dot{s}$, $\ddot{s}$ and f in (5.34)). If we think of excitation as merely the real part of $f(t)$ according to (5.35), therefore $f(t) = \varphi_R(\omega) \cdot \cos(\omega t) - \varphi_I(\omega) \cdot \sin(\omega t)$, then the real part of (5.36) alone is to be seen as the solution.

We now consider any excitation function $f(t)$ being composed of an infinite number of harmonic elementary excitations $\varphi(\omega) \cdot \exp(i\omega t)$. In that case,

$$f(t) = \int_{-\infty}^{\infty} \varphi(\omega) \cdot \exp\{i\omega t\} \, d\omega, \tag{5.39}$$

with the Fourier transform

$$\varphi(\omega) = 1/(2\pi) \int_{-\infty}^{\infty} f(t) \cdot \exp\{-i\omega t\} \, dt, \tag{5.40}$$

for which we want to assume that it exists. Since we can combine any number of particular solutions of a linear differential equation to give other solutions,

$$s(t) = \int_{-\infty}^{\infty} \sigma(\omega) \cdot \exp\{i\omega t\} \, d\omega = \int_{-\infty}^{\infty} H(\omega) \cdot \varphi(\omega) \cdot \exp\{i\omega t\} \, d\omega, \qquad (5.41)$$

is true for the steady-state linear system, with the Fourier transform

$$\sigma(\omega) = 1/(2\pi) \int_{-\infty}^{\infty} s(t) \cdot \exp\{-i\omega t\} \, dt. \qquad (5.42)$$

If we construct the excitation function $f(t)$ from a sequence of consecutive impulses $f(\tau)$ at times $t = \tau$ then

$$f(t) = \int_{-\infty}^{\infty} f(\tau) \cdot \delta(t - \tau) \, d\tau \qquad (5.43)$$

is true for an infinitely sustained excitation. Here, the Dirac–Delta function is $\delta(t - \tau) = 1$ for $t = \tau$; otherwise it is always 0. Since the contribution of every pulse to the response function of the transient state is just the same as $f(\tau) \cdot h(t - \tau)$ with h defined as a complex impulse response function to a unit impulse $f(\tau) = 1$ at time $t = \tau$, then

$$s(t) = \int_{-\infty}^{\infty} f(\tau) \cdot h(t - \tau) \, d\tau \qquad (5.44)$$

applies. Equations (5.40) and (5.41), however, give

$$s(t) = \int_{-\infty}^{\infty} H(\omega) \cdot \left[1/(2\pi) \int_{-\infty}^{\infty} f(\tau) \cdot \exp\{-i\omega\tau\} \, d\tau \right] \cdot \exp\{i\omega t\} \, d\omega,$$

i.e. by comparing the components in the last two equations, with $t := t - \tau$

$$h(t) = 1/(2\pi) \int_{-\infty}^{\infty} H(\omega) \cdot \exp\{i\omega t\} \, d\omega \qquad (5.45)$$

and thus the Fourier Transform

$$H(\omega) = \int_{-\infty}^{\infty} h(t) \cdot \exp\{-i\omega t\} \, dt. \qquad (5.46)$$

By Fourier Transform of both sides of (5.44), it can be shown that (5.37) is true, without alteration, also to the interpretation of $\varphi(\omega), \sigma(\omega)$ and $H(\omega)$, extended by (5.40), (5.42) and (5.46).

We now take $f(t)$ and $s(t)$ as random samples in the random processes $\{F(t)\}$ and $\{S(t)\}$. Then

$$E[S(t)] = \int_{-\infty}^{\infty} E[F(t - \theta)] \cdot h(\theta) \, d\theta \qquad (5.47)$$

is valid as in (5.44), with $\theta = t - \tau$. In stationary random processes, the expected value (mean) $E[F(t - \theta)] = E[F(t)]$ is a constant (see (5.9)). Therefore, with (5.46) and (5.47)

$$E[S(t)] = E[F(t)] \cdot \int_{-\infty}^{\infty} h(\theta)\mathrm{d}\theta = E[F(t)] \cdot H(0) \tag{5.48}$$

applies. The following characteristic is therefore valid for linear systems: if the stationary random excitation process has a zero mean level, then the stationary random process of the system response also has a zero mean level.

We now seek a relation between the autocorrelation functions or the spectra of the two processes $\{F(t)\}$ and $\{S(t)\}$, so that we can determine the corresponding characteristics of the response process $\{S(t)\}$ with our knowledge of the system characteristics $H(\omega)$ or $h(t)$, and of the probabilistic characteristics of the excitation process $\{F(t)\}$. In particular, the statements in Sect. 5.1.2 on up crossing rates for Gaussian excitation processes also apply to the response processes, which themselves are Gaussian.

For $\tau = t - \theta$ in (5.44) we get

$$E[S(t) \cdot S(t + \tau)] = E\left[\int_{-\infty}^{\infty} \int_{-\infty}^{\infty} F(t - \theta_1) \cdot F(t + \tau - \theta_2) \cdot h(\theta_1)h(\theta_2)\mathrm{d}\theta_1\,\mathrm{d}\theta_2 \right],$$
$$\tag{5.49}$$

and then considering (5.10) it follows that

$$R_{SS}(\tau) = \int_{-\infty}^{\infty} \int_{-\infty}^{\infty} R_{FF}(\tau + \theta_1 - \theta_2) \cdot h(\theta_1) \cdot h(\theta_2)\mathrm{d}\theta_1\,\mathrm{d}\theta_2. \tag{5.50}$$

We now multiply both sides of (5.50) by $\exp\{-i\omega t/(2\pi)\}$ and integrate from $-\infty$ to $+\infty$. Then, according to (5.12), we have

$$S_{SS}(\omega) = 1/(2\pi) \int_{-\infty}^{\infty} \exp\{-i\omega\tau\}\mathrm{d}\tau \cdot [\mathrm{RHS}(5.50)] \tag{5.51}$$

with the abbreviation RHS meaning 'right hand side of'. Interchanging the integration sequence and supplementing exp factors whose product is 1 gives

$$S_{SS}(\omega) = \int_{-\infty}^{\infty} h(\theta_1) \cdot \exp\{i\omega\theta_1\}\mathrm{d}\theta_1 \int_{-\infty}^{\infty} h(\theta_2) \cdot \exp\{-i\omega\theta_2\}\mathrm{d}\theta_2$$

$$\int_{-\infty}^{\infty} R_{FF}(\tau + \theta_1 - \theta_2) \cdot \exp\{-i\omega(\tau + \theta_1 - \theta_2)\}\mathrm{d}\tau/(2\pi).$$

Substituting $t = \tau + \theta_1 - \theta_2$, we can interpret the third integral in accordance with (5.12) as the spectral density of the excitation process

$$S_{FF}(\omega) = 1/(2\pi) \int_{-\infty}^{\infty} R_{FF}(t) \cdot \exp\{-i\omega t\}\mathrm{d}t. \tag{5.52}$$

From (5.46), the first two integrals are interpreted as complex conjugate transfer functions $H(-\omega)$ or $H(\omega)$, therefore

$$S_{SS}(\omega) = H(-\omega) \cdot H(\omega) \cdot S_{FF}(\omega) \tag{5.53}$$

is true. Since $H(-\omega)$ and $H(\omega)$ are complex conjugate functions, see (5.38), it then follows that

$$S_{SS}(\omega) = |H(\omega)|^2 \cdot S_{FF}(\omega). \tag{5.54}$$

By using the spectral density of the system response from (5.53) and the expected value from (5.48), a weakly stationary response process $\{S(t)\}$ of a system with one degree of freedom (single mass oscillator) is completely defined. If in addition the $\{F(t)\}$ originates from a stationary Gaussian process, then the initial remark in Sect. 5.1.2 holds good, i.e. the statements in Sect. 5.1.2 are just as valid correspondingly for $\{S(t)\}$ as for $\{F(t)\}$, with both random processes also being stationary.

Of course we may substitute any other linear relation, described by the linear differential equation (5.34), for the random variables F and S dealt with here. For example, with wave elevation Z as input signal for a floating offshore platform and motion S as output signal, we can build up the transfer functions $H(\omega)$ necessary for the important equations (5.37) or (5.53) and (5.54), point-by-point, according to the principles of hydromechanical analysis dealt with in Chap. 3, i.e. for harmonic input signals with a sequence of ω frequencies (see Fig. 2.4 and examples in Figs. 2.6, 2.9 and 2.22). Where necessary, we use indices of output and input signal, e.g. H_{sz} in the previously quoted example.

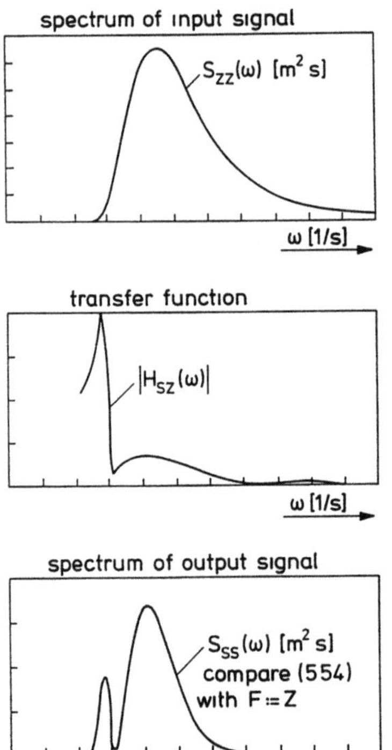

Fig. 5.3. Principle of spectral analysis.

It is on this principle that a major part of the analysis of marine structures, known as spectral analysis, is built. We will be coming back to it more frequently, but at this point we merely wish to represent the basic model of the spectral analysis analogues to (5.54) with $H_{SZ}(\omega)$ schematically in Fig. 5.3. In this case, the phase information between the input and output signal is of course lost.

It should be noted that to simplify matters we have from the outset only been considering steady-state linear systems. For practical applications in marine technology, it is important in this connection to know that for a damping ratio $\delta = 0.05$ a time of approximately four natural periods is sufficient to cause any transient phenomenon affected by an initial state (e.g. displacement from the rest position) to decay. For $\delta = 0.01$, a time of about 20 natural periods is necessary. After this period of time, we may use the term 'steady-state' in the probabilistic sense for a (weakly) stationary response of a linear system.

For the most important applications in evaluating the safety of marine structures, we restrict ourselves to dealing with stationary or weakly stationary excitation or response processes. We indicate in Chap. 6 how we can reduce systems with more than one degree of freedom to the principles developed here.

5.2 Evaluating Stochastic Processes in the Marine Environment

In Sect. 5.2.1 we examine ways of describing a natural seaway probabilistically, i.e. essentially, the classic superposition model, the more modern state space model, as well as several characteristic variables of the seaway derived from these models. In Sect. 5.2.2, these models will be supplemented with statistical models of the seaway related to observations of it, in particular with the presentation of standard spectra and their use for short- and long-term predictions of extreme values of characteristic variables of the seaway, insofar as they are needed for evaluating marine structures. In the final section (5.2.3), several possible methods of presenting the corresponding values of characteristic variables for wind and sea currents are presented.

The marine environment is therefore predominantly described in this section in stochastic (probabilistic–statistical) terms, because this concept seems to us the most suitable for understanding the practical methods of evaluating marine structures in the marine environment. In other fields of research (e.g. geophysics) (in this case, oceanography and marine research), the marine environment is described from the viewpoint of exploration and explanation of certain natural phenomena in any essentially physical way (e.g. in [11]). Some of the geophysical models of the marine environment developed on this basis, and making the most of the increasing possibilities of computers, are also used in the practical evaluation of marine structures in the marine environment, partly in reasonable synthesis with the stochastic models. For example, a number of oil companies engaged in the practical use of offshore resources took part in the North European Storm Study (NESS) project. This involved simulating statistical seaway data and, along with actual statistical weather data, constructing geophysical models of the interactions of the atmosphere and sea for the whole of the North European area in which marine structures are used (see, for example, [12]).

Apart from these parallel developments for a purely stochastic evaluation of marine structures, even a book of the scope of this does not allow complete coverage of the wealth of research results in the field of the purely stochastic approach to the marine environment of marine structures, even only approximately. We must limit ourselves to giving the reader the most important concepts so that, on this basis, he can understand and pursue further developments. For this purpose, we think that the latest report on the subject from the Committee V1 (Environmental Conditions) of the International Ship Structure Congress (ISSC) is particularly suitable [13]. It is published every three years, and gives comprehensive summaries of the latest developments, with relatively detailed comments and references. For the special case of the shallow sea with fetch limited wind effects, which is not considered here, we make special reference to [14].

5.2.1 Probabilistic Description of the Stationary Seaway

If we look at the surface of the natural wind driven seaway (wind-sea), then we can generally identify waves of different height H, length L, and mean direction of propagation μ. We can also recognize in most cases a characteristic of these waves discussed in Chap. 3, namely the higher propagation velocity of longer waves in comparison with shorter waves, which means that the instantaneous deviation (wave elevation) ζ from the still water level (sea level) at any point of the sea surface appears as a result of a random superposition of many elementary waves of different height, length and direction of propagation. In the following, we assume that each of these waves is perfectly regular and infinitely long.

5.2.1.1 *The Superposition Model of the Seaway*

The introductory description of the seaway forms the basis of the so-called 'superposition model' of the seaway, for which the underlying principle has been known a considerable time (see, for example, [15]). To simplify representation of the model, we will examine N stochastically independent elementary waves with the same direction of propagation ($\mu = 0$), measured from an x axis at sea level of an (x, y, z) right hand co-ordinate system, of which the positive z axis points out of the water, and we imagine that each of the elementary waves must only fulfil the conditions of the linear (Airy) wave theory. Then

$$\zeta(t) = \sum_{n=1}^{N} \zeta_{an} \cdot \cos\{-\omega_n t + \varepsilon_n\} = \sum_{n=1}^{N} \zeta_n(t) \qquad (5.55)$$

is true for this so-called 'long-crested seaway' on the whole of the y axis ($x = 0$). Here, $\varepsilon_n = \varepsilon(\omega_n)$ is the phase angle of the elementary waves, i.e. $\zeta_n(0) = \zeta_{an} \cdot \cos \varepsilon_n$ is an elementary wave elevation at time $t = 0$ in $x = 0$, and $\zeta_{an} = \zeta_a(\omega_n)$ is the respective amplitude. We are considering the random character of the process of elementary wave superposition thus described in the classic superposition model by defining the phase angle ε_n as uniformly distributed random phase angles $E_n = E(\omega_n)$. The distribution density function of the phase angles is thus

$$f_{En}(\varepsilon_n) = 1/(2\pi), \quad 0 \leqslant \varepsilon_n < 2\pi. \tag{5.56}$$

The random wave elevation for $x = 0$ is therefore given by

$$Z(t) = \sum_{n=1}^{N} \zeta_{an} \cdot \cos\left\{-\omega_n t + E_n\right\} = \sum_{n=1}^{N} Z_n(t). \tag{5.57}$$

(We use capitals to distinguish random variables and lower case letters to distinguish any realization of these random variables, i.e. here Z is the random wave elevation with the realization ζ, and E is the random phase angle, with realization ε.)

Since we presuppose that none of the elementary waves from which we have formed our natural wind-sea by means of superposition as in (5.57) has a particularly marked effect (equal distribution of random phase angles E), the central limit theorem of probability theory (see details, for example, in [16]) states that the probability density function $f_z(\zeta)$ is a normal distribution for a sufficiently high value of N, with Z according to (5.57) (see (5.18) and (5.19)). We therefore may consider a Gaussian random process $\{Z(t)\}$ to which we can apply our principles developed in Sect. 5.1.2, if this process is stationary (homogeneous in time). We shall return to this point in the following section, but first we must discuss how the amplitudes ζ_{an} of the elementary waves are related to the characteristic seaway parameters, e.g. to its spectrum in the definition from (5.12).

For this we calculate the density function of elementary wave elevations at any time t at the co-ordinate origin $x = 0$. This is on the basis of the transformation equation

$$\zeta_n(t) = \zeta_{an} \cdot \cos\left\{-\omega_n t + \varepsilon_n\right\}, \tag{5.58}$$

which we are using for the monotonic 1-to-1 transformation of the density function $f_{En}(\varepsilon_n)$ into the desired distribution $f_{Zn}(\zeta_n)$ (see (A1.25)).

$$\begin{aligned}
f_{Zn}(\zeta_n) &= f_{En}(\varepsilon_n)/|\mathrm{d}\zeta_n/\mathrm{d}\varepsilon_n| \\
&= f_{En}(\varepsilon_n)/|\zeta_{an} \cdot \sin\left\{-\omega_n t + \varepsilon_n\right\}| \\
&= 2/(2\pi)/\sqrt{\zeta_{an}^2 - \zeta_{an}^2 \cdot \cos^2\left\{-\omega_n t + \varepsilon_n\right\}} \\
&= 1/\pi/\sqrt{\zeta_{an}^2 - \zeta_n^2}, \quad \zeta_n < \zeta_{an}.
\end{aligned}$$

(Because the sine function is not monotonic, we have allowed twice each half monotonic section of the function.) We observe that $f_{Zn}(\zeta_n)$ is symmetrical to $\zeta_n = 0$, i.e. for the expected value

$$E[Z_n] = m_{Zn} = 0. \tag{5.59}$$

The variance of the random variable Z_n is therefore obtained as follows:

$$\mathrm{Var}[Z_n] = \sigma_{Zn}^2 = \int_{-\zeta_{an}}^{\zeta_{an}} \zeta_n^2 \cdot f_{Zn}(\zeta_n)\mathrm{d}\zeta_n$$

$$= 1/\pi \cdot \int_{-\zeta_{an}}^{\zeta_{an}} \zeta_n^2 / \sqrt{\zeta_{an}^2 - \zeta_n^2}\, d\zeta_n$$

$$= 1/\pi \cdot \left[-\zeta_n/2 \cdot \sqrt{\zeta_{an}^2 - \zeta_n^2} + \zeta_{an}^2/2 \cdot \sin^{-1}\{\zeta_n/\zeta_{an}\} \right]_{-\zeta_{an}}^{\zeta_{an}}$$

$$= 1/\pi \cdot \left[\zeta_{an}^2/2 \cdot (\sin^{-1}\{1\} - \sin^{-1}\{-1\}) \right]$$

$$\sigma_{Zn}^2 = 1/\pi \cdot \left[\zeta_{an}^2/2 \cdot (\pi/2 + \pi/2) \right] = \zeta_{an}^2/2. \tag{5.60}$$

The variance in the sum of stochastically independent random variables is equal to the sum of the variances of those variables. Since we have assumed the elementary waves (more accurately, the wave elevations defined by them at any time in any place) to be stochastically independent of each other, then

$$\sigma_Z^2 = \sum_{n=1}^{N} \sigma_{Zn}^2 = \sum_{n=1}^{N} \zeta_{an}^2/2$$

is true for the stationary (Gaussian) seaway process with (5.60). According to (5.14), therefore

$$\sigma_Z^2 = \int_{-\infty}^{\infty} S_{ZZ}(\omega)\, d\omega \doteq \int_{-\infty}^{\infty} \zeta_a^2(\omega)/2, \tag{5.61}$$

is true on condition that an elementary wave of amplitude $\zeta_a(\omega)$ is allocated to every $\omega = \omega_n$ where there is no Riemann integral on the right of (5.61) but a summation $(\sum \rightarrow \int)$. This relationship is certainly satisfied for

$$\zeta_a^2(\omega)/2 \doteq S_{ZZ}(\omega)\, d\omega, \quad \zeta_a(\omega) \doteq \sqrt{2 \cdot S_{ZZ}(\omega)\, d\omega}. \tag{5.62}$$

Thus we have the amplitudes of the different harmonic elementary waves, the superposition of which forms the stationary seaway process $\{Z(t)\}$, in each case interpreted as a strip of infinitesimal width $d\omega$ of the spectrum $S_{ZZ}(\omega)$ of this process: the spectrum defines the amplitudes of the elementary waves by its distribution, and *vice versa*.

Equation (5.62) finally permits a physical interpretation of the spectrum itself: as already outlined in Chap. 3, a wave of length L has the total energy $\rho L \zeta_a^2/2$ (sum of kinetic and potential energy). In relation to the water surface, L is eliminated from this expression, i.e. according to (5.62), so that the whole surface under the spectrum is proportional to the total energy of the seaway. For this reason we also refer to the term energy spectrum $S_{ZZ}(\omega)$ of the seaway. From this we do have something of a problem in obtaining a sensible interpretation in physical terms of the negative wave frequencies ω, to be considered formally as in (5.12). Because of the symmetry of the spectrum, just as of the autocorrelation function in accordance with (5.13), as regards $\omega = 0$, we may limit ourselves to positive (i.e. physically meaningful) frequencies ω if we interpret the seaway spectrum as

$$S_{ZZ}(\omega) \leftarrow S_{ZZ}(-\omega) + S_{ZZ}(\omega), \quad 0 \leqslant \omega < \infty. \tag{5.63}$$

On the subject of the seaway and the effects associated with it on marine structures, in the following we start exclusively from positive wave frequencies without further reference to (5.63) when integrating within the limits $0 \leqslant \omega < \infty$. Thus, from (5.57) and (5.62) with the provisions made for (5.61), we obtain

$$Z(t) = \int_0^\infty \sqrt{2 \cdot S_{ZZ}(\omega) d\omega} \cdot \cos\{-\omega t + E(\omega)\}. \tag{5.64}$$

At the lower integration limit in this equation, it should be recognized that the energy spectrum of the seaway according to (5.63, LHS) is meant. Just as (5.61, RHS) expresses a summation, so does (5.64, RHS), and it must not be interpreted as a Riemann integral.

The seaway superposition model developed thus far has gained outstanding practical significance in relation to the so-called 'spectral analysis', i.e. to the analysis of the seaway and its effects on marine structures in the frequency domain (see Fig. 5.3). For different reasons, which we will go into at the appropriate place, we must, however, switch to simulation techniques from time to time i.e. to observations in the time domain. The seaway superposition model has also for a long time been used for this on the principle of (5.64), yet more recent studies with simulation over longer periods show that certain problems then occur, which we would briefly like to touch on here [17].

We obtain a random sample of the seaway in accordance with (5.64) by means of

$$\zeta(t) = \sum_{n=1}^{N} \sqrt{2S_{ZZ}(\omega_n)\Delta\omega_n} \cos\{-\omega_n t + \varepsilon(\omega_n)\} = \sum_{n=1}^{N} \zeta_n, \tag{5.65}$$

with the spectrum $S_{ZZ}(\omega_n)$ being divided simply into N strips of width $\Delta\omega_n$ at frequency ω_n. From (5.65) we generate a phase angle $0 \leqslant \varepsilon(\omega_n) < 2\pi$ for each of these strips from a uniform density function $f_{En}(\varepsilon_n) = 1/(2\pi)$, and add the elementary wave elevations $\varepsilon_n = \varepsilon(\omega_n)$ represented at each time t, following (5.65). The following questions arise from this:

1. How high should N be?
2. Should $\Delta\omega_n = \omega_{max}/N = \Delta\omega$ be constant?
3. How should $\Delta\omega_n$ be selected if the answer to question 2 is negative?

It is shown in [17] that, for constant $\Delta\omega$, we obtain a random sample $\zeta(t)$ which is periodic over a time interval

$$T = 2\pi N/\omega_{max} = 2\pi/\Delta\omega. \tag{5.66}$$

Assuming real seaway processes are practically stationary over 10 minutes, then $\Delta\omega = 2\pi/600 = 0.0105/s$, i.e. $N = \omega_{max}/\Delta\omega = 300$ with $\omega_{max} = \pi/s$. Figure 5.4 shows a seaway spectrum $S_{ZZ}(\omega)$ as a continuous line, from which the random sample of the seaway presented in Fig. 5.5 was simulated for these data by means of (5.65).

Despite several numerically-based fluctuations, we recognize from the random sample the periodicity of about 600 s expected of a seaway simulated in such a way, i.e. the seaway is realistically represented for only 600 s. For seaway effects which are characterized by periods evidently longer than those of the waves of the seaway (e.g. mooring forces), this may be of great significance. The situation could be inproved by increasing N or, even more effective, by selecting variable

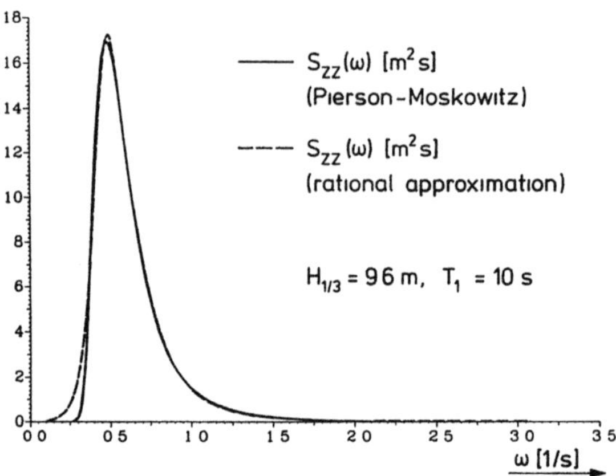

Fig. 5.4. Spectrum of rough seas in two analytical representations according to [17].

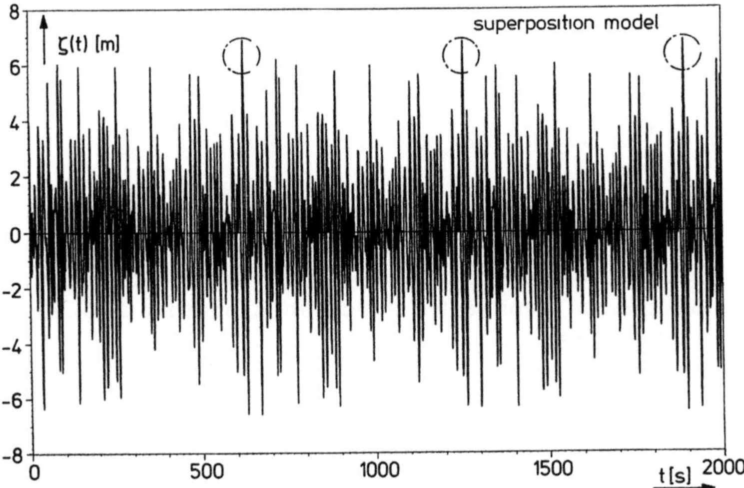

Fig. 5.5. Random sample of rough seas with the spectrum according to Fig. 5.4.

values for $\Delta\omega_n$ such that the N intervals $\Delta\omega_n$ are not rationally related to each other, e.g. by selecting

$$\Delta\omega_n = \sqrt{n/(n+1)}\cdot\omega_{max} \left/ \sum_{n=1}^{N} \sqrt{n/(n+1)}, \right. \tag{5.67}$$

with the autocorrelation function then also losing (gradually) its periodic qualities with increasing simulation time [17].

Nevertheless, this type of simulation is expensive and unsatisfactory for long-term simulations, which we will discuss again in Chap. 6 in connection with mooring problems. In the following, therefore, we will establish the most important principles of an alternative model to the superposition model of the seaway.

5.2.1.2 The State Space Model of the Seaway

With this model we are considering the seaway as the response of a linear system to an input signal which originates from white noise with the spectrum $S_{RR}(\omega)$. We therefore know both the excitation and response spectrum $S_{ZZ}(\omega)$ of the seaway so that our task is to identify the pertinent linear system, i.e. to determine its complex transfer function $H_{ZR}(\omega)$. We will do this exercise formally with reference to (5.53) as follows:

$$S_{ZZ}(\omega) = H_{ZR}(-\omega)\cdot H_{ZR}(\omega)\cdot S_{RR}(\omega) = H_{ZR}(-\omega)\cdot H_{ZR}(\omega), \qquad (5.68)$$

for the uniform/spectral density $S_{RR}(\omega)$ of white noise may be fixed at unity without restricting the generality of this equation. (Index R indicates white noise.) The transfer function $H_{ZR}(\omega)$ is now developed into a rational polynomial, the coefficient of which we can determine from the given spectrum because of (5.68) [17–19],

$$H_{ZR}(\omega) = \sum_{k=0}^{K} b_k\cdot(i\omega)^k \Big/ \sum_{j=0}^{J} a_j\cdot(i\omega)^j, \quad K \leqslant J, a_J = 1. \qquad (5.69)$$

If we designate the Fourier transforms of the random samples of white noise $r(t)$ or of the seaway $\zeta(t)$ by $r_F(\omega)$ and $\zeta_F(\omega)$, then from (5.69)

$$\sum_{j=0}^{J} a_j\cdot(i\omega)^j\cdot\zeta_F(\omega) = \sum_{k=0}^{K} b_k\cdot(i\omega)^k\cdot r_F(\omega).$$

From this, after a Fourier transform of both sides [7], we obtain

$$\sum_{j=0}^{J} a_j\cdot(i\omega)^j\cdot\partial^j\zeta(t)/\partial t^j = \sum_{k=0}^{K} b_k\cdot(i\omega)^k\cdot\partial^k r(t)/\partial t^k,$$

from which the recursive differential equation system for all states $s_j(t)$ is derived

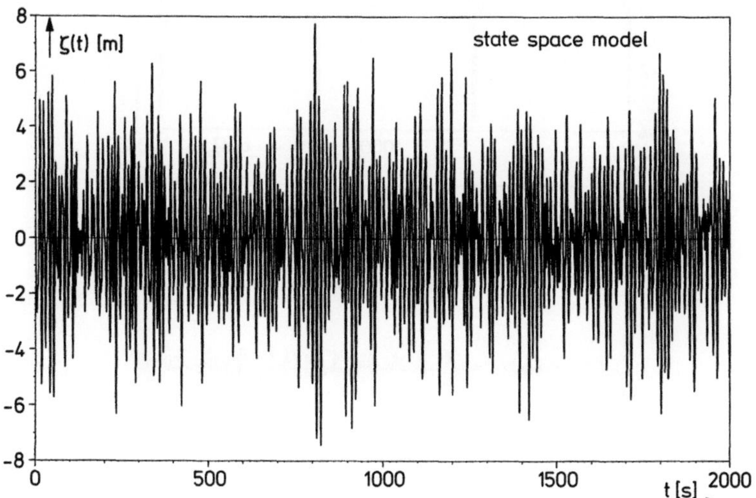

Fig. 5.6. Random sample of rough seas with the spectrum according to Fig. 5.4.

as a state space model [20]

$$\dot{s}_{J-1-j}(t) = s_{J-j}(t) - a_j \cdot s_0(t) + b_j \cdot r(t), \quad j = 0, 1, 2, \ldots, J-1, \tag{5.70}$$

with $s_0(t) = \zeta(t)$, $s_J(t) \equiv 0$, and with the initial condition at time $t = 0$: $s_{J-1-j}(t) = s_{J-1-j}(0)$, where $b_j = 0$ for $j > k$. For any input signal $r(t)$, the output signal $\zeta(t)$ is calculated from (5.70) by known numerical methods (see [17]). With coefficients a_j and b_k being in (5.69), at first glance this model of the seaway, which is perhaps involved mathematically, is numerically very effective and realistic in terms of the periodicity of the classic superposition model, recognizable in Fig. 5.5. Figure 5.6 represents a random sample $\zeta(t)$ for the seaway of the spectrum from Fig. 5.4 (dotted line) approximated rationally from (5.69). This random sample has no periodicity, and the relevant autocorrelation function $R_{ZZ}(\tau)$ shown in Fig. 5.7 declines rapidly to levels which are numerical inaccuracies of the noise generator.

It is true of both methods that from a simulated random sample, seaway parameter are only obtained as statistical, not probabilistic values, so we now turn again to the spectral analysis of a stationary seaway, coming first of all to a purely probabilistic description of those seaway parameters.

5.2.1.3 Probabilistic Seaway Parameters

We first look at the seaway parameters, which we can already define under the conditions valid for stationary Gaussian random processes, namely a characteristic

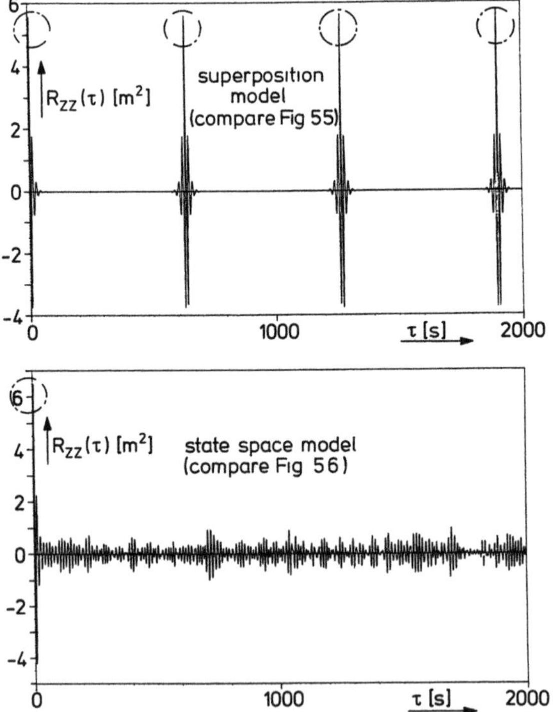

Fig. 5.7. Autocorrelation functions of random samples of Figs. 5.5 and 5.6

wave period as a reciprocal of the rate of zero up-crossing $v_0^+ = v_z^+(0)$ in the seaway process regarded as narrowband, therefore

$$T_0 = 1/v_0^+, \tag{5.71}$$

which can be calculated very simply from the seaway spectrum $S_{ZZ}(\omega)$ according to (5.20) and (5.63), and the significant wave height, which we define as the expected value of the 1/3-highest waves, and could thus be estimated statistically by sorting the wave heights of a measured or calculated (simulated) wave train. The exact value is obtained for a narrowband stationary Gaussian seaway process in accordance with this definition as the centre of gravity of the upper 1/3 area under the Rayleigh density function according to (5.23), (5.24) and (5.2) from the three equations which follow:

$$P[Z > \zeta_{200/3}] = \exp\left\{-(\zeta_{200/3})^2/(2\sigma_Z^2)\right\} = 1/3, \tag{5.72}$$

$$\zeta_{1/3} = 3 \cdot \int_{\zeta_{200/3}}^{\infty} (\zeta^2/\sigma_Z^2) \cdot \exp\left\{-\zeta^2/(2\sigma_Z^2)\right\} d\zeta, \tag{5.73}$$

$$H_{1/3} = 2\zeta_{1/3}. \tag{5.74}$$

Equation (5.72) gives the lower limit $\zeta_{200/3}$ of the integral in (5.73), which we denote as a $200/3\% = 66.7\%$ fractile of the wave amplitudes. (A $p\%$ fractile is the value which is not exceeded with probability p.) Therefore

$$\zeta_{200/3} = \sigma_Z \cdot \sqrt{-2\ln\{1 - 2/3\}} = \sigma_Z \cdot \sqrt{2\ln 3}, \tag{5.75}$$

applies, for which we write

$$\sigma_Z^2 = \int_0^{\infty} S_{ZZ}(\omega) d\omega. \tag{5.76}$$

from (5.14) and (5.63). When considering the spectral density $S_{ZZ}(\omega)$ of the seaway, the significant wave height $H_{1/3}$ may thus be determined from (5.73) to (5.76): we obtain, by substituting $\zeta = \sqrt{2} \cdot \sigma_Z \cdot z$ and $UG = \sqrt{\ln 3}$

$$\zeta_{1/3} = 3 \cdot 2\sqrt{2}\sigma_Z \cdot \int_{UG}^{\infty} z \cdot z \cdot \exp\left\{-z^2\right\} dz = 3 \cdot 2\sqrt{2}\sigma_Z \cdot I,$$

$$I = \left[-0.5 \cdot z \cdot \exp\{-z^2\}\right]_{UG}^{\infty} + 0.5 \cdot \int_{UG}^{\infty} \exp\left\{-z^2\right\} dz,$$

$$I = 0.5\,UG/3 + 0.5\sqrt{\pi}/2 \cdot (1 - \operatorname{erf}\{UG\}).$$

The error function erf is tabulated (e.g. [21]), and we then obtain

$$\zeta_{1/3} = \sigma_Z\left[\sqrt{2\ln 3} + 3\sqrt{\pi}/2(1 - \operatorname{erf}\{\sqrt{\ln 3}\})\right] \simeq 2\sigma_Z. \tag{5.77}$$

If we apply the definition of the i-th moments of the seaway spectral density according to (5.27) and (5.63), namely

$$m_i = \int_0^{\infty} \omega^i \cdot S_{ZZ}(\omega) d\omega, \tag{5.78}$$

then

$$T_0 = 2\pi \cdot \sqrt{m_0/m_2} \tag{5.79}$$

applies; taking into account (5.20) and (5.71) and with (5.74), (5.76) and (5.77) it then follows that

$$H_{1/3} = 4 \cdot \sqrt{m_0} = 4\sigma_Z. \tag{5.80}$$

We go into more detail of the particular significance of the seaway parameters T_0 and $H_{1/3}$ for the practical concerns of marine technology in Sect. 5.2.2. Before that we return once more to (5.72) to (5.74): the centres of gravity of other upper reference surfaces of area $1/n$ of the Rayleigh distribution density function can be calculated in analogue form and style, i.e. in our interpretation mean values of the $1/n$ highest wave amplitudes or heights can be specified. Some of the results can be found in Table 5.1.

Also in this table, we find other characteristics of wave heights which are interesting for practical applications: if, for example, we are looking for an amplitude value ζ_N, which is exceeded just once in N waves, then from (5.72) we get

$$\exp\left\{ -\frac{1}{2}\left(\frac{\zeta_N}{\sigma_Z}\right)^2 \right\} = \frac{1}{N}.$$

From this it follows that

$$\zeta_N = \sqrt{2\sigma_Z^2 \cdot \ln\{N\}}, \tag{5.81}$$

and with (5.76), (5.78) and (5.80) we obtain

$$H_N = H_{1/3} \cdot \sqrt{\ln\{N\}/2}, \tag{5.82}$$

for $H_N = 2\zeta_N$. With $N = 1000$ there would be, for example, one wave exceeding a height $H_{1000} = 1.86 \cdot H_{1/3}$ (see Table 5.1), in which the analogous interpretation is also given as a $p\%$ fractile.

If N is relatively high, then H_N is an extreme value of the natural seaway, regarded as stationary, and (5.82) therefore represents a very important relationship between an extreme and the significant wave height; it would also be very useful for practical purposes to determine a similar relationship for wave periods. We

Table 5.1. Wave-height parameters (wind-sea, deep water)

$1/n$	$H_{1/n}/H_{1/3}$	N	$H_N/H_{1/3}$	p in %	$H_{p\text{-}\%}/H_{1/3}$
1/1	0.63	2	0.59	50.0	0.59
1/3	1.00	7.4	1.00	86.5	1.00
1/10	1.27	500	1.76	99.8	1.76
1/100	1.67	1000	1.86	99.9	1.86

$H_{1/n}$ mean of the $1/n$-highest wave-heights
H_N excess of H_N in N waves
$H_{p\text{-}\%}$ p-% fractile of all wave-heights

return to this in the following section on the basis of statistical seaway information; the purely probabilistic analysis of the seaway developed up to now is still not adequate.

5.2.2 Statistical Analysis of the Seaway

It should be noted at the outset that in this section only a limited selection of the most relevant research results are being selected from a very extensive special subject of seaway research. The possibilities of and limits on statistical principles for the analysis of marine structures will therefore be explained, insofar as the most practically important points of stochastic evaluation methods based on them are recognizable.

5.2.2.1 Short-term Statistics

After explanation of the stationary seaway in the previous section, we must first attempt an analytical representation of the seaway spectrum $S_{ZZ}(\omega)$ in order to apply the results obtained.

From (5.10) and (5.12) we can describe the spectrum mathematically as

$$S_{ZZ}(\omega) = 1/(2\pi) \int_{-\infty}^{\infty} E[Z(t) \cdot Z(t + \tau)] \cdot \exp\{-i\omega\tau\} d\tau. \tag{5.83}$$

Here, $Z(t)$ is the wave elevation related to sea level. For a fully developed stationary wind-sea, it is assumed that the spectrum does not change in its essential shape over a relatively short period of time at higher frequencies (shorter waves), and is therefore in a type of equilibrium state. This equilibrium state is, of course, determined by the same physical parameters that also determine the (random) wave elevation $Z(t)$, and that from the superposition model of the previous section, is visualized as (randomly) composed of many elementary waves of all possible frequencies ω and amplitudes $\zeta_a(\omega)$. For all higher-frequency elementary waves, it is assumed that the sea surface is such that no water at all is released from the wave crests, for which, therefore, the local downward acceleration of the water particles in the vicinity of the wave crest is equal to acceleration due to gravity g. On this assumption, the relevant part of the seaway spectrum is likewise determined by g and, of course, by ω. Because all other physical parameters for higher frequencies are small in comparison with g, a simple dimensional approach to the spectrum, defined in (5.83), gives (see [22]),

$$[S_{ZZ}(\omega)] = [\text{length}^2 \cdot \text{time}] = [g^2 \cdot \omega^{-5}]. \tag{5.84}$$

We therefore find the statement

$$S_{ZZ}(\omega) = \alpha \cdot g^2 \cdot \omega^{-5} \tag{5.85}$$

for higher frequencies. This includes α as a scale factor (Phillips constant) which must be determined from observation for certain seaway or wind conditions. These observations can simultaneously be used to decide where that range of lower frequencies begins, which is not represented by (5.85).

Developing (5.85) further, Pierson and Moskowitz [23] have compared different statements, already suggested previously, for an analytical representation of the seaway spectrum from data, and in so doing have derived a model frequently used in marine technology, the original form of which was

$$S_{ZZ}(\omega) = \frac{\alpha g^2}{\omega^5} \exp\left\{ -\beta \left(\frac{\omega_0}{\omega} \right)^4 \right\}, \quad \omega \geqslant 0. \tag{5.86}$$

An exponential function therefore determines the shape of the spectrum in the lower frequency range (longer waves). For this, $\alpha = 0.0081$, $\beta = 0.74$, $\omega_0 = g/w$, with wind velocity w measured at a height of 19.5 m above sea level.

With (5.78) to (5.80) we obtain, by means of (5.86), a relationship between the wind velocity w on the one hand, and on the other, the seaway parameters $H_{1/3}$ and T_0

$$H_{1/3} = 2\frac{w^2}{g}\sqrt{\frac{\alpha}{\beta}}, \quad T_0 = 2\pi\frac{w}{g}\sqrt[4]{\frac{1}{\beta\pi}}, \tag{5.87}$$

as the necessary integrations may be carried out in closed form. We can thus bring the so-called Pierson–Moskowitz spectrum from (5.86) into one of the more commonly used forms

$$S_{ZZ}(\omega) = 4\pi^3 \frac{H_{1/3}^2}{T_0^4} \cdot \frac{1}{\omega^5} \exp\left\{ -16\frac{\pi^3}{T_0^4} \cdot \frac{1}{\omega^4} \right\}. \tag{5.88}$$

Wind velocity can be eliminated from (5.87), so that the following relationship between seaway parameters $H_{1/3}$ and T_0, which is only true for the Pierson–Moskowitz spectrum, is obtained

$$H_{1/3} = k \cdot \alpha \cdot g \cdot T_0^2, \quad k = 0.99771. \tag{5.89}$$

If, instead of the mean period between consecutive zero upward crossings of wave elevation $Z(t)$ (i.e. instead of $T_0 = 2\pi\sqrt{m_1/m_2}$) we introduce in analogue form a mean period $T_1 = 2\pi m_0/m_1$ to the representation of the Pierson–Moskowitz spectrum, $\omega_1 = 2\pi/T_1$ defines the centre of gravity of the area under the spectrum, and we then obtain a second commonly used form (see Fig. 5.4)

$$S_{ZZ}(\omega) = 5.57\pi^3 \frac{H_{1/3}^2}{T_1^4} \cdot \frac{1}{\omega^5} \exp\left\{ -22.28\frac{\pi^3}{T_1^4} \cdot \frac{1}{\omega^4} \right\}. \tag{5.90}$$

In the case where we apply (5.88) as a two-parameter spectrum for independent values $(H_{1/3}, T_0)$ without regard for the secondary condition given by (5.89), then this in general signifies a modification of the original single-parameter Pierson–Moskowitz spectrum in accordancce with (5.86). The same is true for seaway parameters $(H_{1/3}, T_1)$ in (5.90). Correspondingly, the spectrum according to (5.88) or (5.90) is also more frequently termed a modified Pierson–Moskowitz spectrum.

This seaway spectrum has proved to be a realistic representation in general terms of the fully developed wind-sea in deep water, and for unlimited wind effect area (fetch), but it is not unreservedly valid for shallow water, and in extreme seaway conditions. Extensive measurements were therefore taken off the German North Sea coast in what was called the JONSWAP (Joint North Sea Wave Project) project, on the basis of which the homogeneous JONSWAP spectrum was formulated as follows [24]:

$$S_{ZZ}(\omega) = \alpha g^2 \cdot \frac{1}{\omega^5} \exp\left\{-1.25\left(\frac{\omega_M}{\omega}\right)^4\right\} \cdot \gamma^p,$$

$$p = \exp\left\{-\frac{(\omega - \omega_M)^2}{2b^2 - \omega_M^2}\right\}$$

If $\gamma^p = 1$ is applied, this spectrum represents a third alternative formulation of the modified Pierson–Moskowitz spectrum. The factor γ^p is a function which increases the original Pierson–Moskowitz spectrum in the region of the modal frequency ω_M, and is therefore termed the 'peak enhancement' factor. For the part of the North Sea coast in which the JONSWAP measurements were carried out, the enhancement at ω_M is given by $\gamma = 3.3$, and for the constant b, which controls the width of the enhancement, $b = 0.07$ is then true for $\omega < \omega_M$ and $b = 0.09$ for $\omega \geqslant \omega_m$.

For a Pierson–Moskowitz spectrum, the modal frequency (frequency at the maximum value of the spectrum) is $\omega_M = 4.44/T_0$.

A necessary condition for the alternative representation of Pierson–Moskowitz and JONSWAP spectra for the same seaway is that the total energy, or the variable m_0, as in (5.78), must be equal for both standard formulations. Under this condition, a formulation of the JONSWAP spectrum suitable for practical applications was developed [25], and this formulation allows us to start from the seaway parameters $(H_{1/3}, T_1)$ as in the application of (5.90):

$$S_{ZZ}(\omega) = 5.32\pi^3 \frac{H_{1/3}^2}{T_1^4} \cdot \frac{1}{\omega^5} \exp\left\{-32.29\frac{\pi^3}{T_1^4} \cdot \frac{1}{\omega^4}\right\} \cdot \gamma^p, \tag{5.91}$$

$$p = \exp\left\{-(\omega \cdot T_1/5.32 - 1)^2/(2b^2)\right\}, \gamma = 3.3$$

$$b = 0.07 \text{ for } \omega < 5.32/T_1; b = 0.09 \text{ for } \omega \geqslant 5.32/T_1.$$

The JONSWAP spectrum formulation as in (5.91) gives a modal frequency ω_M (frequency of the highest level of S_{ZZ}), which is not equal to the modal frequency of the Pierson–Moskowitz spectrum of equal energy.

If we want to distribute the seaway energy in different directions μ, then we can use a product statement; for example,

$$S_{ZZ}(\omega, \mu) = S_{ZZ}(\omega) \cdot 2/\pi \cdot \cos^2\{\mu_0 - \mu\}, \quad -\pi/2 < \mu < \pi/2.$$

We may visualize the effect of the $\cos^2$ spreading function by applying (5.64) and (5.90) for $t = 0$ at variable x, y-values (see Figs. 5.8 and 5.9).

$S_{ZZ}(\omega, \mu)$ is termed a 'directional spectrum' in this context, the relevant seaway is then known as short-crested, and μ_0 is the main direction of the sea waves. Other details will be found in, for example, [26], and a series of other standard spectra will be found in [27]. Apart from the more specialized problems of research and development, we generally get by in practice with the aforementioned alternatives.

Fig. 5.8. Natural seaway representation by the superposition model (equations (5.64) and (5.90) with a $\cos^2$-spreading function).

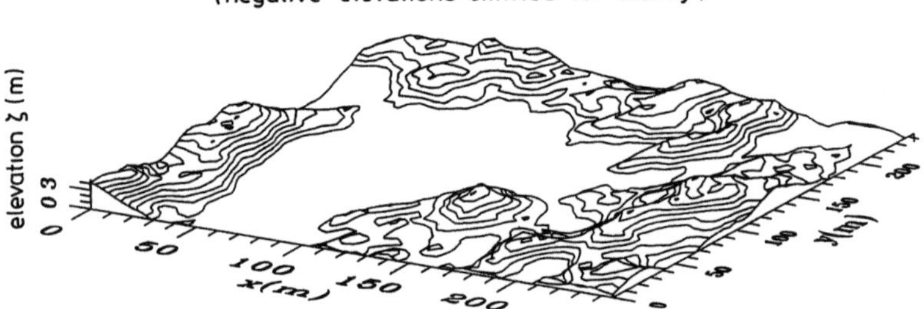

Fig. 5.9. Three-dimensional topography of the natural seaway of Fig. 5.8.

If the parameters $H_{1/3}$, T_0 (and if necessary μ_0) are interpreted as so-called 'seaway class parameters', and if they are therefore allocated values in the range, e.g. $H_{1/3} = 7.0$ to 7.49 m or $T_0 = 5.0$ to 6.99 s, then we can allocate a representative seaway spectrum, e.g. with representative values $(H_{1/3}, T_0) = (7.25\,\text{m}, 6.0\,\text{s})$ for all levels in the selected example. This possibility is of the greatest significance in practical applications as the $(H_{1/3}, T_0$ or $T_1)$ parameters may be estimated extremely well by visual observation of the seaway. This will be demonstrated in Fig. 5.10 with linear regression lines, obtained from numerous different visual observations (H_V) and measurements $(H_{1/3})$, taken from [28].

We recognize from Fig. 5.10 that the following simple relationship between the significant wave height $H_{1/3}$ and the statistical (regressive) mean of the observed wave height, H_V, i.e.

$$H_{1/3} = 4\sqrt{m_0} = 4\sigma_z \simeq H_V, \tag{5.92}$$

is to some extent realistic for the range of moderate to higher waves. We must make it clear in this connection that we are relating statistical parameters with (5.92), for, as a regression value, H_V represents both higher and lower observations, and $H_{1/3}$ is – as has just been explained – used to designate a seaway class, and therefore represents higher and lower values alike. As explained at the beginning, because of their statistical origin, analytical descriptions of a seaway spectrum may not, in any case, accomplish more than an accurate prediction, on average,

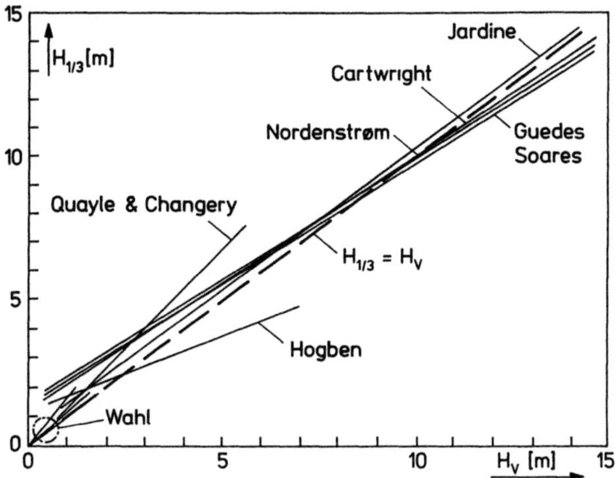

Fig. 5.10. Correlation between H_V and $H_{1/3}$ according to [28].

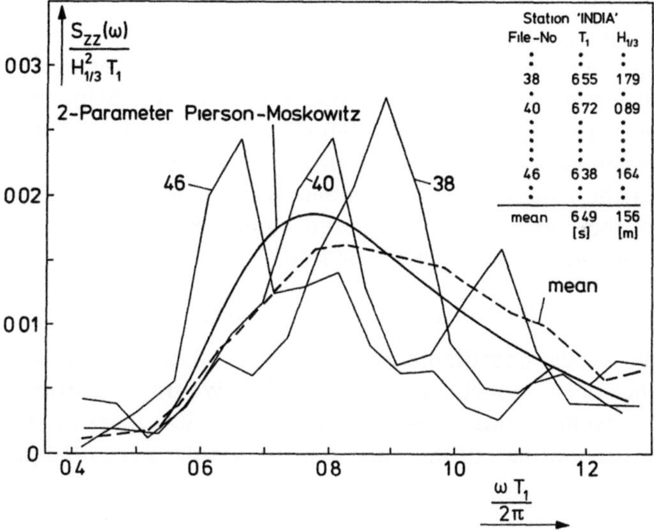

Fig. 5.11. Measured seaway spectra of approximately equal parameters $H_{1/3}$ and T_1.

of the general trend of many random samples from one class of seaway, but measured at different times in any location in the sea, e.g. Fig. 5.11 from [29], where several measured results for the seaway class $(H_{1/3}, T_0) = (1.56\,\mathrm{m}, 6.49\,\mathrm{s})$ are compared. The Pierson–Moskowitz spectrum here represents an approximation for the mean value of the spectra of current random samples. With $H_{1/3} = 0.89$ to $1.96\,\mathrm{m}$ and $T_1 = 6.27$ to $6.74\,\mathrm{s}$, we obtain class widths of $H_{1/3} = 1.07$ and $\Delta T_1 = 0.47\,\mathrm{s}$.

A relationship analogous to (5.92) for the correlation between the mean period T_1 and the mean of the visually-observed periods T_V is just as common in practice. In particular,

$$T_1 = 2\pi \cdot m_0/m_1 \simeq T_V, \quad T_0 = 0.92\,T_1 \tag{5.93}$$

is true for the Pierson–Moskowitz spectrum. The equations (5.88) and (5.90), together with (5.92) and (5.93), pass into the standard spectrum of the fully developed wind-sea, recommended by the ITTC (International Towing Tank Conference), 1969, and the ISSC (International Ship Structure Congress) as early as 1964.

Other seaway parameters may now be calculated according to the probabilistic procedure explained in the previous section, (see, for example, Table 5.1). The results for this are based on statistical estimations of the seaway parameters or of the seaway spectrum, and they are, therefore, themselves statistical point estimations of a random value with confidence limits, not specified in detail, which, as shown qualitatively in Fig. 5.11, may turn out to be relatively wide.

If, for example, we formulate a Rayleigh distribution function analogous to (5.72) for the wave amplitudes,

$$1 - F_Z(\zeta) = P[Z > \zeta] = \exp\{-\zeta^2/(2m_0)\}, \tag{5.94}$$

and in it determine m_0 from a standard spectrum which is representative of one class of seaway, then this distribution is to be interpreted for every value ζ as a statistical point estimation of the exceedance probability. Before we go into the background of this interpretation of (5.94) we would like to present a distribution function (representative of one class of seaway as well) for the wave period:

In Fig. 5.12a the values of the non-dimensional wave period T^* are given on the abscissa of the coordinate system, and values of the non-dimensional wave amplitudes ζ^* on the ordinate [30]. For this

$$T^* = T/T_1 = T/(2\pi \cdot m_0/m_1), \quad \zeta^* = \zeta/\sqrt{2m_0}, \tag{5.95}$$

and every point in the coordinate system in Fig. 5.12a indicates an observation of the occurrence of the associated values (ζ^*, T^*). We notice a stochastic relationship between wave amplitudes and periods which is less marked at higher levels. Figure 5.12b gives contours of the same distribution density of the pair of variates

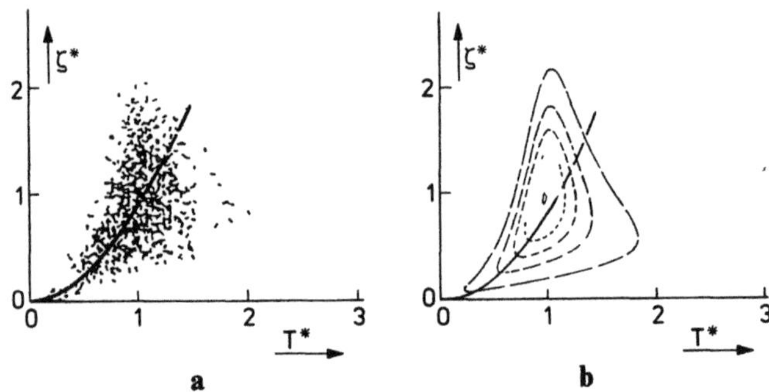

Fig. 5.12. Observed values of ζ^* and T^* according to [30].

$(\zeta^*,\ T^*)$, and it can be seen that the plotted function

$$T^* = (1/0.675^{1/4})\cdot\sqrt{\zeta^*} \tag{5.96}$$

approximately connects the modal values of the conditional distribution densities $f(T^*|\zeta^*)$ of the wave periods T^* [31]. Thus we may interpret (5.96) as a (statistical) estimate of the non-dimensional wave period at a given non-dimensional wave amplitude. If we regard the marginal density function of the non-dimensional wave amplitudes, i.e. the integration of the joint density function of ζ^* and T^* over dT^*, according to (5.94), as a Rayleigh distribution, then

$$f_Z(\zeta^*) = 2\zeta^*\cdot\exp\left\{-\zeta^{*2}\right\} \tag{5.97}$$

is true. By means of monotonic one-to-one transform of this density function in accordance with (A1.25) we obtain from it with the (monotonically increasing) transformation equation (5.96) a marginal density function for the non-dimensional wave periods, representative of one class of the seaway, which is termed a 'non-dimensional Bretschneider distribution density' [32],

$$f_T(T^*) = 2.7(T^*)^3\cdot\exp\left\{-0.675(T^*)^4\right\}. \tag{5.98}$$

The complementary Bretschneider distribution function for dimensional wave periods is then

$$\bar{F}_T(\tau) = 1 - F_T(\tau) = P[T > \tau] = \exp\left\{-0.675\tau^4/T_1^4\right\}$$
$$= \exp\left\{-0.484\tau^4/T_0^4\right\}. \tag{5.99}$$

For the case in which application of joint distributions of wave heights (amplitudes) and periods is necessary, the reader is again referred to [30].

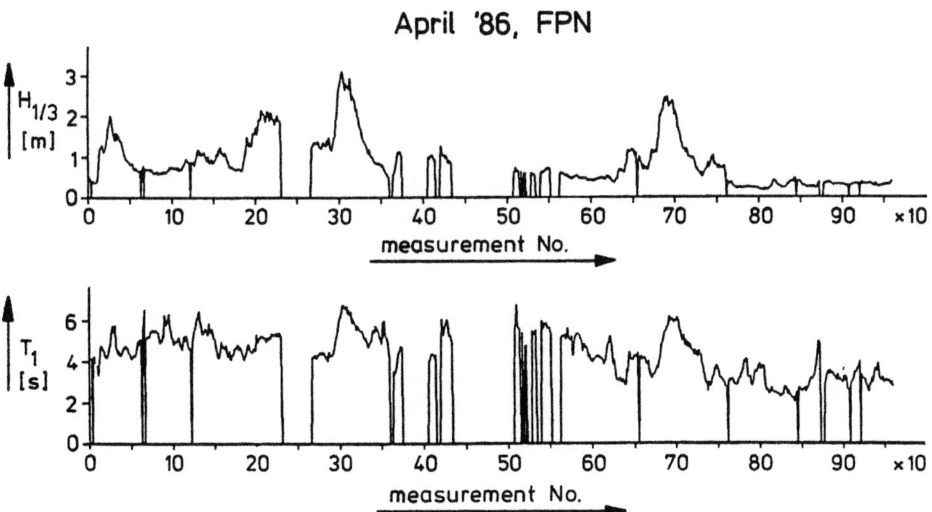

Fig. 5.13. Long-term development of seaway parameters at the FPN research platform according to [34].

5.2.2.2 Long-term Statistics

Until now, we have assumed a stationary seaway, i.e. invariability over time of the representative spectrum for a brief but not closely specified duration. We would now like to consider the seaway statistics over longer periods, for which we cannot assume one stationary state. To illustrate this we will look at characteristic seaway parameters $H_{1/3} = 4\sqrt{m_0}$ according to (5.80) and $T_1 = 2\pi \cdot m_0/m$, according to (5.93), as obtained from measurements at the North Sea Research Platform (see Fig. 6.16 [33], and Fig. 5.13 [34]). Here we have measurements for which a 10-minute sample of wave elevation $\zeta(t)$ was recorded every 3/4 h.

As expected, the premise of seaway parameters representative of the steady-state is only true for some few hours according to how wide or how narrow the class widths ($\Delta H_{1/3}$, ΔT_1) are selected for the various seaway classes represented by the parameters ($H_{1/3}, T_1$). This becomes particularly clear if we determine, from the variation of seaway parameters against time (as in Fig. 5.13), the frequencies with which seaways fall into the classes given (Figs. 5.14 and 5.15) for different class widths.

We see that with the increase in class widths,

$$\Delta H_{1/3} \simeq 0.25\,\text{m} \rightarrow 0.5\,\text{m}, \quad \text{Fig. 5.14,}$$
$$\Delta T_1 \simeq 0.50\,\text{s} \rightarrow 1.0\,\text{s}, \quad \text{Fig. 5.15,}$$

these histograms are smoothed out, because then more measured seaway parameters fall into fewer classes. This means that seaways from classes of greater width also last longer in that class, on average, than seaways from classes of smaller widths. We must therefore interpret the stationary concept here in relation to the class widths considered. In this sense, we define the quasi-stationary concept to describe seaway developments which only take place within a given seaway class. If the class width tends towards zero, the assumption of a stationary seaway still holds good on average for the 10-minutes recording of the seaway sample.

We now wish to look at long-term seaway developments as an ergodic time sequence of randomly arranged seaway classes, in which the seaway may be considered quasistationary. We see from the example in Fig. 5.13 that the seaway parameters $H_{1/3}$ and T_1 do not develop completely independently of each other, so the separate classification as in Figs. 5.14 and 5.15 still does not contain all the statistical information. Usually, therefore, we classify according to the joint occurrence of the class parameters, i.e. according to seaway classes ($H_{1/3i}, T_{ij}$). For the larger of the class widths selected in the sample in Fig. 5.13, the data set out in Table 5.2 are obtained, now converted to relative frequencies in the usual manner, i.e. the number of measurements of seaway parameters in each class was divided by the total number of measurements (in the example 744).

In this connection we also use the term 'scatter diagram' for a statistical representation (long-term statistics) of the local and seasonal wave climate. Large numbers of scatter diagrams, for areas of significance to shipping and marine technology, have been published (e.g. [35–38]); in particular cases, the Hydrographic Institutes responsible for one area may produce these diagrams upon request.

```
WAVE HEIGHT FREQUENCY ANALYSIS
-- ------------------------------------

    744 DATA ITEMS PROCESSED

HISTOGRAM PARAMETERS        DELTA         MIN        CLASSES
                            0.2500       0.0000         15

     CLASS       COUNT
  -------------------
     0.125        54      #*******
     0.375       192      #*************************
     0.625       151      #********************
     0.875       109      #**************
     1.125        90      #***********
     1.375        47      #*******
     1.625        19      #**
     1.875        29      #***
     2.125        20      #**
     2.375        16      #**
     2.625         9      #*
     2.875         7      #
     3.125         1      #
     3.375         0      #
     3.625         0      #


HISTOGRAM PARAMETERS        DELTA         MIN        CLASSES
                            0.5000       0.0000         10

     CLASS       COUNT
  -------------------
     0.250       246      #***********************************
     0.750       260      #*************************************
     1.250       137      #******************
     1.750        48      #*******
     2.250        36      #*****
     2.750        16      #**
     3.250         1      #
     3.750         0      #
     4.250         0      #
     4.750         0      #
```

Fig. 5.14. Statistics of $H_{1/3}$ for the seaway at the FPN research platform, April 1986 according to [34].

Below, we describe two methods suitable to determine long-term extreme wave height values on the basis of data as given in Table 5.2. The methods are illustrated with an example using data from Table 5.2.

Method 1

As in (A1.22), we start from the general definition of the mean value $E[N]$ of a discrete random number N

$$E[N] = \sum_{(n)} n \cdot P[N = n], \quad n = 1, 2, \ldots . \tag{5.100}$$

For this, we may interpret $P[N = n]$ as the marginal mass distribution of a two-

```
WAVE PERIOD FREQUENCY ANALYSIS
-----------------------------------------

    744 DATA ITEMS PROCESSED

HISTOGRAM PARAMETERS          DELTA          MIN        CLASSES
                              0.5000       0.0000         15

      CLASS        COUNT
   -------------------------
      0.250          0      #
      0.750          0      #
      1.250          0      #
      1.750          0      #
      2.250         22      #**
      2.750         67      #*********
      3.250        107      #**************
      3.750         56      #********
      4.250        144      #*********************
      4.750        104      #*************
      5.250        131      #******************
      5.750         78      #***********
      6.250         27      #****
      6.750          8      #*
      7.250          0      #


HISTOGRAM PARAMETERS          DELTA          MIN        CLASSES
                              1.0000       0.0000         10

      CLASS        COUNT
   -------------------------
      0.500          0      #
      1.500          0      #
      2.500         89      #************
      3.500        163      #***********************
      4.500        248      #**********************************************
      5.500        209      #****************************
      6.500         35      #****
      7.500          0      #
      8.500          0      #
      9.500          0      #
```

Fig. 5.15. Statistics of T_1 for the seaway at the FPN research platform, April 1986 according to [34].

Table 5.2. Relative frequencies for FPN (Research Platform, North Sea) April 1986

T_1 in s	$H_{1/3}$ in m							
	0.250	0.750	1.250	1 750	2.250	2 750	3.250	
2.5	0.101	0.019						0.120
3.5	0.156	0.051	0.012					0.219
4.5	0.062	0.122	0.130	0.016	0.003			0.333
5.5	0.012	0.149	0.035	0.048	0.036			0.281
6.5		0 008	0.007		0.009	0.022	0.001	0 047
Σ	0.331	0.349	0.184	0.065	0.048	0.022	0.001	1.000

dimensional probability distribution $P[N = n \cap X_i]$, in which a random variable X may assume values x from the classes X_i (see (A1.9)), i.e. as in (A1.17)

$$P[N = n] = \sum_{i=1}^{I} P[N = n \cap X_i]. \tag{5.101}$$

Likewise, as in (A1.17), (5.100) gives us

$$E[N] = \sum_{(n)} n \cdot \sum_{i=1}^{I} P[N = n \cap X_i] = \sum_{i=1}^{I} \sum_{(n)} n \cdot P[N = n | X_i] \cdot P[X_i]$$

$$= \sum_{i=1}^{I} E[N | X_i] \cdot P[X_i] = \sum_{i=1}^{I} E[N_i] \cdot P[X_i]. \tag{5.102}$$

With reference to (5.100), we see that in (5.102) a conditional mean value of N is defined which we have here abbreviated to $E[N_i]$. If we had now interpreted $P[N = n]$ as a marginal distribution of a three-dimensional probability distribution $P[N = n \cap X_i \cap Y_j]$ then, in the same way, we would have derived the following equation instead of (5.102) with the shortened version $P[X_i \cap Y_j] =: P[X_i, Y_j]$

$$E[N] = \sum_{i=1}^{I} \sum_{j=1}^{J} E[N_{ij}] \cdot P[X_i, Y_j]. \tag{5.103}$$

We now apply this generally valid equation to the special problem of calculating a mean (positive) up crossing rate v_B^+ of a value ζ_B in the long-term seaway process $\{Z(t)\}$ observed over a period of time T_B. This occurs by substituting for X_i in (5.103) the seaway class with the visually observed wave height $H_{Vi} = H_{1/3i}$, and substituting for Y_j the wave period $T_{Vj} = T_{ij}$. We thus obtain

$$v_B^+ = \sum_{i-1}^{I} \sum_{j=1}^{J} v_{Bij}^+ \cdot P[H_{Vi}, T_{Vj}] \tag{5.104}$$

for all more rarely expected events, which we would regard as events in a stationary Poisson process using $v_B^+ = E[N(T_B)]/T_B$, as in (5.28) (see also (5.116)).

We may construe $v_{Bij}^+ = v_B^+(H_{Vi}, T_{Vj})$ as the mean up crossing rate of level ζ_B in the quasi-stationary seaways defined by (H_{Vi}, T_{Vj}), of which the probability of occurrence $P[H_{Vi}, T_{Vj}]$ within T_B is substituted by the relative frequency with which these seaways were observed (see Table 5.2). With (5.25) we obtain, for any stationary seaway with class parameters, (H_{Vi}, T_{Vj}), which we regard as a Gaussian process,

$$v_{Bij}^+ = v_{0ij}^+ \cdot P[Z_{ij} > \zeta_B], \tag{5.105}$$

where

$$v_{0ij}^+ = 1/(2\pi \sqrt{m_{0ij}/m_{2ij}}), \ P[Z_{ij} > \zeta_B] = \exp\left\{ -\zeta_B^2/(2m_{0ij}) \right\}$$

applies, as in (5.24), (5.80), (5.20) and (5.27), and m_{0ij} and m_{2ij} from, e.g. the Pierson–Moskowitz or JONSWAP spectrum, as in (5.90) or (5.91), may be calculated with $H_{1/3} = H_{Vi}$ and $T_i = T_{Vj}$. From (5.104), it thus follows that with

(5.92)

$$n_B^+ = T_B \cdot \sum_{i=1}^{I} \sum_{j=1}^{J} v_{0ij}^+ \cdot \exp\{-2H_B^2/H_{Vi}^2\} \, P_{ij} \qquad (5.106)$$

for the number of up-crossings $n_B^+ = v_B^+ \cdot T_B$ of $H_B = 2\zeta_B$ within a time period T_B. Here we have abbreviated $P_{ij} = P[H_{Vj}, T_{Vj}]$. To simplify, assuming a constant value $v_{0ij}^+ \to v_{0m}^+$ with $v_{0m}^+ = 1/T_{0m}$ (see (5.31)) for all seaways, we may transform (5.106) to and approximate form in frequent practical use

$$n_B^+ \simeq T_B/T_{0m} \cdot \sum_{i=1}^{I} \sum_{j=1}^{J} \exp\{-2H_B^2/H_{Vi}^2\} \cdot P_{ij}, \qquad (5.107)$$

$$n_B^+ \simeq T_B/T_{0m} \cdot \sum_{i=1}^{I} \exp\{-2H_B^2/H_{Vi}^2\} \cdot P_{i}, \qquad (5.108)$$

where the second equation is applied to the marginal histogram $P_i = P[H_{Vi}]$.

If we insert, for example, $T_B = 100$ years and $n_B^+ = 1$, then we obtain an estimate of the 100-year value for wave height with an estimated value for T_B/T_{0m} (order of magnitude $5 \cdot 10^8$ for $T_B = 100$ years) for an iterative evaluation of this equation. We therefore obtain a value $H_B = \zeta_B$, which in a sea area is exceeded on average once in 100 years, with the long-term wave climate of that sea area being described by probabilities P_{ij} or P_i. Such extreme values of wave elevation we define as design values, and correspondingly T_B is the relevant design period of time.

The advantage of this method of establishing design values is based on the fact that when calculating the design value, which, theoretically, may occur in any seaway class, each seaway class is considered according to its probability of occurrence. A disadvantage to be noted is the possibility that extreme seaway conditions are not taken into account when applying seaway statistics which were only provided for a relatively short period of time – as, e.g. in Fig. 5.13. It would therefore not be very sensible to apply Method 1 directly to the data given here for one month only when calculating a 100-year design value of the month of April at the North Sea Research Platform.

It is therefore necessary to try to include non-observed, but possible extremes of the characteristic seaway parameters by statistical estimations of their probability of occurrence. This happens when seaway observations are approximated by means of a suitable stochastic model, i.e. by a suitable distribution function, the parameters of which are estimated on the basis of the observed values (random samples). The Weibull distribution, which has been proven for these purposes in broad areas of science and technology, when linked with visual observations of the significant wave height, has shown itself to be the model best suited for this purpose (see, for example, [39] and [40], where also other distributions are supplied alongside the Weibull distribution):

$$1 - F_{H1/3}(h_{1/3}) = P[H_{1/3} > h_{1/3}] = \exp\{-[(h_{1/3} - h_0)/d]^b\}. \qquad (5.109)$$

Here, the location parameter h_0 is a lower limit for $H_{1/3}$ (for marginal distributions of the significant wave heights generally zero), and d and b are scale or form

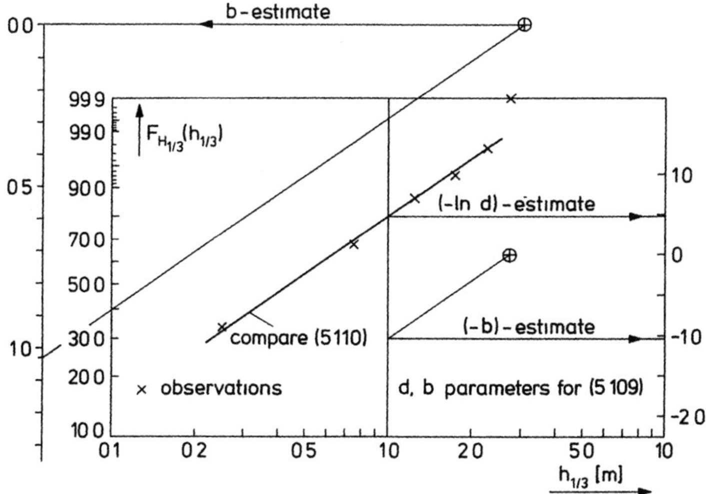

Fig. 5.16. Weibull distribution of $H_{1/3}$ for the seaway at the FPN research platform with data from Fig 5.14 ($\Delta = 0.5$).

parameters which may be determined from the data available, if, when plotted on Weibull paper, a straight line is approximated.

Although there is little data available for a very limited period of time in the example in Fig. 5.14 we can demonstrate the real suitability of the Weibull distribution as a statistical model of the representative characteristic wave height (Fig. 5.16).

Except for the statistically uncertain maximum significant wave height representative of the observed classes, we are therefore able to get a good approximation of histogram values by means of the following Weibull (cumulative) distribution,

$$P[H_{1/3} > h_{1/3}] = \exp\left\{-[h_{1/3}/0.618]^{1\ 03}\right\}. \tag{5.110}$$

We now compare the histogram values of the $H_{1/3}$ marginal distribution, as in Table 5.2, with the values estimated from this Weibull distribution (see Table 5.3 as an illustration).

Method 1 will be applied to the values in Tables 5.3 defined by the Weibull distribution. For $n_B^+ = 1$ and $T_{0m} = 4.5\,\text{s}$ (estimated from Fig. 5.15), with $T_B = 100/12$ years ($= 100$ 'April' months), i.e. $T_B/T_{0m} = 0.58 \times 10^8$, it follows from

Table 5.3. Long-term statistics for FPN, April, with sample data from Fig. 5.11

$H_{1/3}$ in m									
0.25	0 75	1.25	1.75	2.25	2.75	3.25	3 75	4.25	>4 5
$P[H_{1/3}]^a$ 0.331	0.349	0.184	0 065	0 048	0.022	0.001	—	—	—
$P[H_{1/3}]^b$ 0.325	0.380	0.168	0.073	0.031	0 013	0.006	0.002	0.001	0.001

[a]see Table 5.2
[b]from Weibull distribution (5.110)

(5.108) that

$$1.7 \cdot 10^{-8} = \sum_{i=1}^{10} \exp\left\{-2H_B^2/H_{1/3i}^2\right\} \cdot P[H_{1/3i}].$$

An iterative evaluation of this equation gives $H_B \simeq 11.15\,\text{m}$ as the crossed level about once in 100 years in the month of April, with $H_{1/3} > 4.5\,\text{m}$ in the calculations instead of $H_{1/3} = 4.75\,\text{m}$. Without modifying Method 1 using the stochastic model as in (5.110), H_B would be too low.

Method 2

Here, we are seeking the significant wave height $H_{1/3B}$ which is exceeded just once in n_B years, i.e. the design seaway. For this, we first estimate the associated probability $P[H_{1/3} > H_{1/3B}]$ based on an assumption of the duration of that seaway, i.e. for storm duration T_S

$$P[H_{1/3} > H_{1/3B}] := T_S/T_B. \tag{5.111}$$

Here again, T_B is the design period. We now find the associated design value for the significant wave height $H_{1/3B}$ according to (5.110)

$$H_{1/3B} = d \cdot \left[-\ln\left\{T_S/T_B\right\}\right]^{1/b}, \tag{5.112}$$

and we can estimate the mean wave period of the seaway, (e.g. from (5.89))

$$T_{0B} \simeq 3.55 \cdot \sqrt{H_{1/3B}}, \quad (H_{1/3B} \text{ in m, } T_{0B} \text{ in s}). \tag{5.113}$$

With $N = T_S/T_{0B}$ as the number of waves in the seaway, with $H_{1/3B}$, an estimated value, is obtained from (5.82) for the largest individual wave H_{BN} which is therefore exceeded, on average, once in this seaway, i.e. also once in the design period T_B, or

$$H_{BN} \simeq H_{1/3B} \cdot \sqrt{\ln\{N\}/2}. \tag{5.114}$$

Method 2 is illustrated in the example shown in Fig. 5.13 with two assumed times for the storm duration ($T_{S1} = 1\,\text{h}$ and $T_{S2} = 5\,\text{h}$) for a design period $T_B = 100 \times 30 \times 24\,\text{h}$ (100-year value for calculating the 100-year design wave height in the month of April at the North Sea Research Platform). The results are presented in Table 5.4.

It can be seen that the estimate of storm duration T_S has no significant effect on the result, yet it should not have too high a value, as comparison with the result of modified Method 1 shows ($H_B = 11.15\,\text{m}$).

Table 5.4. Design value H_{BN} for FPN, April, with sample data from Fig. 5 11

i	T_{Si} in h	$H_{1/3\,Bi}$ in m	T_{0Bi} in s	N_i	H_{BNi} in m
1	1	6.44	9.01	400	11.15
2	5	5.54	8.36	2154	10.85

($d = 0.618$, $b = 1.03$, see (5.94) and Fig. 5.13)

For practical applications, to evaluate marine structures in relatively deep water and for unlimited area of wind effects (fetch), this detail is, however, of secondary importance, and this is also the case for a whole range of modifications to Methods 1 and 2, details of which we will not go into here. But the question of practical interest is the probability of a design value H_B being exceeded at certain times t within or outside the design period T_B. The following is a comment on this.

For k independent samples with the random result $N_1, N_2, \ldots N_k$ from the same basic population of random numbers N (all random samples are distributed identically), for which there is a valid stochastic model $F_N(n) = P[N \leqslant n]$, we find the stochastic model for the results of all the samples being lower than n from

$$
\begin{aligned}
P_{nk} &= P[N_1 \leqslant n, N_2 \leqslant n, \ldots, N_k \leqslant n] \\
&= P[N_1 \leqslant n] \cdot P[N_2 \leqslant n] \cdot \ldots \cdot P[N_k \leqslant n] \\
&= (P[N \leqslant n])^k = (F_N(n))^k.
\end{aligned}
\tag{5.115}
$$

Now $P[N > n] = 1 - P[N \leqslant n] = 1/n$ is the probability of a value which is exceeded once in n random samples if N is uniformly distributed. We should regard this condition as being fulfilled for one random sample per year from the seaway event in k consecutive years for the n-year design value $H_{Bn}, n > 1$. We thus obtain the probability of H_{Bn} being exceeded in time $T_k = k \cdot T_1, T_1 = 1$ year as

$$
\begin{aligned}
P_{nk} &= P[H > H_{Bn}] = 1 - (1 - 1/n)^k \\
&\simeq 1 - \exp\{-k/n\}, \quad 1/n < 0.1 \\
&= 1 - \exp\{-T_k/T_n\} \\
&= 1 - \exp\{-v_n \cdot T_k\} = P[T \leqslant T_k],
\end{aligned}
\tag{5.116}
$$

where $T_n = n \cdot T_1$ and $v_n = 1/T_n$. Equation (5.29) also gives this result, from which it is possible to make the important inference in connection with (5.104) that the exceedance of an n-year reference value H_{Bn} with low crossing rates v_n may also be interpreted as a stationary Poisson process. In particular, the probability of the design value being encountered annually increases exponentially: for $k = n$ years after exceeding an H_{Bn} design value, it follows that, for example from (5.116), this design value will be exceeded again with a probability $1 - \exp\{-1\} = 0.63$.

In this section we have discussed several fundamental, statistical features of the natural seaway, important for evaluating marine structures, and have also given some indication of confidence limits in these features. Empirical seaway research is now concentrating particularly intensively on confidence limits in the statistical features of the seaway, e.g. in relation to stochastic models and their parameters, as there is a possibility of considering statistical uncertainties quantitatively, to a certain degree, when evaluating marine structures. This is against a background of modern reliability analysis methods, more details of which can be found in Sect. 6.3. Within this book, we restrict ourselves to indicating to interested readers further technical literature (e.g. [41], [42] (especially), or [43] as a more recent investigation of the specialized subject of verification (qualified falsification) of seaway observations as a Gaussian process).

5.2.3 Wind and Sea Currents

We have already mentioned that wind and sea currents in general have relatively less effect on marine structures in comparison with the effects of the seaway. There are, however, a few practical aspects involved in allowing for these environmental influences, which should not be overlooked when evaluating marine structures from a design or safety technology viewpoint. We will go into this briefly in the following.

5.2.3.1 Wind

If we analyse the air movements near the surface of the land or sea, defined as wind, then we detect a mean speed fluctuating relatively slowly (over several hours), this also being known as the 'continuous wind speed'. Relatively rapid, randomly occurring fluctuations, known as gusts or turbulence, are added to the continuous wind speed. The cause of both wind phenomena is first and foremost variations in atmospheric pressure which give rise to relatively steady, so-called 'geostrophic winds' at great heights, where boundary layer effects are negligible, the directions of the winds being determined for a quasi-stationary weather situation by the lines of equal air pressure (isobars). The gradient speed of the geostrophic wind is reduced by boundary layer effects near the earth's surface (between 300 and 600 m above ground), and the previously mentioned turbulence occurs due to momentum exchange between different boundary layers, with the mean wind speed still being subject to a general change in direction near the ground.

For an evaluation of marine structures we are interested in wind in the direct vicinity of the surface of the sea up to a height of about 100 m, i.e. in the region of the lower boundary layer, which is described most effectively within the framework of practical engineering applications by stochastic (statistical–probabilistic) parameters. These may be applied with a relatively modest claim to accuracy in design formulae for the reasons outlined earlier.

The principal stochastic parameters for the wind speed, W defined as a random variable are the mean value $\bar{w}$ and the standard deviation σ_W. When determining $\bar{w}$ from wind observations, the problem of stipulating a sensible length of random sample arises, from which the mean should be taken. This is a similar problem to that involving a sensible class width when evaluating seaway observations, which was covered in more detail in Sect. 5.2.2. Because of the natural characteristics of wind, random samples of the order of 1 h, i.e. representing the 60-min mean values $\bar{w}_{60}$, would seem to be the most sensible of all the possibilities [44, 45], but from the point of view of evaluating marine structures, 10-min mean values $\bar{w}_{10}$ are safer, for the highest 10-min mean is of course higher than the highest 60-min mean, $\bar{w}_{10max} > \bar{w}_{60max}$, so that (at least) $\bar{w}_{10max}$ is regarded as standard for evaluating structural strength under wind load.

The standard deviation σ_W must, of course, be suitably determined for the mean value observed at that time, i.e. we calculate with either the stochastic parameters $(\bar{w}_{60}, \sigma_{W60})$ or $(\bar{w}_{10}, \sigma_{W10})$. If, on this basis, we are dealing with representation of speed peaks as fractiles of the probability distribution $F_W(w)$ (e.g. with the level p which belongs to the highest 3-s mean wind speed, that is, with the $p\%$ fractile belonging to the so-called 'gust' wind speed), then the results, each belonging to one parameter combination ($\bar{w}$, σ_w) and to the probability distribution defined by it, must be equal in spite of different $p\%$ values. For practical reasons, the parameter combination (w_{60}, σ_{w60}) is selected here, as the following empirical relationship can then be applied to determining σ_{w60}, [46],

$$\sigma_{w60} = 2.58\sqrt{\kappa} \cdot \bar{w}_{60}(z_{10}), \quad z_{10} = 10\,\text{m}. \tag{5.117}$$

Here, we apply $\bar{w}_{60}$ to the generally standard observation height $z = 10\,\text{m}$ above sea level, κ is a drag coefficient which we can set at 0.001 for rough sea (or 0.002 for very rough sea).

Data for the highest 60-min mean value $\bar{w}_{60\text{max}}$ over a 50-year period of measurement are given in [47], so it is possible to calculate $\sigma_{w60\text{max}}$ over the same period of time from (5.117). We may thus determine the parameters of the Type 1 extreme value distribution recommended as a stochastic model for wind speeds (e.g. in [48])

$$F_w(w) = \exp\left\{-\exp\left\{-\delta(w-\mu)\right\}\right\}, \tag{5.118}$$

which is also known as a Gumbel distribution, for

$$\delta = 1.282/\sigma_w; \quad \mu = \bar{w} - 0.577/\delta = \bar{w} - 0.45\sigma_w \tag{5.119}$$

is true for the parameters δ (dispersion value) and μ (modal value). If, for example, we use the value $\bar{w}_{60\text{max}} = 34$ m/s as in [47] for the North Sea, just off the coast at $z_{10} = 10\,\text{m}$, then from (5.117), $\sigma_{w60\text{max}} = 2.77$ m/s ($\kappa = 0.001$), and from (5.118) and (5.119)

$$F_w(w) = \exp\left\{-\exp\left\{-0.462(w-32.75)\right\}\right\}$$

applies. In the same way, in [47], a mean value of over 3 s (0.05 minutes) is recommended for determining a peak wind speed, the magnitude of this mean value being given by

$$\bar{w}_{0\,05\text{max}} = 1.37\bar{w}_{60\text{max}} \tag{5.120}$$

over open sea, i.e. in the example $\bar{w}_{0.05\text{max}} = 46.6$ m/s. (Factors higher than 1.37 apply over land.) We obtain $F_w(\bar{w}_{0\,05\text{max}}) = 0.9983$ for the 3-s mean value calculated from (5.120) with the above numerical values, so the gust wind speed defined can thus be interpreted as a 99.8% fractile: about 0.2% of all wind speeds within an hour, in which the highest 60-min mean value $\bar{w}_{60\text{max}}$ expected in 50 years occurs, are higher than $\bar{w}_{0\,05\text{max}}$. This result allows us to define, by means of (5.118) and (5.119), permissible design values $\bar{w}_{0.05\text{max}}$ as 99.8% fractiles on the basis of statistical data if available, with the design values thus obtained being approximately equivalent to the values given in [47]. For this, converting the 50-year time period to any other period of time (k times the 50-year period) may also be represented using (5.115) from [48]

$$F_{wk}(w) = (F_w(w))^k = \exp\left\{-k\cdot\exp\left\{-\delta(w-\mu)\right\}\right\}, \tag{5.121}$$

with the mean value changing, not the standard deviation, in contrast to the initial distributions, as in (5.118):

$$\bar{w}_k = \bar{w} + 1.8\sigma_{wk}\cdot\log\{k\}; \quad \sigma_{wk} = \sigma_w. \tag{5.122}$$

If we have observed, for example, wind speeds $\bar{w}_{60}$ over a period of n years, then we know estimated values for the moments $\bar{w}_{60max}(n)$, and, from (5.117) $\sigma_{60max}(n)$, and from (5.119) the parameters δ_n and μ_n for the associated Gumbel distribution (5.118). On this basis we can estimate, from (5.121), a realistic design value $\bar{w}_{0,05max}$ for the design period of N years

$$\bar{w}_{0\,05max}(N\text{ years}) = \mu_n - \ln\{-n/N\cdot\ln 0.998\}\delta_n. \tag{5.123}$$

In the example quoted above, $n = 50$ years, and with the data we previously used we obtain (e.g. the 100-year design value with $n/N = 50/100$) at $w_{0\,05max}$ (100 years) $= 47.7\,\text{m s}$, i.e. a reference value slightly higher than that of $46.2\,\text{m s}$ ($n = N = 50$) for the 50-year period. This difference is insignificant for applications in marine structures, and any discussion of the question about the correct design period is superfluous against the background of the generally subordinate significance of wind loading in comparison with sea loading, as well as other statistical uncertainties connected with estimating μ_n or δ_n.

Equation (5.123) has a practical use in determining local design loads (e.g. on helicopter decks, etc.) in all the cases for which we have observations for only a relative short period (e.g. for one year or several years ($n = 1,2,\ldots$)), but a 50-year or 100-year design value is of interest. This equation is only valid under the conditions of (5.120); for ratios other than $\bar{w}_{0\,05max}/\bar{w}_{60max} = 1.37$, readers may now develop equations analogous to (5.123) for themselves.

From [47] we can then determine the 10-min mean wind speed applicable to global loading on marine structures from the following relationship, which applies to open sea conditions

$$\bar{w}_{10max}(N\text{ Years}) = \bar{w}_{0\,05max}(N\text{ Years})/1.3, \tag{5.124}$$

where we use the reference value $\bar{w}_{0\,05max}$, which is calculated as described previously. If we eliminate $\bar{w}_{0\,05max}$ from (5.120) and (5.124), we finally obtain the 60-min mean value associated with it

$$\bar{w}_{60max}(N\text{ Years}) = \bar{w}_{10max}(N\text{ Years})/1.054. \tag{5.125}$$

The method described from (5.123) to (5.125) is not the only possible one, but it does have the advantage that the design wind speed values needed for design purposes can be represented relatively simply and generally accurately enough if initial data $\bar{w}_{60}$ are known for a relatively short observation period (e.g. 1 year).

Until now, we have looked at wind conditions only at $z_{10} \simeq 10\,\text{m}$ above sea level. For the range up to $100\,\text{m}$ above sea level, which is of interest to evaluate marine structures, the standard deviation σ_w is practically independent of z [45], i.e. the result from (5.117) is applicable for any z. We identify the coefficient of variation $\sigma_w/\bar{w}(z)$ in this connection also as a gust or turbulence intensity. The latter decreases as the height z increases, for $\bar{w}(z)$ increases within the boundary layer, even with a constant standard deviation, according to an exponential law in the form

$$\bar{w}(z) = \bar{w}(z_{10})\cdot(z/z_{10})^{1/r}, \quad 8 < r < 10. \tag{5.126}$$

If we formalize statement (5.120) as

$$\bar{w}_{0.05\,max} = [1 + \varphi] \cdot \bar{w}_{60\,max}, \quad \varphi = 0.37, \tag{5.127}$$

then it is true that φ is also independent of z because of the independence of σ_w from the height within $0 < z < 100\,m$ above sea level, for φ in this formula does take into account stochastic deviations in wind speed from the mean value due to turbulence. Since July 1984, variable values have been proposed for φ in [47] (Proposed Revisions) at any rate as a function of $\bar{w}_{60\,max}$: from this, for wind speeds over $31\,m/s$, we obtain $\varphi > 0.37$ (e.g. $\varphi = 0.38$ for $\varphi = 34\,m/s$ and $\varphi = 0.43$ for $\bar{w}_{60\,max} = 42\,m/s$). We obtain, independent of the actual values of φ, a simple design formula from (5.126) for determining the highest 3-s mean wind speed effective at different heights z above sea level

$$\bar{w}_{0.05\,max}(z) = [(z/z_{10})^{1/r} + \varphi] \cdot \bar{w}_{60\,max}(z_{10}). \tag{5.128}$$

If we regard $\bar{w}_{60\,max}(z_{10})$ as the cause of a wind-sea with a design significant wave height $H_{1/3B}$, that is as the mean wind speed of the design storm, then we may also start from (5.87) ($z = 19.5\,m$ with (5.126) converted to z_{10} with $r = 8$) for design values, so that the highest 3-s mean wind speed, in accordance with (5.128), is approximately equal to the value

$$\bar{w}_{0.05\,max}(z)\,[m/s] := 6.3[(z/z_{10})^{1/8} + \varphi] \cdot \sqrt{H_{1/3B}[m]}. \tag{5.129}$$

We calculate $H_{1/3B}$, for example, as detailed in Sect. 5.2.2.2 with (5.111) and (5.112). An almost identical equation was also suggested by the IMO (Intergovernmental Maritime Organization) to the Ship Design & Equipment Sub-Committee, item 5 on the agenda for the 30th meeting in March 1987, for the case of simultaneous extreme wind and wave conditions for unlimited wind effect area (fetch) and (quasi) infinite depth of water, otherwise see [14]. It is obviously not valid for particular weather conditions, e.g. for hurricane wind speeds, which are much greater in relation to the significant wave height developing at the same time.

For more detailed evaluations of marine structures using the spectral analysis technique, of which we shall give more details in Chap. 6, the standard deviation σ_w is also calculated more accurately by means of (5.14) instead of (5.117), if we know the wind spectra $S_{ww}(\omega)$ under condition $\bar{w}$. From [49] we obtain

$$S_{ww}(\omega) = 4\kappa\bar{w}^2 \cdot \frac{\zeta^2}{\omega(1 + \zeta^2)^{4/3}}, \tag{5.130}$$

but this can also be calculated, according to [46], from

$$S_{ww}(\omega) = 4\kappa\bar{w}^2 \cdot \frac{\zeta}{\omega(2 + \zeta^2)^{5/3}}, \tag{5.131}$$

with ω being the angular frequency, and $\zeta = \omega\Lambda/(2\pi\bar{w})$ with the empirically defined value $\Lambda = 100\,m$ in (5.130) and $\Lambda = 1800\,m$ in (5.131). The values $\kappa = 0.001$ are again introduced for wind over a rough sea, and thus $\bar{w} = \bar{w}(Z_{10})$

Another standard wind spectrum, in which the effect of height z is included, is described in [50], but such fine details are not in general significant in practice when applied to marine structures, on which the seaway produces much greater effects than the wind.

5.2.3.2 Sea Currents

We give the name 'sea currents' to water movements which transport a body of water of considerable dimensions in a certain direction at measurable speed. For this, we differentiate three types of current, essentially by cause:

1. Thermosaline convection currents which are the result of different temperatures and/or salt contents in the sea.
2. Tidal currents which are essentially caused by the gravitational forces of the moon in its orbit about the earth.
3. Wind-induced sea currents which arise when wind energy is converted to sea-water flow energy, in particular by the complicated process of the wind producing sea waves.

All types of sea currents are influenced by topographical irregularities of the seabed and coasts, and partly by effects of the earth's rotation. These effects are significant when evaluating marine structures, and are described in general as deterministic parameters for the purposes of practical engineering based on observations. Of these deterministic parameters, we are particularly interested in the design value of the velocity of flow u at the sea surface, including all the factors affecting it, and the distribution of u against depth of water z, the so-called 'velocity profile' $u(z)$.

Relevant details on local phenomena connected with thermosaline and tidal currents may be supplied by hydrographic institutes (e.g. [51]). Data on wind-induced surface currents and current profiles are difficult to measure, and yet models suitable for numerical simulation of surface currents are now available (e.g. [52]). When calculating the current loading on marine structures, we are interested in what is termed the Euler mean current, i.e. the (vector) mean velocity over several wave cycles at a fixed location. The values in Table 5.5 are (indirectly) calculated Euler mean velocities of the wind-induced surface current $u(z)$ as a function of wind speed w, as in [52]. They give us a realistic idea of the order of magnitude of natural values.

Table 5.5. Euler mean values for wind-driven oceanic currents $u(z)$ according to [52]

w in m/s	z in m	
	0.0	-2.00
5	1.3[a]	0.42
10	1 2	0 91
20	1 2	0.93
30	1.0	0.86

[a]u in % of w

By approximation we may use $w = \bar{w}_{60}(z_{10})$ when applying percentiles, as in Table 5.5.

The reverse effect (i.e. the seaway affected by a thermosaline or tidal current) may be presented in a (non-linear) expansion of the relationship developed deterministically in Chap. 3 on the basis of the linear wave theory, between wave height and velocity of flow, principally by means of an alteration in the original Pierson–Moskowitz seaway spectrum, as in (5.86) due to sea current [53]

$$S_{Z_u Z_u}(\omega) = 4 S_{ZZ}(\omega)/[\chi \cdot (1 + \chi)^2], \quad \chi = \sqrt{1 + 4u\omega/g}. \tag{5.132}$$

This result implies that a wind speed w (here measured 19.5 m above the surface) converts less energy into wave energy if w and u are acting in the same direction than if the situation were reversed. In practice, it is usual to use spectra without considering the effect of velocity of flow. For this it is consistent if the effect of the current is left out of the calculation of transfer functions of the seaway effect when using the spectral analysis technique, which we presented in principle in Fig. 5.3. The effect of current is only superimposed subsequently on the calculated seaway effect.

The simplest and most usual presentation of the velocity profile $u(z)$ for evaluating a structure affected by sea currents is a result of superimposing the profile $u_p(z)$ of a submarine current, which we define as including thermosaline as well as tidal currents, or both together, and the velocity profile $u_w(z)$ of the wind-induced current

$$u(z) = u_p(0) \cdot [(z + d)/d]^{1/7} + u_w(0) \cdot (z + d)/d, \quad 0 \geqslant z \geqslant -d, \tag{5.133}$$

with z as the vertical coordinate axis, whose origin is at sea level, and d as the depth of water. We obtain a reference value $u_B(z)$ e.g. with $u_{Bp}(0)$ as the statistical maximum of a series of observations of submarine currents over a long period of time, and with $u_{BW}(0) = 0.012 \bar{w}_{60\,max}(z)_{10}$ (see Table 5.5); extra components from the waves themselves (Stokes drift) are also to be added if necessary. See [47] (Proposed Revisions) for other possibilities.

If the current in the seabed area is of particular significance such as, for example, when evaluating safe supports for underwater pipelines, then it is sensible to modify (5.133) (see e.g. [54]). When calculating seaway effects taking into account the surface contour of the sea waves, the velocity profile of the sea current in question must also in consequence be extended right to this surface contour [55]. In practice, however, these issues are of marginal importance, as forces due to sea currents are in general relatively small compared with those due to the seaway.

A design value $u_B(z)$, as in (5.133), is of course dependent to a considerable degree on a realistic definition of reference water depth d_B, if the submarine current component u_{Bp} is determined partly or wholly by a tidal current, for of course water depth can alter noticeably with the tides. Even local atmospheric pressure can have a marked effect on water depth.

In this connection, it is usual to use the term 'sea-level' for the elevation of the surface of the sea above a standard water depth. Sea-level maxima at the location, which were observed over along period of time, must be added to the standard water depth for fixed-location marine structures.

Following a recommendation from the International Hydrographic Conference, the standard value, the sea chart datum (*CD*), should be set at a depth below which the sea level rarely falls. The *CD* is therefore set at the lowest astronomical tide (LAT) taken over a longer period of time in most sea areas, and it is sensible to establish the design water depth d_B as the sum of the highest astronomical tide (HAT) and the depth of water at the sea chart datum d_{CD}, i.e.

$$d_B \geqslant d_{CD} + \text{MHAT}. \qquad (5.134)$$

We use the $\geqslant$ sign because the astronomical tides considered when finding the mean of the highest astronomical tide (MHAT) are subject to astronomical conditions (moon/sun/constellations), not to weather characteristics (atmospheric pressure and wind). For an accurate definition of d_B, particularly in coastal areas, sea-level increases due to weather conditions should be taken into consideration. These, as with all the influencing variables mentioned previously, may be supplied in general by the relevant Hydrographic Institutes for the location of a fixed-location marine structure (see [51] and comments in [47]).

5.3 List of Symbols

This list contains mainly symbols used in different sections of this chapter to signify the same thing. Some are symbols generally found in the technical literature, and some need further clarification by the definition given here. See derived symbols (e.g. $E[A], m_a$ etc.) under X (e.g. $E[X], m_X$).

$\text{Cov}[X_1, X_2]$	Covariance of random values X_1 and X_2
C	Covariance matrix
D	Random number with realisation d_i to characterize discrete states
E	Random phase angle
E_n	Random phase angle of the n-th wave component
$E[X]$	Expected value of $X(=m_X)$
$E[\boldsymbol{X}]$	Vector of expected values
F	Stochastic excitation
$F_X(x)$	Distribution function $P[X \leqslant x]$
$\bar{F}_X(x)$	Complementary distribution function $P[X > x] = 1 - F_X(x)$
H	Wave height
H_B	Wave height design value for T_B
H_V	Observed wave height
H_{V_i}	Observed wave height representative of the i-th seaway class
$H_{XY}(w)$	Complex transfer function between the excitation process $\{Y(t)\}$ and the response processes $\{X(t)\}$
$H_{1/3}$	Significant wave height (mean of 1/3 highest wave heights)
$H_{1/3B}$	Design significant wave height for T_B
$H_{1/3i}$	Significant wave height representative of i-th seaway class
I	Maximum value of i, e.g. number of seaway classes $H_{1/3i}$ or H_{V_i}
J	Maximum value of j, e.g. number of seaway classes T_{ij} or T_{V_j}

K	Maximum value of k, e.g. number of stochastic variables X_k
L	Wave length
N	Maximum value of n, e.g. number of harmonic components $\zeta_n(t)$ to simulate the natural seaway
N	Integer random number (event number)
$\{N/(t)\}$	Random process of integer random number N (counting process)
$\boldsymbol{P}$	Matrix of probabilities p_{ij} for transition from state d_i to d_j
$P[X]$	Probability $P[x < X \leqslant x + \mathrm{d}x] = f_X(x)\mathrm{d}x$
$P[X_i]$	Probability $P[x \in X_i]$
P_{ij}	Probability $P[H_{Vi}, T_{Vj}]$ of seaway observations H_V and T_V from the seaway class with parameters H_{Vi} and T_{Vj}
$R_{XX}(\tau)$	Autocorrelation function of stationary random process $\{X(t)\}$
R_{XY}	Cross-correlation function of stationary random processes $\{X(t)\}$ and $\{Y(t)\}$
$R_{ZZ}(\tau)$	Autocorrelation function of the seaway process $\{Z(t)\}$
S	Stochastic variable of motion
$S_{WW}(\omega)$	Wind speed spectrum
$S_{XX}(\omega)$	Autospectrum (spectrum) of the stationary random process $\{X(t)\}$
$S_{XY}(\omega)$	Cross-spectrum of the stationary random processes $\{X(t)\}$ and $\{Y(t)\}$
$S_{ZZ}(\omega)$	Seaway spectrum (autospectrum) of the stationary random process $\{Z(t)\}$
T	Period
T	Period of harmonic oscillation $2\pi/\omega$
T	'Transpose' (if in the exponent of vectors or matrices)
T_B	Design (operation) period (order of magnitude: years/decades)
T_S	Storm duration (order of magnitude: hours)
T_V	Observed wave period
T_{Vj}	Observed wave period representative of the j-th seaway class
T_0	Mean time between consecutive zero up crossings in the seaway process $\{Z(t)\}$
T_0	'Characteristic' wave period ($= 2\pi/\omega_0$)
T_{0B}	Characteristic wave period of the design seaway
T_{0m}	Long-term mean of T_0 (over several years/decades)
T_1	Mean period ($= 2\pi/\omega_1$)
T_{1j}	Mean period T_1 representative of the j-th seaway class
U	Standardized random variable $(X - m_X)/\sigma_X$
$\mathrm{Var}[X]$	Variance of random variable $X(= \sigma_X^2)$
W	Random wind speed
X	Random variable, also representing E, F, S, W, Z for derived parameters in this nomenclature
X	Random vector: $(X_1, X_2, \ldots, X_K)T$
$\{X(t)\}$	Random process of random variable X
$\{Z(t)\}$	Stochastic seaway process (random process of wave elevation)
$Z_n(t)$	Random variable of wave elevation of the n-th harmonic wave component at time t

b	Weibull distribution form parameter
d	Damping coefficient
d	Water depth
d	Weibull distribution scale parameter
d_i	Realization of random discrete states D
det	Determinant
f	Excitation force
$f_X(x)$	Probability density function of random variable X
g	Acceleration due to gravity $(=9.81\,\mathrm{m/s^2})$
$h(t)$	Complex momentum response function
$h_{1/3}$	Realization of $H_{1/3}$ (if $H_{1/3}$ is a random variable)
i	Counting index $(i = 1, 2, \ldots, I)$, e.g. states d_i; class index for wave heights $H_{1/3i}$ or H_{Vi}
i	$\sqrt{-1}$
j	Counting index $(j = 1, 2, \ldots, J)$, e.g. states d_j; class index for wave periods T_{ij} or T_{Vj}
k	Random sample number
k	Coefficient of restoration
m	Mass
m_X	Mean value of random variable $X(=E[X])$
m_i	i-th moments of the spectrum
n	Counting index $(n = 1, 2, \ldots, N)$ e.g. to characterize the harmonic components of the natural seaway
n	Realization of the integer random value N
n_B^+	Abbreviated form $n_X^+(x_B)$: up-crossings of value x_B in the reference time period T_B
p_{ij}	Probability of transition from state d_i to d_j
s	Motion variable $\dot{s} = \partial s/\partial t$, $\ddot{s} = \partial^2/\partial t^2$
t	Time variable
u	Realization of the standardized random variable U
u	Velocity of sea current
w	Wind speed ($\bar{w}$: mean value)
x	Realization of random variable X
x	Horizontal coordinate axis, perpendicular to y, z
x_i	Realization of random variable X_i
x	Vector $(x_1, x_2, \ldots, x_K)^T$
(x, y, z)	Cartesian coordinate system (right-hand system)
y	Horizontal coordinate axis, perpendicular x, z
z	Vertical coordinate axis, perpendicular to x, y
$\Delta H_{1/3}$	Class width of the seaway class represented by $H_{1/3}$
ΔT_1	Class width of the seaway class represented by T_1
$\Delta \omega_n$	Width of a frequency interval in the spectrum
α	Width parameter of the spectrum (regularity factor)
α	Phillips constant
δ	Damping measure (damping ratio)

$\delta(t-\tau)$	Dirac–Delta function
ε	Realization of the random phase angle E
ζ	Wave elevation (deviation from sea level)
$\zeta(t)$	Random sample of the seaway process $\{Z(t)\}$
ζ_B	Design value of wave elevation in T_B
$\zeta_a(\omega)$	Amplitude of an elementary wave of frequency ω
$\zeta_n(t)$	Realization of $Z_n(t)$
μ	Direction of wave progress
μ_0	Main direction of waves of natural seaway
ν	Event rate
ν	Frequency of harmonic oscillation $1/T$
ν_B^+	Up-crossing rate of ζ_B or H_B
ν_{Bij}^+	Up-crossing rate ν_B^+ in the seaway class with parameters H_{Vi} and T_{Vj}
ν_0^+	Zero up-crossing rate $(=\nu_X^+(x=0))$
ν_{0ij}^+	ν_0^+ in the seaway class with parameters H_{Vi} and T_{Vj}
π	$=3.1415926535$
$\pi_X(i)$	Probability of being in state d_i at time $t=x$
π_X	Vector of state probabilities at time x
ρ	Density
σ_X	Standard deviation of the random process $\{X(t)\}$
$\sigma(\omega)$	Complex amplitude of harmonic motion, Fourier transform of a motion function $s(t)$
τ	Time interval t_2-t_1, realisation of T, if T is regarded as a random number
$\varphi(u)$	Standard normal density function (standardized Gaussian (normal) density function)
$\varphi(\omega)$	Complex amplitude of an harmonic excitation force, Fourier transform of an excitation function $f(t)$
ω	Circular frequency of harmonic oscillations $2\pi/T$
ω_M	Modal value of circular frequency of a seaway spectrum
ω_1	Circular frequency at the centre of gravity of a seaway spectrum
ω_0	Natural circular frequency of an oscillating system, zero up-crossing circular frequency

References

1 Papoulis A. Probability, random variables, and stochastic processes. 2nd ed. New York: McGraw-Hill (1984)

2 Crandall SH, Mark WD. Random vibration in mechanical systems. New York: Academic Press (1973)

3 Newland DE. Random vibrations and spectral analysis. London: Longman (1975)

4 Lin YK. Probabilistic theory of structural dynamics. (Reprint). New York: Krieger (1976)

5 Ross SM. Introduction to probability models. New York: Academic Press (1980)

6 Benjamin JR, Cornell, CA. Probability, statistics, and decision for civil engineers. New York: McGraw-Hill (1970)

7 Bracewell RN. The Fourier transform and its application. 2nd ed. New York: McGraw-Hill (1978)

8   Rice SO. Mathematical analysis of random noise. Bell Syst. Tech. J (1944) 23: 228–332; (1944) 25: 46–156

9   Cartwright DE, Longuet-Higgins MS. The statistical distribution of the maxima of a random function. Proc. R. Soc. A (1956) 237. 212–232

10   Bogdanoff JL, Kozin F. Probabilistic models of cumulative damage New York: Wiley (1985) 19–30

11   Kamenkovich VM Fundamentals of ocean dynamics. Oceanographic Series, Vol. 16. Amsterdam: Elsevier (1977)

12   Francis PE. The north European storm study (NESS). Modelling the offshore environment. Proc Soc. Underwater Technol., London (1987)

13   Committee V1 Reports. Proc 1 to 10. Int. Ship Struct. Cong. (ISSC) (1958, 1961,..., 1988)

14   Shore protection manual Vol. I. Washington D.C.. Dept. of the Army, Coastal Eng. Res. Center US Government Printing Office (1984) 3–42 to 3–66

15   Pierson WJ, Neuman G, James RW. Practical methods for observing and forecasting ocean waves. U.S. Navy Hydrographic Office Publ No. 603 (1955) 23–25

16   Heinhold J, Gaede KW. Ingenieur-Statistik (Engineering Statistics). Munchen: Oldenburg (1972) 156–164

17   Jiang T Rationale Beschreibung des Seegangs in der Zeitebene (Rational description of the seaway in the time domain). Hamburg: Germanischer Lloyd Report (1987)

18   Flower JO, Vijeh N. A note on ratio-of-polynomials curve-fitting of seaway spectra. J. Shipbuilding Prog. (1983) 30

19   Flower JO, Vijeh N. Further consideration of the ratio-of-polynomial form-fit of seawave spectra J. Shipbuilding Prog. (1985) 32 2–5

20   Schmiechen M. On state space models and their application to hydrodynamic systems Univ. of Tokyo. Naut. Rep. 5002 (1973)

21   Abramowitz M, Stegun IA Handbook of mathematical functions. New York: Dover (1970) 295–329

22   Phillips OM. The equilibrium range in the spectrum of wind-generated waves. J. Fluid Mech. (1958) 4: 426–434

23   Pierson WJ; Moskowitz L. A proposed spectral form for fully developed wind seas based on the similarity theory of S. A Kitaigorodskii. J. Geophys Res. (1964) 69: 181–190

24   Hasselmann et al Measurements of wind-wave growth and swell decay during the joint North Sea wave project (JONSWAP). Deutsches Hydrographisches Institut, Series A (8°) (1973) No 12

25   Soding H. Lastannahmen bei der direkten Dimensionierung (Design loads for direct dimensioning). Continuing Education Course 22, Institute of Naval Architecture (IFS), Hamburg (1986)

26   Wiegel RL. (Ed.) Directional wave spectra applications New York: ASCE 1981

27   Chakrabarti SK. Hydrodynamics of offshore structures Southampton, Berlin· Computational Mechanics Publication and Springer 1987

28   Guedes Soares, C. Assessment of the uncertainty in visual observations of wave height Ocean Eng 13 (1986) 37–56

29   Hoffman D, Lewis EV Analysis and interpretation of full scale data on midship bending stresses of dry cargo ship. Ship Struct. Comm. Rep. SSC-196 (1969)

30   Longuett-Higgins. MS. On the joint distribution of wave periods and amplitudes in a random wave field Proc R Soc. London A 389 (1983) 241–258

31   Östergaard C. Written discussion of committee I.1-Rep. Int. Ship Struct Cong 1 (1985) 171–174

32   Bretschneider CL. On the determination of the design ocean wave spectrum. Look Lab/Hawaii 7 (1977)

33   Rathlev J. et al Darstellung kennzeichnender Werte für FPN-Seegangsdaten (Expression of significant values for seaway data from the North-Sea Research Platform). Data records IAP, Univ. Kiel 1986

34   Koch T, Östergaard C. Seegangsprozesse FPN (Seaway processes at the North-Sea Research Platform). Hamburg. Germanischer Lloyd Report STB-1370 (1987)

35   Hogben N, Dacunha NM, Olliver GF (primary contributors) Global wave statistics. Brit. Maritime Technol. (1986)

36   Yoshifumi T, Tsugio M, Shigeo O. Winds and waves of the north Pacific Ocean, 1964–1973. Ship Res. Inst. Tokyo (1980)

37   Schiffsregister der UdSSR: The winds and waves at oceans and seas. Leningrad: Izdatelstwo 'Transport' (1974)

38   Secretary of the Navy. Oceanographic atlas of the north Atlantic Ocean. Sec IV Sea and Swell. U.S Naval Oceanographic Office Reprint (1970)

39   Nordenström N. A method of predicting long term distributions of waves and wave induced motions and loads on ships and other floating structures. Det Norske Veritas (1973) 31

40   Dept. of Energy: Waves recorded at ocean weather station Lima; Dec. 1975–Nov 1981. Inst. Oceanogr. Sc Taunton OTH 84204 (1985)

41   Spanos PD. (Ed.). Probabilistic offshore mechanics. Southampton CML Publ. (1985)

42   Haver S, Moan T. On some uncertainties related to the short term stochastic modelling of ocean waves. In: Spanos PD (Ed). Probabilistic offshore mechanics. Southampton: CML Publ. (1985) 1–16

43   Spidsoe N, Karunakaran D Statistical and directional properties of measured ocean waves Offshore Technol Conf. OTC 5414 (1987) 469–480

44   Van der Hoven I. Power spectrum of horizontal wind speed in the frequency range from 0.0007 to 900 cycles per hour. J. Meteorol. (1957) 14

45   Harris RI. The nature of the wind. Mod. Des. Wind Sensitive Struct CIRIA (1971) 29–55

46   Harris RI. On the spectrum and autocorrelation function of gustiness in high winds. E R.A. Report 5273 (1968)

47   Dept. of Energy Offshore Installations Guidance Notes. Part II. Sect. 2 (1984), as well as Proposed Revisions of Part II. Sect. 2. (1987). London· HMSO.

48   Rackwitz R. et al. Wind velocities in western Europe First order reliability concepts for design codes basic notes on actions. A–07. Bull D'Information No 112. Joint Committee CEB/CECM/ CIB/FIP/IABSE on Structural Safety (1976) A–00.6

49   Davenport AG. The spectrum of horizontal gustiness near the ground in high winds. Q J R Soc. (1961) 87: 194–211

50   Kaimal JC, Wyngaard JC, Izumi Y, Cote OR. Spectral characteristics of surface layer turbulence. Q.J.R. Met. Soc. (1972) 98: 563–589

51   Atlas der Gezeitenstrome für die Nordsee, den Kanal und die Britischen Gewasser (Atlas of tidal currents for the North Sea, the Channel, and British Waters). Hamburg: Deutsches Hydrographisches Inst. (1963)

52   Jenkins AD. A dynamically consistant model for simulating near surface ocean currents in the presence of waves. Modelling the offshore environment. Proc. Soc Underwater Technol., London (1987)

53   Hunag NE, Davidson TC, Tung CC, Smith J. Steady non-uniform currents and gravity waves with applications for current measurements. J. Phys Oceanography (1972) 2: 421–431

54   Deigaard R, Jacobson V, Bryndum MB A critical review of near bottom boundary layer models with special attention to stability of marine pipelines. Proc. 4th Int OMAE Symp 1. New York: ASME (1985) 542–549

55   Eastwood JW, Townend IH, Watson CJH. The modelling of wave current velocity profiles in the offshore design process. Modelling the offshore environment. Proc. Soc. Underwater Technol., London (1987)

6. Evaluation of Marine Structures

As mentioned in the introduction to Chap. 5, we will concentrate mainly on evaluating the characteristics of marine structures determined by the sea, and on some effects of wind and currents. While in Chap. 5 we dealt with stochastic analysis of the marine environment itself represented by these influences, in this chapter we analyse its effects on marine structures. Because the seaway could be considered with classical stochastic (probabilistic–statistical) methods within the framework of short- or long-term analysis (see Sect. 5.2.2.1, 5.2.2.2), we will also use this concept here for analysing the seaway effects, which can be considered specifically as decisive for the dimensioning of marine structures.

In Sect. 6.1 we first describe the short-term analysis of various marine structures, for which specific methods of analysis have been developed over the years. In Sect. 6.2 we then deal with long-term analysis, which is based on the former, on the one hand for representing design values which permit safe dimensioning of the structure for extreme seaway conditions, and on the other hand for representing fatigue and serviceability criteria that ensure the use of the structure in all seaway conditions to be expected during service. We thus lay the basis for an analysis and evaluation concept for marine structures which can be regarded as classical, while highlighting some specific aspects of the methods associated with the multiplicity of types of work platforms in the sea, whether free floating, flexibly moored and anchored or fixed to the sea bottom.

In Sect. 6.3 we finally describe additional analysis methods based on modern principles of reliability analysis, which now permit more extensive treatment of some design and safety problems of marine structures. In the whole field of civil engineering, these principles are a prerequisite for understanding the latest design regulations, which we deal with in Sect. 7.4.1. In this chapter our first aim is to introduce the reader to modern developments from which marine technology expects progress in the direction of rational design and operation strategies.

The whole evaluation concept, whether classical or modern, is based on stochastic evidence as an essential component of any decision which must be taken during the development and operation planning on a marine structure. This means for the decision maker, whether designer or user, taking risks, knowing their magnitude, causes and effects (if they occur), in order to control them. Therefore, in our opinion, it is more important in this chapter to offer as wide as possible a view of various practical aspects of the stochastic analysis of marine structures, rather than to examine one or other area of a problem in the greatest possible detail.

6.1 Classical Methods of Short-Term Evaluation

In Sect. 6.1.1 we present the concept of the spectral analysis method applied to motions of a semisubmersible, and describe some possibilities for practical design or operation decisions. The analysis of offshore mooring or anchoring of semisubmersibles and other floating structures leads to a whole series of special problems, which we subsequently deal with in Sect. 6.1.2, where we compare the results of sample calculations made on the basis of the spectral analysis method as well as time simulation. In Sect. 6.1.3 we deal finally with specific aspects which must be considered in the analysis of fixed structures.

The methods presented here essentially serve as the basis for the long-term analysis of marine structures dealt with in Sect. 6.2. Short-term analyses are, when considered alone, of only marginal practical importance, so we restrict ourselves to discussing the most important aspects. We regard these methods as classical because they now constitute a self-contained evaluation concept, well proven in practice, which has been used for a long time without fundamental alterations.

6.1.1 Floating Structures

We refer to the concept designated as a spectral analysis method in Sect. 5.1.4, which is particularly well suited to the analysis of marine structures in the marine environment, because seaway and wind, the two most important stochastic excitations for marine structures, can be modelled by Gaussian random processes, the spectra of which can be realistically represented within certain confidence limits (see Sect. 5.2.2, 5.2.3), and because the system characteristics of marine structures which are most important for actual service can often be modelled linearly (see Chap. 3 and 4). There are two reasons for only dealing initially with floating structures: first, this accords with the historical development of marine technology, because the exploration of crude oil resources in deep waters led first to the development of suitable floating structures, called semisubmersibles, and second, the spectral analysis method can be illustrated particularly well with rigid body motions of floating structures.

We consider as an example the heave motion s_3 which is particularly important for successful drilling operations of a semisubmersible. For this we have used the value of the transfer function $|H_{S3Z}(\omega)|$ in Fig. 6.1b based on the example of a special semisubmersible design [1]. We can find the value of the transfer function most easily from the (real) amplitude ratio s_{3a}/ζ_a for various frequencies ω, for

$$s_a(\omega)/\zeta_a(\omega) = \sqrt{s_a^2(\omega)/\zeta_a^2(\omega)}.$$
$$= \sqrt{2S_{SS}(\omega)d\omega/(2S_{ZZ}(\omega)d\omega)}$$
$$= |H_{SZ}(\omega)|, \tag{6.1}$$

applies, generalizing (5.54) and (5.62) for a random response process of the motion s of a linear system, excited by the (Gaussian) seaway process, with the amplitude $s_a(\omega) = \sqrt{\sigma^2(\omega)} = \sqrt{\sigma_R^2(\omega) + \sigma_I^2(\omega)}$ (see (5.36)). By calculating the amplitudes s_{a_i} of the motion components s_i $(i = 1, 2, \ldots, 6)$ of a semisubmersible in elementary

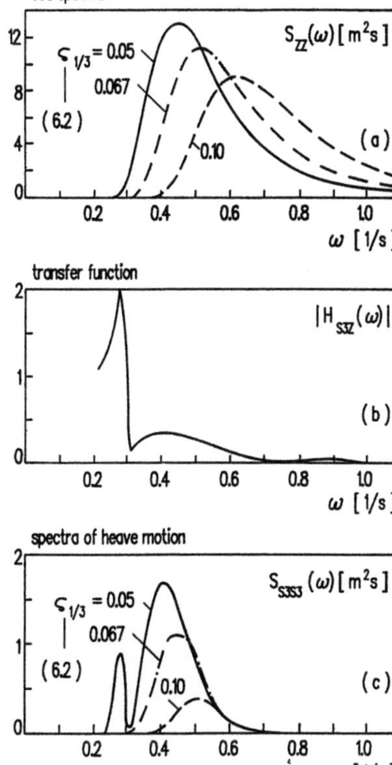

Fig. 6.1. Spectral analysis of heaving motion of semi-submersible in a seaway [1].

(harmonic) waves of amplitude ζ_a (cf. details in Chap. 3), we obtain each relevant transfer function point by point, i.e. for selected values of ω, e.g. the transfer function $|H_{S3Z}(\omega)|$ of the heave motion $s_3(\omega)$. Here it is possible to take into consideration all cross-couplings with other rigid-body motion values s_i ($i = 1, 2, 4, 5$ and 6 in the example with the heave motion s_3) which may occur simultaneously. In connection with $s_a(\omega)/\zeta_a(\omega)$ we also refer to the term Response Amplitude Operator (RAO) at frequency ω.

To demonstrate the effects of various seaways we have used the Pierson–Moskowitz Spectrum, as in (5.88), defined by the significant wave steepness

$$\varsigma_{1/3} = 2\pi \cdot H_{1/3}/(g T_0^2) \tag{6.2}$$

as a parameter in Fig. 6.1a. The significant wave height $H_{1/3}$ was chosen such that all seaways have approximately the same energy in a different distribution, and we can see that the effects of the seaways thus defined, represented in Fig. 6.1c as spectra $S_{S3S3}(\omega)$ of the heave motion, are clearly different, despite their relation to almost the same total wave energy. This is due to the influence of the transfer function $H_{S3Z}(\omega)$ as a seaway response characteristic of the semisubmersible being considered.

In principle, representations like those in Fig. 6.1 are satisfactory for fully characterising a marine structure's behaviour in natural seaways, and for analysing its seaworthiness with regard to any particular response to the action of stationary seaways. In practice, however, another representation has proved more suitable in the (rare) cases where decisions have to be taken on the basis of such short-term characteristics of floating structures in a seaway: if we relate the significant seaway effect, which we can, for example, represent by the significant double amplitude of the motion s for linear systems subject to Gaussian excitation as a generalization of (5.14) analogously to (5.80)

$$2s_{1/3} = 4 \sqrt{\int_0^\infty S_{ss}(\omega)d\omega,} \tag{6.3}$$

to the significant wave height $H_{1/3}$, we obtain, depending on the characteristic wave period T_0, which we can vary in the Pierson–Moskowitz Spectrum in (5.88), a significant transfer function $2s_{1/3}/H_{1/3}$, which we have shown in Fig. 6.2 for the leave motion s_3, using the example in Fig. 6.1 [1].

With a significant transfer function such as that in Fig. 6.2, we can immediately read off the associated characteristic double amplitude $2s_{1/3}$ of any relevant seaway effect for any seaway class $(H_V, T_V) \simeq (H_{1/3}, 1.087T_0)$. (cf. (5.92) and (5.93)), and can draw conclusions from that for extreme values in a stationary seaway, generalizing from (5.82) to the motion value considered. This concept is used in practice, for instance, in operation decisions for offshore cranes [2, 3] and barges for laying undersea pipe lines [4].

Of course, the spectral analysis method described in Fig. 6.1 using the example of a motion value can be extended to all seaway effects which have a linear relationship with the seaway. For example, it can be applied to stresses and deformations of the structure, provided the latter can be considered linearly elastic. It has already been shown [5] that for such cases, from a practical point of view, it can be useful to apply the unit loads concept, i.e. to initially use unit loads and unit accelerations at the centre of gravity instead of the actual hydrodynamic loads and accelerations for the strength analysis of a floating structure, and then, for all elementary waves under consideration, to multiply the calculated hydrodynamic loads and accelerations by the previously determined effects of the unit loads and

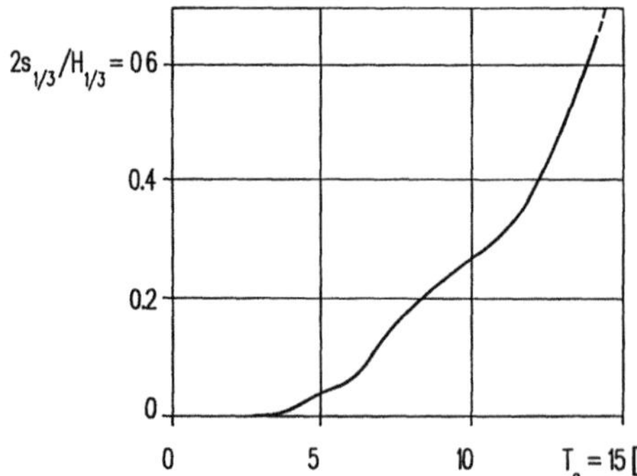

Fig. 6.2. Significant transfer function of heaving motion of semisubmersible [1]

accelerations to obtain the actual stress or deformation values under wave action. In this way, transfer functions of stresses and deformations are obtained, as well as stress and deformation spectra using the spectral method of analysis. Hence, all statistical–probabilistic, i.e. stochastic, characteristics of interest can be defined.

For all further considerations which are based on calculated transfer functions of the response to seaway excitation in question, we must point out from the start that not only statistical uncertainties of the seaway spectrum itself, discussed in more detail in Chap. 5, are of significance, but also model uncertainties.

As an illustration of model uncertainties we consider a systematic investigation of the 17th International Towing Tank Conference (ITTC) of 1984 [6], which concerned comparative calculations for a model semisubmersible structure, the motions of which were tested in a towing tank so that the model-test results served simultaneously as a test of the quality of the analysis programs and of the quality of the theories behind them, and their assumptions. The model tests were carried out for two wave heights: $2\zeta_a = 0.046\,\mathrm{m}(\hat{=} H = 4.1\,\mathrm{m})$ and $2\zeta_a = 0.16\,\mathrm{m}(\hat{=} H = 10.24\,\mathrm{m})$, with ζ_a the wave amplitude in the model test and H the wave height in full-scale. Twenty-eight institutes from ten countries took part in the comparative analysis, all using linear motion analysis. Here we are interested mainly in comparing the calculated results, and we again want to consider the transfer functions of the heave motions (Fig. 6.3).

We see first that linear motion behaviour (independence of wave height) is confirmed in the experiments for this semisubmersible in the range below the period with the relative minimum of the transfer function (cancellation period T_C): $T_C \simeq 2.75\,\mathrm{s}$ on scale 1/64, i.e. $T_C \simeq 22.0\,\mathrm{s}$ at full scale (Froude scaling law). We can also see, however, that this structure which is apparently hydrodynamically simple leads to definite differences in the transfer functions coded with different letters,

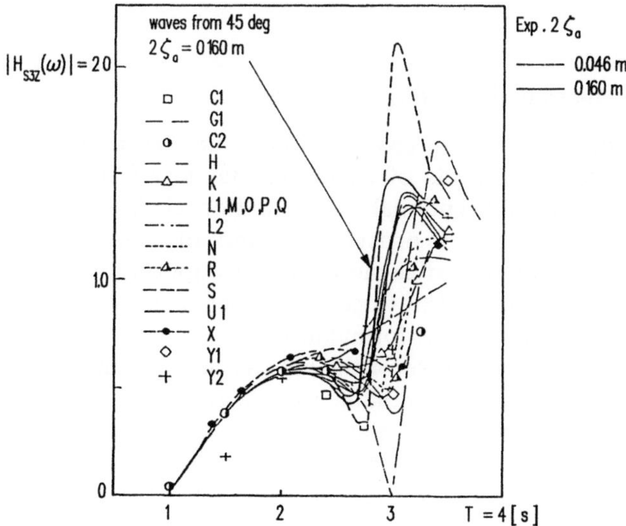

Fig. 6.3. ITTC comparative analysis [6].

as calculated by 28 internationally renowned institutes. The extent of the scatter of the calculated results for this one semisubmersible is almost as great as the scatter in Fig. 2.24 of corresponding transfer functions of six semisubmersibles with widely differing structures.

There are many causes for varying analysis results. They are due partly to the theory used, and are partly rooted in the numerics, and also in the idealization (modelling) of the geometry of the structure used for the calculation. Details of this are to be found in the comparative study [6], and in a more detailed discussion of this subject [7].

At the moment we are interested mainly in the fact of the scatter of the calculated results, not so much in its possible causes, because this fact shows how important it is to include all causes of possible uncertainty in a rational analysis concept: this need not only be statistical uncertainty, which we have already discussed fully in connection with the analytical representation of the seaway in Sect. 5.2.2.1 (see Fig. 5.11), but also, or especially, mechanical model uncertainty, which is very clear in Fig. 6.3 in the results of a motion analysis. Both types of uncertainty are noticeable from the application of the basic equation (5.54) of the spectral method of analysis in the responses of the system, as illustrated in Fig. 6.1c for the spectra of the heave motion, and in all calculations based on them. Insofar as these calculations are of a probabilistic nature, we also have to cope generally with stochastic model uncertainty, i.e. with the question of whether the populations of the random values considered behave as we would expect by choosing a distribution function (cf. the example of the Weibull distribution of the characteristic wave height $H_{1/3}$ as in (5.109), (5.110) and Fig. 5.16).

In Sect. 6.3 we discuss a modern analysis concept of load-bearing structures which in principle allows all uncertainties to be considered rationally. Until then we continue to rely on the classical evaluation concept, assuming the validity of mechanical models (transfer functions, in this case) and their correct use.

6.1.2 Flexible Mooring of Floating Structures

By flexible moorings we mean any positioning devices which allow the marine structure to reduce the effects of environmental forces to a certain extent by movement, including conventional cable and/or chain anchoring of floating structures as well as mooring to articulated towers and similar modern positioning concepts. When evaluating flexible moorings it is important to include the random characteristics of horizontal drift forces, for which a simple theory is used which we will develop here in its basic formulation.

We imagine a random sample of the seaway as a chronologically continuous sequence of single waves, i.e. the random sample as a juxtaposition of single waves. We use the term single wave for a wave between two successive zero up-crossings of wave elevation. With the elementary wave concept, we have already defined the infinitely long uniform wave train in Sect. 5.2.1. We can add formally here that an elementary wave consists of continuous juxtapositions of single waves of the same frequency and amplitude.

The single waves of the random sample of the natural seaway have different periods T (times between two successive zero up-crossings of wave elevation) and mean frequencies $\omega = 2\pi/T$ and mean amplitudes $\zeta_a = (\zeta_{max} + |\zeta_{min}|)/2$ within T. Thus we can consider each of these single waves, by way of approximation, as a part (strip) of different elementary waves of constant amplitude and frequency, for which we can calculate the associated drift forces with second-order hydrodynamic theories (see Chap. 3, Sect. 3.4.4 of Vol. I and [8–10]). From the juxtaposition of the different single waves to form a random sample of the natural seaway, we obtain the juxtaposition of different drift forces, which we can now consider as a random sample of the drift force $f^{(2)}(t)$ of a marine structure depending on the amplitude fluctuation $a(t)$ of the random sample of the seaway in a very simple form, which is, however, generally satisfactory for practical applications:

$$f^{(2)}(t) = \varrho/2 \cdot gL \cdot \alpha_0^2 \cdot a^2(t). \tag{6.4}$$

Here α_0^2 is a mean drift force coefficient, which we define below (see (6.6)), and we define $a(t)$ as the envelope of the random seaway sample which thus shall correspond to the amplitude fluctuation of the random seaway sample. The factor L is a characteristic length of the structure under consideration. Two methods are used in practice for calculating stochastic characteristics of drift force and the associated mooring (anchoring) forces on the basis of (6.4), which we examine briefly below.

Method I: Spectral Analysis
The stationary random process $\{F^{(2)}(t)\}$ related to the drift force has, according to (6.4), at any time t a mean value

$$E[F^{(2)}] = \bar{F}^{(2)} = \varrho/2 \cdot gL \cdot \alpha_0^2 \cdot E[A^2]. \tag{6.5}$$

We define the mean drift force coefficient α_0^2 in (6.5), like the mean ω_0^2 in (5.20), as a mean weighted to the seaway spectrum

$$\alpha_0^2 = \int_0^\infty \alpha^2(\omega) \cdot S_{ZZ}(\omega) d\omega / \int_0^\infty S_{ZZ}(\omega) d\omega. \tag{6.6}$$

We have values for $\alpha^2(\omega)$ from the hydrodynamic calculations of the mean horizontal drift force $f^{(2)}(\omega)$, which is calculated as the effect of elementary waves of amplitude $\zeta_a (= a(t) = a = \text{const.}$ (see Fig. 6.4b) and frequency ω:

$$\alpha^2(\omega) = f^{(2)}(\omega)/(\varrho/2 \cdot gL \cdot \zeta_a^2). \tag{6.7}$$

Results of such calculations for an example dealt with in [11] are reproduced in Fig. 6.5, the relevant floating structure (a pontoon) being shown in Fig. 6.6. (For use of lower and upper case letters in (6.4), (6.5) and the following equations see introduction to Chap. 5.)

To determine the expectation $E[A^2]$ in (6.5) we first consider a random sample $\zeta(t)$ of the seaway process $\{Z(t)\}$ of wave elevation, and define the envelope in the time domain of Fig. 6.4a as a radius vector $r(t)$ of the corresponding points in the phase domain, which is given by the coordinates ζ and $(d\zeta/dt)/\omega_0$, ω_0 being regarded initially as a freely selectable reference frequency. Figure 6.4b serves to

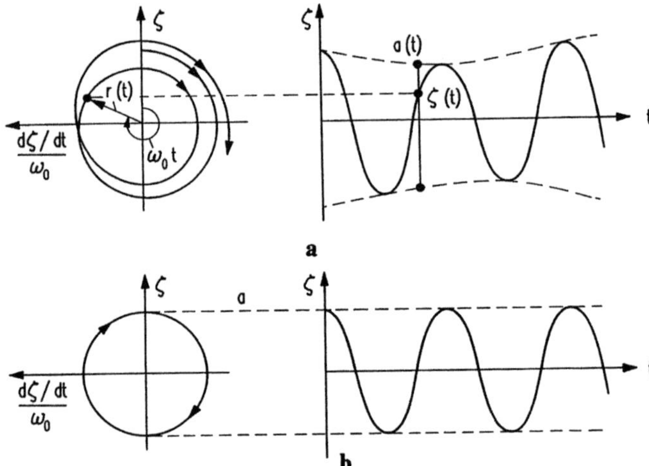

Fig. 6.4a,b. Random sample functions with envelopes.

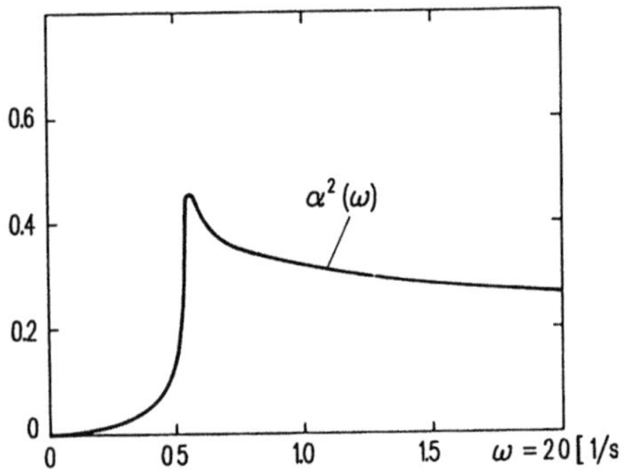

Fig. 6.5. Drift force coefficient for pontoon in Fig. 6.6.

clarify the simple case of an elementary wave with $\zeta = \zeta_a \cdot \cos\{\omega_0 t\}$, i.e. $(d\zeta/dt)/\omega_0 = -\zeta_a \sin\{\omega_0 t\}$ and $r^2(t) = \zeta_a^2$, thus $a^2(t) = \zeta_a^2 = \text{const.}$

In a purely amplitude-modulated representation ($\omega_0 = \text{const}$), which we use here to simplify the subsequent derivations, without the results thereby becoming unrealistic:

$$\zeta(t) = a(t) \cdot \cos\{\omega_0 t\} \tag{6.8}$$

applies more generally in accordance with Fig. 6.4a. From this we obtain the time average of the random sample over a long period of time $T \to \infty$ as

$$\langle \zeta^2(t) \rangle = 1/T \cdot \int_0^T a^2(t) \cos^2\{\omega_0 t\}\,dt$$

$$\stackrel{\scriptscriptstyle\triangle}{=} \langle a^2(t) \rangle \cdot 1/T \cdot \int_0^T \cos^2\{\omega_0 t\}\,dt$$

$$= \langle a^2(t) \rangle \cdot [2\omega_0 + \sin\{2\omega_0 T\}/T]/[4\omega_0]$$

$$= \langle a^2(t) \rangle \cdot 1/2.$$

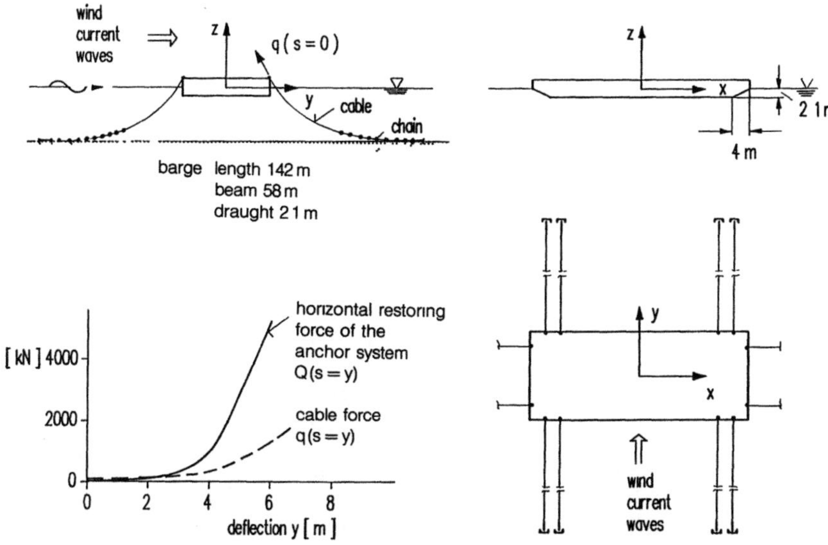

Fig. 6.6. Pontoon anchorage and characteristic functions $Q(y)$ and $q(y)$ [11].

If we consider $\zeta(t)$ as a random sample of an ergodic seaway process (cf. (5.7) and (5.8)), at any time t

$$E[Z^2] = E[A^2]/2,$$

$$E[A^2] = 2E[Z^2] = 2\sigma_Z^2 = 2\cdot \int_0^\infty S_{ZZ}(\omega)\mathrm{d}\omega = E[2Z^2], \tag{6.9}$$

applies for the random process values A and Z. Taking (6.5) and (6.6) into account, we obtain from this

$$\bar{F}^{(2)} = \varrho\cdot gL\cdot \int_0^\infty \alpha^2(\omega)\cdot S_{ZZ}(\omega)\mathrm{d}\omega \tag{6.10}$$

for the expected $\bar{F}^{(2)}$ of the random process $\{F^{(2)}(t)\}$ at any time t. In the next step we want to determine the spectrum of the drift force, as from that we obtain its standard deviation in accordance with (5.14).

With ergodicity, similarly to (5.7) and (5.8) with (6.8),

$$\langle \zeta^2(t)\cdot \zeta^2(t+\tau)\rangle$$

$$= 1/T\cdot \int_0^T a^2(t)\cos^2\{\omega_0 t\}\cdot a^2(t+\tau)\cos^2\{\omega_0(t+\tau)\}\,\mathrm{d}t$$

$$\triangleq \langle a^2(t)\cdot a^2(t+\tau)\rangle\cdot 1/T\cdot \int_0^T \cos^2\{\omega_0 t\}\cdot \cos^2\{\omega_0(t+\tau)\}\,\mathrm{d}t$$

$$= \langle a^2(t)\cdot a^2(t+\tau)\rangle\cdot (2 + \cos\{2\omega_0\tau\})/8.$$

applies for a time average value of the random sample $\zeta^2(t)$ corresponding to auto-correlation with $T \to \infty$.

Due to the ergodicity of the random process under consideration, by observing (5.8) and (5.10) we obtain a relation between the autocorrelation functions of the random processes $\{Z^2(t)\}$ and $\{A^2(t)\}$

$$4R_{Z2Z2}(\tau) = R_{A2A2}(\tau) + R_{A2A2}(\tau)\cos\{2\omega_0\tau\}/2. \tag{6.11}$$

The Fourier Transform of the two sides of this equation gives, with

$$\cos\{2\omega_0\tau\} = \exp\{i2\omega_0\tau\}/2 + \exp\{-i2\omega_0\tau\}/2,$$

the result

$$4S_{Z2Z2}(\omega) = S_{A2A2}(\omega) + [S_{A2A2}(\omega - 2\omega_0) + S_{A2A2}(\omega + 2\omega_0)]/4 \tag{6.12}$$

(see also the modulation theorem of the Fourier Transforms, e.g. [12]). The last two terms can be interpreted as the spectrum of the envelope of the autocorrelation function $R_{A2A2}(\tau)$ modulated with $\cos\{2\omega_0\tau\}$. We can see that an initial spectrum $S_{A2A2}(\omega)$ distributed about $\omega = 0$ occurs in a considerably weakened form about $\omega = \pm 2\omega_0$, if we assume a narrowband spectrum $S_{ZZ}(\omega)$. These weakened mappings are not important for the mooring problem we are considering, because only the low frequencies contribute significant components to the drift motion. In the following we therefore consider only the first term in (6.11) and (6.12) and use (exclusively for the mooring problem)

$$R_{A2A2}(\tau) \simeq 4R_{Z2Z2}(\tau) \quad \text{or} \quad S_{A2A2}(\omega) \simeq 4S_{Z2Z2}(\omega). \tag{6.13}$$

We now want to show a relationship between $S_{Z2Z2}(\omega)$ and $S_{ZZ}(\omega)$, the seaway spectrum. For this we assume that $\{Z(t)\}$ is a Gaussian random process with the mean value zero. According to [13],

$$E[Z(t_1)Z(t_2)Z(t_3)Z(t_4)] = E[Z_1Z_2Z_3Z_4]$$
$$= E[Z_1Z_2]E[Z_3Z_4] + E[Z_2Z_3]E[Z_1Z_4] + E[Z_1Z_3]E[Z_2Z_4].$$

This equation applies also to $t_1 = t_3$ and $t_2 = t_4$, i.e.

$$E[Z_1^2Z_2^2] = E[Z_1^2]E[Z_2^2] + 2E^2[Z_1Z_2].$$

With $t_1 = t$ and $t_2 = t + \tau$, we obtain in accordance with (5.10) and (5.11)

$$R_{Z2Z2}(\tau) = R_{ZZ}^2(0) + 2R_{ZZ}(\tau) \cdot R_{ZZ}(\tau). \tag{6.14}$$

According to the convolution theorem of Fourier Transforms, the Fourier Transform of a product is equal to the convolution of the Fourier Transform of the factors. So, by Fourier Transform of $R_{Z2Z2}(\tau)$

$$S_{Z2Z2}(\omega) = 2\pi \cdot R_{ZZ}^2(0)\delta\omega + 2S_{ZZ}(\omega)*S_{ZZ}(\omega).$$

The first term with the Dirac–Delta function $\delta\omega$ has a value other than zero only at $\omega = 0$, and is of no further interest for the mooring problem, for the wave drift force is always zero at $\omega = 0$ (see Fig. 6.5 and (6.4)). We use the second term in

(6.13), and find (exclusively for the mooring problem)

$$S_{A2A2}(\omega) = 8S_{ZZ}(\omega)*S_{ZZ}(\omega) = 8 \cdot \int_{-\infty}^{\infty} S_{ZZ}(\mu) \cdot S_{ZZ}(\mu - \omega) d\mu$$

$$= 8 \cdot \int_{-\infty}^{\infty} S_{ZZ}(\mu) \cdot S_{ZZ}(\mu + \omega) d\mu.$$

We can verify the last statement with the substitution $\mu = -\mu$ due to the symmetry of the seaway spectrum $S_{ZZ}(\mu)$ to $\mu = 0$. Using alternatively one of the last two statements for $\mu \geqslant 0$, we find

$$S_{A2A2}(\omega) = 8 \cdot \int_{-\infty}^{0} S_{ZZ}(-\mu) \cdot S_{ZZ}(-\mu - \omega) d\mu$$

$$+ 8 \cdot \int_{0}^{\infty} S_{ZZ}(\mu) \cdot S_{ZZ}(\mu + \omega) d\mu.$$

The integrands of the two integrals are symmetrical functions to $\mu = 0$. So by considering exclusively frequencies $\mu \geqslant 0$ when applying (5.63), the sum of the two integrands is equal to the second integrand, so that finally

$$S_{A2A2}(\omega) = 8 \cdot \int_{0}^{\infty} S_{ZZ}(\mu) \cdot S_{ZZ}(\mu + \omega) d\mu. \tag{6.15}$$

In accordance with (6.4) we write for the stochastic processes $\{F^{(2)}(t)\}$ and $\{A^2(t)\}$

$$F^{(2)}(t) = \varrho/2 \cdot gL \cdot \alpha_0^2 \cdot A^2(t).$$

By forming expectations on both sides of this equation, we obtain, observing (6.13)

$$R_{F(2)F(2)}(\tau) = (\varrho/2 \cdot gL)^2 \cdot \alpha_0^4 \cdot R_{A2A2}(\tau)$$

$$= (\varrho gL)^2 \cdot \alpha_0^4 \cdot R_{Z2Z2}(\tau)$$

$$= 2(\varrho gL)^2 \cdot \alpha_0^4 \cdot R_{ZZ}(\tau) R_{ZZ}(\tau).$$

Observing the restriction to $\mu \geqslant 0$ already mentioned for (6.15), the Fourier Transform of both sides finally gives the drift force spectrum

$$S_{F(2)F(2)}(\omega) \simeq 2\varrho^2 \cdot g^2 L^2 \cdot \alpha_0^4 \cdot \int_{0}^{\infty} S_{ZZ}(\mu) \cdot S_{ZZ}(\mu + \omega) d\mu \tag{6.16}$$

in a form of approximation which is generally sufficiently accurate for practical applications. We cite for comparison a slightly modified form of a widely used approximation, which was (partially) derived in [14] in another way:

$$S_{F(2)F(2)}(\omega) \simeq 2\varrho^2 \cdot g^2 L^2 \cdot \int_{0}^{\infty} \alpha^4(\mu + \omega/2) \cdot S_{ZZ}(\mu) \cdot S_{ZZ}(\mu + \omega) d\mu. \tag{6.17}$$

Although the juxtaposition principles of the drift forces behind (6.4) simplify reality quite significantly, and cannot be realistic for single waves which are short compared with the characteristic length L of the marine structures considered (i.e. they have relatively high frequencies), this formulation has proved itself in connection with the evaluation of flexible moorings, for the essential components of the drift force must be assigned to very low frequencies of the wave spectrum, which is clearly verified by the convolution in (6.16) or (6.17). We can now deal with the application of the drift force spectrum to the evaluation of forces and motions of flexible moorings.

We are essentially interested in the forces acting in the mooring or anchoring elements (chains, cables, etc.) associated with the horizontal motion s. For this we solve the linear differential equation of motion s

$$(m + m_{\mathrm{Hs}})\ddot{s}^{(2)}(t) + d_s\dot{s}^{(2)}(t) + k_s s^{(2)}(t) = f_s^{(2)}(t). \tag{6.18}$$

Here m_{Hs} is the hydrodynamic mass of the marine structure under consideration in the direction of the motion s, d_s is the hydrodynamic damping coefficient of the structure and k_s is the stiffness coefficient of the mooring system in direction s. The hydrodynamic mass m_{Hs} is generally frequency-dependent, but we have just seen that only small frequencies make a significant contribution to the drift force, so the assumption of constant hydrodynamic mass over the relatively narrow range of frequencies of practical interest in the mooring problem is justified. The same applies of course to the damping coefficient d_s.

The assumption of a constant stiffness coefficient k_s is somewhat more problematical, since observation of the system's restoring characteristic $Q(s = y)$ or a cable/chain restoring characteristic $q(y)$ for the transverse motion y of the simple pontoon in Fig. 6.6 reveals behaviour which is extremely non-linear in parts. (For the determination of such anchoring characteristics and the principles to be observed when designing anchoring systems, see [15].)

We overcome this non-linearity with the definition of a quasi-linear stiffness coefficient, given by the tangent at a point on the system curve, which is determined by the mean drift force as in (6.10)

$$\bar{s} = s(Q = \bar{F}^{(2)} + f_{\mathrm{WS}}), \tag{6.19}$$

where f_{WS} is the sum of wind and current forces which is regarded as constant with time. The notation $s(Q)$ defines here the inverse function of the curve $Q(s)$ of the anchoring characteristics of a pontoon as in Fig. 6.6. Thus the anchoring system can be regarded as linear over a limited range of fluctuation of $s^{(2)}$, and we find the spectrum of the drift motion as in (5.54) to be

$$S_{S(2)S(2)}(\omega) = |H_{S(2)F(2)}(\omega)|^2 \cdot S_{F(2)F(2)}(\omega). \tag{6.20}$$

The drift force spectrum $S_{F(2)F(2)}(\omega)$ can be calculated in accordance with (6.16) or (6.17); we find the square of the value of the transfer function $H_{S(2)F(2)}$ from (5.38), taking (6.18) into account, to be

$$k_s^2 \cdot |H_{S(2)F(2)}(\omega)|^2 = 1/[\{1 - (\omega/\omega_0)^2\}^2 + 4\delta_s^2(\omega/\omega_0)^2],$$

$$\omega_0^2 = k_s/(m + m_{\mathrm{Hs}}), \ \delta_s = d_s/[2\sqrt{k_s(m + m_{\mathrm{Hs}})}]. \tag{6.21}$$

From this we obtain the variance of the random process $\{S^{(2)}(t)\}$ of the horizontal excursion, similarly to (5.14),

$$\sigma^2_{S(2)} = \int_0^\infty S_{S(2)S(2)}(\omega)d\omega. \tag{6.22}$$

Furthermore, adding the squares of the standard deviations of the random processes of first and second order motions which are regarded as stochastically independent of one another, gives the total variance of the sum of the two processes

$$\sigma^2_S = \sigma^2_{S(2)} + \sigma^2_{S(1)}. \tag{6.23}$$

Indeed, we cannot be certain that the mean value $\bar{s}$ as in (6.19) and the standard deviation σ_s of the composite motion as in (6.23) are characteristics of a Gaussian random process, as we have only made a linear approximation to the stiffness coefficients, but if our assumptions hold we can determine all probabilistic parameters, e.g. $p\%$ fractiles s_p, in precisely the same way as the wave elevation in (5.77) to (5.82), in order to calculate design values for the forces in the anchoring elements from the associated curves $q(s)$ by means of

$$q_p = q(\bar{s} + s_p).$$

It is therefore of great practical importance either to underpin the assumption of a Gaussian process for the combined motion, or to avoid it completely by using another procedure, if this is more practical.

Method II: Simulation
We obtain the random sample of the seaway process, for example, in accordance with (5.65) as

$$\zeta(t) = \sum_{n=1}^N \sqrt{2S_{ZZ}(\omega_n)\Delta\omega_n}\cos\{-\omega_n t + \varepsilon(\omega_n)\}, \tag{6.24}$$

where the spectrum $S_{ZZ}(\omega_n)$, which is defined, for example, as in (5.88) for specified class parameters $(H_{1/3}, T_0)$, is divided into N strips of width $\Delta\omega_n$. (We have discussed the special problems of this procedure and an alternative in more detail in Sect. 5.2.1.1 and 5.2.1.2.) For each of these strips we generate a phase angle $0 \leqslant \varepsilon_n < 2\pi$ from a uniform distribution $f_{En}(\varepsilon_n) = 1/(2\pi), \varepsilon_n = \varepsilon(\omega_n)$. Thus we also find a random sample $a^2(t)$ of the random process $\{A^2(t)\}$, as in Fig. 6.4, to be

$$a^2(t) = \zeta^2(t) + (d\zeta/dt)^2/\omega_0^2. \tag{6.25}$$

Here $\omega_0 = 2\pi/T_0$ can be defined by the parameter T_0 of the specified seaway class $(H_{1/3}, T_0)$.

If we now describe a random sample $f^{(1)}(t)$ of the random process $\{F^{(1)}(t)\}$ using (6.24) at time t by superposition of N first-order harmonic seaway forces, which we can calculate with a procedure developed in Chap. 3, and obtain a random sample $f^{(2)}(t)$ of the random process $\{F^{(2)}(t)\}$ of the drift force with (6.25) by using (6.4), it would be possible to superpose the two forces in correct phase to form a sample of the whole wave force. However, as the first- and second-order motions $s^{(1)}$ and $s^{(2)}$ are practically independent of one another – the processes belong to separate frequency domains which hardly coincide – this is not necessary. We can simulate the first- and second-order forces or motions independently of one another, and superpose the results without considering the correct phasing.

Thus, simulating the first-order wave forces $f^{(1)}(t)$ and the associated motions $s^{(1)}(t)$ becomes much simpler: the simulation procedure consists firstly of representing the input signal $\zeta(t)$ formed as in (6.24) against ω by means of Fourier Transform, then multiplying it, for example, as in (5.37) with the relevant transfer function $H_{F(1)Z}(\omega)$ or $H_{S(1)Z}(\omega)$, and converting the result by means of a new Fourier Transform back into a random sample of the random process $\{F^{(1)}(t)\}$ or $\{S^{(1)}(t)\}$. This method can easily be carried out with electronic data processing by means of Fast Fourier Transforms.

The random samples $f^{(1)}(t)$ and $s^{(1)}(t)$ thus determined permit two possibilities for the further treatment of the anchorage problem with regard to the further simulation of the drift forces for $f^{(2)}(t)$ and the second-order motions $s^{(2)}(t)$, which were tried in [11]. We briefly describe these possibilities under the headings Simulation I and Simulation II, and compare them with the aid of the sample results presented in [11] and expanded here.

Simulation I:
1. Calculation of the wind current forces f_{WS}, which, by way of approximation, are considered to be constant with time.
2. Calculation of a random sample of the first order wave force $f^{(1)}(t)$ by means of $H_{F(1)Z}(\omega)$ as in (5.37) described above.
3. Calculation of a random sample of wave drift force $f^{(2)}(t)$ with (6.4), (6.6), (6.7), (6.24) and (6.25), $\alpha^2(\omega)$ (see Fig. 6.5).
4. Superposition $f(t) = f_{WS} + f^{(1)}(t) + f^{(2)}(t)$.
5. Solution of the following equations of motion with $f_s(t) = f(t)$ as in step 4 for the relevant direction of motion s

$$(m + m_{Hs})\ddot{s}(t) + d_s\dot{s}(t) + k_s(s)s(t) = f_s(t). \tag{6.26}$$

We can see that in the simulation, a stiffness coefficient $k_s(s)$ can be used as a tangent to $Q(s)$ for any excursion s. Of course, with this method the system must have reached a stationary state before the results can be interpreted (statistically evaluated), since the exciting seaway process is assumed stationary.
6. Determination of the tensile load $q(t)$ which can be read off for any excursion $s(t)$ from the curve $q(s)$ of the elements of the mooring system.

Simulation II:
1. As simulation I, point 1.
2. Calculation of a random sample of the first-order motions $s^{(1)}(t)$ by means of $H_{S(1)Z}(\omega)$ as in (5.37) described above.
3. As simulation I, point 3.
4. Superposition $f(t) = f_{WS} + f^{(2)}(t)$.
5. Solution of (6.26) with $f_s(t) = f(t)$ as in step 4 for the relevant direction of motion and superposition of the solution $s(t)$ with $s^{(1)}(t)$ as in step 2.
6. As simulation I, step 6, with $s(t) := s(t) + s^{(1)}(t)$ as in step 5.

To illustrate this simulation we consider the anchoring of the simple pontoon shown in Fig. 6.6. The relevant motion is $s = y$, i.e. beam waves act on the pontoon. Figures 6.7 and 6.8 show short time sections of 300 s of the random samples of wave elevation $\zeta(t)$, transverse motion $y(t)$ and tensile force in the luff-side (upwind) mooring elements $q(t)$ determined by the simulation described.

Evaluation of the results for the purpose of design decisions (alterations to configuration or dimensions) demands the statistical evaluation of the whole length of these random samples (order of magnitude 10^4 s), e.g. by determining a probability density distribution $f_Y(y)$ of the calculated values y, which is shown in Fig. 6.9 for both simulation of Method II and for the spectral analysis of Method I.

We can see that Simulations I and II give quite different distributions which, however, coincide relatively accurately in the region, which alone is important for determining design values of low exceeding probability, i.e. the region of large values of y, which seldom occur. From this point of view, both simulations can be used in practice. Interestingly, the deviation of the results from a Gaussian probability density distribution in the case of Simulation I, where we superpose separately calculated first- and second-order forces, is greater than with Simulation II, where we superpose separately calculated first- and second-order motions. Simulation II corresponds better to the purely spectral analysis method presented as Method I, so the results reproduced here do not contradict the assumption of a Gaussian process for the combined first- and second-order motion values.

Both methods imply relatively high physical model uncertainties due to the simplified conceptual model of drift force analysis in the natural seaway (see (6.4)), which could not be completely removed by model tests [16]. In Method II the anchoring characteristic is accurately represented by Simulation I, but the frequency dependence of the hydrodynamic mass and damping, which are significant for first-order motion, cannot be dealt with satisfactorily. Simulation II encompasses somewhat too small an area of the anchoring characteristic, but first-order motion takes sufficient account of the frequency dependences mentioned. The methods presented can nevertheless by used as a realistic basis for comparative evaluations of mooring or anchorage configurations in the design phase.

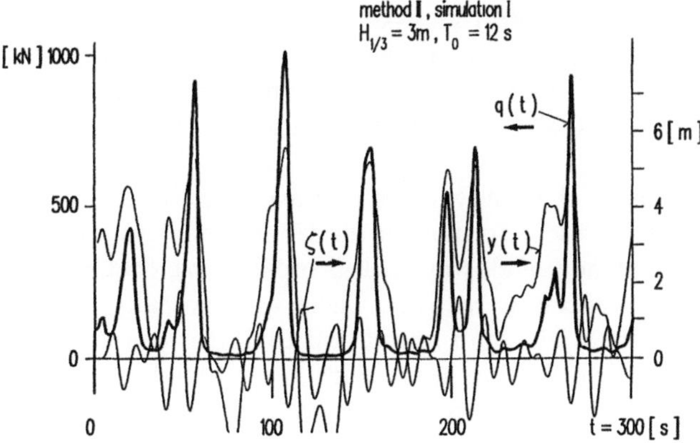

Fig. 6.7. Simulation I for anchorage in Fig. 6.6.

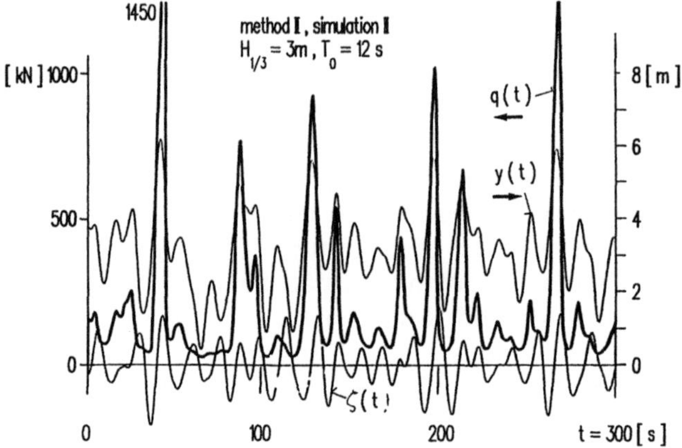

Fig. 6.8. Simulation II for anchorage in Fig. 6.6.

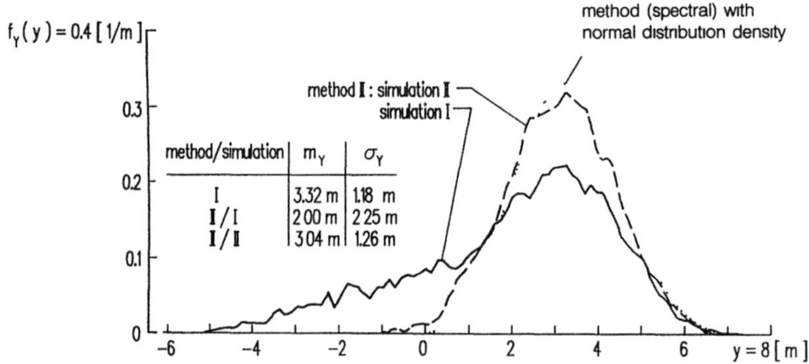

method/simulation	m_Y	σ_Y
I	3.32 m	1.18 m
I/I	2.00 m	2.25 m
I/I	3.04 m	1.26 m

Fig. 6.9. Histograms of transverse motion of pontoon in Fig. 6.6.

6.1.3 Fixed Structures

The application of the spectral analysis method to fixed marine structures does not differ fundamentally from its application to floating structures. The fact that we are nevertheless devoting a separate section to fixed structures is mainly due to the existence of peculiarities of the non-static (i.e. dynamic) behaviour of these structures.

We talk here and in the following of a quasi-static behaviour of a structure, if the frequency of its elastic deformations under the influence of wave forces lies so far outside the frequency of elastic natural oscillations, that a static treatment of the relevant seaway effects is realistic. These are generally structures which are used in water which is less than about 250 to 300 m deep. For such structures with linearly elastic deformation behaviour, the unit load concept of [5] described in Sect. 6.1.1 can be regarded a most effective possibility for rational analysis.

We speak of dynamic structures in contrast to quasi-static structures if the lowest elastic natural frequencies can be excited by the seaway. These are mainly the,

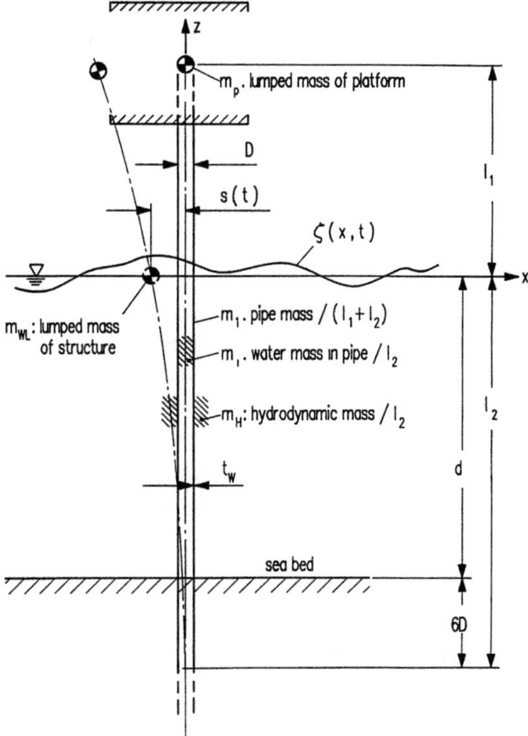

Fig. 6.10. Monopod platform schematic

now rare, deep-water platforms, but shallow-water platforms in particular can also demonstrate dynamic behaviour such as, for example, monopod platforms, which we study as an introductory example for illustrating the specific characteristics of methods for analysing dynamic structures in a stationary seaway.

6.1.3.1 Monopod Platforms in a Stationary Seaway

We consider a platform supported by only one pylon (see Fig. 6.10). This type of platform is often used for unmanned lighthouses or measuring installations in the area of river mouths, where guiding beacons as well as continuous information on water levels and wave heights are very important for ship traffic in restricted waters.

The wave force acts on the wetted part of the pylon according to the Morison formula discussed in detail in Chap. 3

$$f(t) = \int_{-d}^{\infty} [k_m \cdot \dot{u}(z,t) + k_d \cdot u(z,t)|u(z,t)|]\,dz. \tag{6.27}$$

Here z is the vertical coordinate axis with its origin in the still water line, $k_m = C_m \cdot \rho \cdot \pi D^2/4$ and $k_d = C_d \cdot \rho/2 \cdot D$ are defined by the drag and inertia coefficients

C_d and C_m introduced in Chap. 3, and D is the diameter of the pylon. For deep water, the horizontal wave velocities as in Table 3.1 and (3.40) are given by

$$u(z, t) = \omega \cdot \exp \{\omega^2 \cdot z/g\} \cdot \zeta(t). \tag{6.28}$$

If we consider the stochastic seaway process $\{Z(t)\}$, we again use capital letters for the formal characterisation of random values, i.e.

$$F(t) = \int_{-d}^{\infty} [k_m \cdot \dot{U}(z, t) + k_d \cdot U(z, t)|U(z, t)|] dz. \tag{6.29}$$

(We consider k_m and k_d as deterministic values, and point out from practical experience that here the scatter of the wave velocity has a much greater effect on the scatter of the load calculated with the Morison formula than scatter of coefficients k_m and k_d [17].)

If we imagine a Fourier Transform of (6.28), we can describe the relationship between wave velocity and wave elevation, similarly to (5.37), as an input–output relation of a linear system, in which the transfer function for deep water is defined as

$$|H_{UZ}(\omega, z)| = \omega \cdot \exp \{\omega^2 \cdot z/g\}. \tag{6.30}$$

According to (5.54) and (5.14), therefore,

$$S_{UU}(\omega, z) = |H_{UZ}(\omega, z)|^2 \cdot S_{ZZ}(\omega), \tag{6.31}$$

$$\sigma_U(z) = \sqrt{\int_0^{\infty} S_{UU}(\omega, z) d\omega}. \tag{6.32}$$

With (6.29) in accordance with (5.10) the autocorrelation function $R_{FF}(\tau)$ can be formed and in accordance with (5.12) by Fourier Transform the associated spectrum $S_{FF}(\omega)$ of wave forces can be calculated. This method is too complex for the scope of this book, so here we make reference to the detailed derivation in [18] and an example in [19], and consider only the special result for our example in Fig. 6.10, i.e. a support pylon in relatively deep water

$$S_{FF}(\omega) = |H_{FZ}(\omega)|^2 \cdot S_{ZZ}(\omega),$$

$$|H_{FZ}(\omega)|^2 = (k_m \cdot \omega^2/k_0)^2 + (8/\pi) \cdot \left(\int_{-d}^{0} k_d \cdot \sigma_U(z) \cdot |H_{UZ}(\omega, z)| dz \right)^2. \tag{6.33}$$

Here k_0 is the wave number. We can see from Table 3.1 that the first term assumes the asymptotic form $(k_m g)^2$ in deep water or with high wave frequencies, i.e. it has no amplification characteristic restricted to a specific frequency range.

With a given wave spectrum $S_{ZZ}(\omega)$, we can calculate the wave force spectrum $S_{FF}(\omega)$ for the pylon of a monopod platform somewhat more effectively with (6.30) to (6.32) in accordance with (6.33) than with the concept shown in Fig. 5.2 or Fig. 6.1. The latter would, of course, also be applicable here if we constructed point-by-point a transfer function f_a/ζ_a by applying (6.27) to various waves of amplitude ζ_a with frequencies ω; we could count on, for example, $f_a \simeq (f_{max} + |f_{min}|)/2$ from one wave cycle with $\tanh \{kd\} \to 1$.

To determine whether we are dealing with a dynamic system in the case of the monopod platform under consideration, we are primarily interested in the first natural frequency of the bending vibration of the structure, as the next highest natural frequencies are often so high that they are not excited by the seaway. This is due particularly to the fact that the exciting forces of the seaway act upon the vibration nodes of the higher vibration modes.

To determine the first natural frequency, the calculation process can be simplified by transforming the platform structure, which has almost continuously distributed masses and thus theoretically an infinite number of natural frequencies and vibration modes, into a single-mass oscillator, such that the natural frequency of this single-mass oscillator coincides with the first natural frequency of the original system. The transformed equivalent system of the example shown schematically in Fig. 6.10 represents a cantilever beam of length l_2, at the end of which the wave force acts. The equivalent mass is also there

$$m_{\mathrm{WL}} = k_{\mathrm{WL}}/\omega_0^2. \tag{6.34}$$

Here the bending stiffness is given by $k_{\mathrm{WL}} = 3E \cdot I/l_2^3$, E being the modulus of elasticity of the pylon material, I the moment of inertia of the pylon's cross-section, and ω_0 is the lowest natural frequency of the original system obtained from finite element calculations described in Sect. 4.5. (Simpler estimation formulae can also be used here [20].)

In the example we consider the random process $\{S(t)\}$ of the horizontal deflection of the pylon at the water line. By applying (5.54) we obtain for the spectrum of this random process

$$S_{\mathrm{SS}}(\omega) = |H_{\mathrm{SF}}(\omega)|^2 \cdot S_{\mathrm{FF}}(\omega), \tag{6.35}$$

where we calculate $|H_{\mathrm{SF}}(\omega)|$ from (5.38) by means of

$$k_{\mathrm{WL}}^2 \cdot |H_{\mathrm{SF}}(\omega)|^2 = 1/[\{1 - (\omega/\omega_0)^2\}^2 + 4\delta^2(\omega/\omega_0)^2] \tag{6.36}$$

and $S_{\mathrm{FF}}(\omega)$ according to (6.33). We use the Pierson–Moskowitz spectrum $S_{\mathrm{ZZ}}(\omega)$ in the example in the form in (5.90), assuming a very moderate seaway such as is often to be expected in relatively sheltered estuaries (see Table 6.1). We may therefore also assume that the chosen water depth d justifies the assumption of 'infinite' water depth for higher frequencies (here $\omega \geqslant 1.75/s$), so (6.30) to (6.32) can be used for calculating the standard deviation $\sigma_{\mathrm{U}}(z)$ of the wave velocities U in dependence on the vertical coordinate z. Table 6.1 contains the remaining data used for the example, and Fig. 6.11 gives the most important results.

Table 6.1.

$c_{\mathrm{d}}=0.8$	$C_{\mathrm{m}}=2.0$	$g=9.81\,\mathrm{m/s^2}$	$\rho=1025\,\mathrm{Ns^2/m^4}$	$\delta=0.005$
$d=10\,\mathrm{m}$	$D=0.5\,\mathrm{m}$	$t_{\mathrm{w}}=0.012\,\mathrm{m}$	$l_1=7.0\,\mathrm{m}$	$l_2=d+6D=13\,\mathrm{m}$
$E=2.1\cdot10^{11}\,\mathrm{N/m^2}$		$I=5.48\cdot10^{-4}\,\mathrm{m^4}$	$k_{\mathrm{WL}}=166\,490\,\mathrm{N/m}$	$\omega_0=5.48/\mathrm{s}$
$m_1=143.5\,\mathrm{kg/m}$		$m_{\mathrm{i}}=191.7\,\mathrm{kg/m}$	$m_{\mathrm{H}}=201.3\,\mathrm{kg/m}$	$m_{\mathrm{P}}=500\,\mathrm{kg}$
$H_{1/3}=0.75\,\mathrm{m}$			$T_1=2.5\,\mathrm{s}$	
(since wave height)			(mean wave period)	

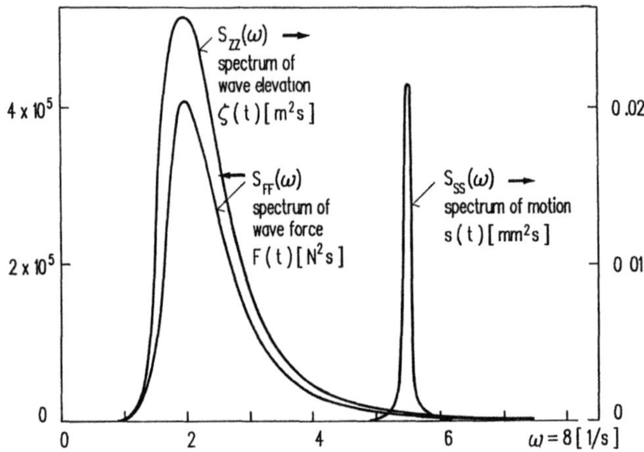

Fig. 6.11. Spectra for the example in Fig 6.10.

We can see from Fig. 6.11 that the wave spectrum $S_{ZZ}(\omega)$ and the horizontal wave force spectrum $S_{FF}(\omega)$ are similarly distributed in the example. This is not always the case: in particular, a second relative maximum of $S_{FF}(\omega)$ can occur at higher frequencies, if the non-linear viscous component of the horizontal force of certain (rarely occurring) elementary waves becomes apparent at triple wave frequency (see Chap. 3 of Vol. I and [7]). (This effect can be very significant for the fatigue strength of deep water platforms.)

We can also see that the response spectrum $S_{SS}(\omega)$ of the horizontal deflection at the water line is very pronounced, although the energy of the seaway in the region of the natural frequency is relatively small. This is the particular characteristic of a dynamic structure which can be important in the evaluation of the fatigue strength of marine structures. (We come back to this in Sect. 6.2.3.) If the response spectrum is narrow, however, we can calculate quite precisely all statistical characteristics of the random process of the seaway effect $\{S(t)\}$ under consideration which are of importance for short-term statistics, for all formulae for determining probabilistic wave height characteristics, e.g. those derived in Sect. 5.2.1.3 for narrowband Gaussian seaway processes are correspondingly valid for the responses of the linearly elastic structure. With larger dynamic systems with many degrees of freedom the response spectra can be relatively wide, and can have several relative maxima, so the comments on the distribution of the maxima of wideband random processes given in connection with Fig. 5.1 should be heeded.

6.1.3.2 Linearly Elastic Structures with more than One Degree of Freedom

The differential equations for describing the equilibrium of the inertia, damping, restoring and excitation forces valid for each of the N degrees of freedom considered are, in matrix notation,

$$\boldsymbol{m} \cdot \ddot{\boldsymbol{s}}(t) + \boldsymbol{d} \cdot \dot{\boldsymbol{s}}(t) + \boldsymbol{k} \cdot \boldsymbol{s}(t) = \boldsymbol{f}(t). \tag{6.37}$$

Here the components of the vector $f(t) = f_a \cdot \exp\{i\omega t\}$ indicate the forces acting at different nodes of the system, the components of the vector $s(t)$ are the associated deformations (deflection from rest), the matrix m contains the masses assumed to be concentrated in the nodes, the symmetrical matrix d contains the damping coefficients, and the symmetrical matrix k contains the stiffness coefficients.

By way of illustration we can imagine the monopod platform dealt with in Sect. 6.1.3.1 with two concentrated masses, or with one mass in two degrees of freedom, so the system of differential equations suitable for determining the first and second natural frequency and the related vibration mode (eigen form) in accordance with (6.37), disregarding the damping, would be

$$\begin{bmatrix} m_1 & 0 \\ 0 & m_2 \end{bmatrix} \cdot \begin{bmatrix} \ddot{s}_1 \\ \ddot{s}_2 \end{bmatrix} + \begin{bmatrix} k_1 + k_2 & -k_2 \\ -k_2 & k_2 \end{bmatrix} \cdot \begin{bmatrix} s_1 \\ s_2 \end{bmatrix} = \begin{bmatrix} f_1(t) \\ f_2(t) \end{bmatrix}. \tag{6.38}$$

There are several ways of solving the system of equations (6.37) for determining the deflections and associated effects. We would like to develop further the method used most frequently in marine engineering, namely modal analysis (dealt with deterministically in Appendix 2) from the point of view of stochastic excitations by the seaway.

If the damping in the system is negligibly small, or if we can assume that the matrix d can be accurately approximated by a linear combination of m and k (i.e. $d = \alpha \cdot m + \gamma \cdot k$ with α and γ as proportionality constants), each of the N uncoupled differential equations can be dealt with according to the principles of the spectral analysis method already illustrated for one degree of freedom. For this, we first determine the natural frequencies ω_i and a transformation matrix T of the normalized eigenvectors, so by means of transformed coefficient matrices I_m, Δ and Ω or excitation forces $v(t)$, i.e.

$$I_m = T^T m T, \quad \Delta = T^T d T, \quad \Omega = T^T k T, \quad v(t) = T^T f(t), \tag{6.39}$$

we obtain an uncoupled system of differential equations

$$I_m \cdot \ddot{w}(t) + \Delta \cdot \dot{w}(t) + \Omega \cdot w(t) = v(t). \tag{6.40}$$

where I_m is the unit mass matrix, Δ is a diagonal matrix (e.g. $\alpha I_m + \gamma \Omega$), and Ω is the diagonal matrix of the squared eigenvalues ω_i (see Appendix 2). The inverse transformations

$$s = T w, \quad s_n = \sum_{i=1}^{N} T_{ni} \cdot w_i \tag{6.41}$$

give the deflections.

If we again understand each excitation of our system as a stochastic process $\{F(t)\}$ or $\{V(t)\}$, each response is also a stochastic process $\{S(t)\}$ or $\{W(t)\}$, the autocorrelation function of which for the n-th degree of freedom of the system can be calculated from the above transformation equation (6.41) as in (5.10)

$$R_{S_n S_n}(\tau) = E\left[\sum_{i=1}^{N} T_{ni} \cdot w_i(t) \sum_{j=1}^{N} T_{nj} \cdot w_j(t+\tau) \right] = \sum_{i=1}^{N} \sum_{j=1}^{N} T_{ni} \cdot T_{nj} \cdot R_{W_i W_j}(\tau).$$

$$\tag{6.42}$$

By Fourier Transform of both sides we obtain the spectrum

$$S_{S_n S_n}(\omega) = \sum_{i=1}^{N} \sum_{j=1}^{N} T_{ni} \cdot T_{nj} \cdot S_{W_i W_j}(\omega) \tag{6.43}$$

for each degree of freedom $n = 1, 2, \ldots, N$. Hence, in addition to autospectra of one and the same random process, we also require cross-spectra of different, simultaneously occurring random processes of the generalized (modal) deflections $\{W_1(t)\}$, $\{W_2(t)\}$, etc., in all degrees of freedom. Similar to (5.49) to (5.53), they can be traced back to the corresponding cross-spectra of the generalised (modal) wave excitations $S_{V_i V_j}(\omega)$

$$S_{W_i W_j}(\omega) = H_{WV_i}(-\omega) \cdot H_{WV_j}(\omega) \cdot S_{V_i V_j}(\omega) = \hat{H}_{WV_i}(\omega) \cdot H_{WV_j}(\omega) \cdot S_{V_i V_j}(\omega). \tag{6.44}$$

(In the following, the $\hat{}$ symbol indicates complex conjugate functions with a more general structure than $H_{WV_j}(\omega)$). Similar to (5.38), we find here, for example, with $2\delta_j \omega_j = \alpha + \gamma \omega_j^2$ and l_m as the diagonal element of the matrix I_m

$$H_{WV_j}(\omega) = 1/[l_m \cdot (\omega_j^2 - \omega^2 + 2i \cdot \delta_j \cdot \omega_j \cdot \omega)]. \tag{6.45}$$

On the basis of the Morison formula and expanding the possibilities developed in [18] for representing cross-spectra of wave velocities and accelerations, which were also used for deriving the wave force as in (6.33), the calculation of the cross-spectra $S_{V_i V_j}(\omega)$ was described in principle in [21]. In [22] this problem is dealt with thoroughly, so we will dispense with the details here. In the following we may assume that (for example, as in [22]) the (complex) transfer functions $H_{V_j Z}(\omega)$ between $v_j(t)$ and the wave elevation $\zeta(t)$ can be calculated, on the basis of which the relationships

$$S_{V_i V_j}(\omega) = \hat{H}_{V_i Z}(\omega) \cdot H_{V_j Z}(\omega) \cdot S_{ZZ}(\omega) \tag{6.46}$$

can be evaluated. Thus the spectrum $S_{S_n S_n}(\omega)$ of the deflections $s(t)$ can be calculated for each degree of freedom n by means of (6.43) to (6.46) on the basis of a standard spectrum $S_{ZZ}(\omega)$ of the seaway, e.g. (5.90). If we create further linear transformations formally as in (6.41), e.g. for the internal sectional load l by means of

$$l_n = \sum_{i=1}^{N} L_{ni} \cdot w_i, \tag{6.47}$$

we obtain, similar to (6.42) and (6.43), the spectrum of the internal sectional loads as

$$S_{L_n L_n}(\omega) = \sum_{i=1}^{N} \sum_{j=1}^{N} L_{ni} \cdot L_{nj} \cdot S_{W_i W_j}(\omega), \tag{6.48}$$

which can also be calculated from the seaway spectrum S_{ZZ} by means of $S_{W_i W_j}(\omega)$ as in (6.44) to (6.46). We would proceed in the same way if we were studying stresses instead of sectional loads. Due to the special role which the cross-spectra of the generalized (modal) deformations $S_{W_i W_j}(\omega)$ play in this procedure, they are distinguished at times by the designation participation factors $S_{ij} := S_{W_i W_j}(\omega)$.

All characteristics of a random process which are of interest for short-term statistics again follow from the autospectra calculated with (6.43) or (6.48), hence, the evaluation of dynamic structures is being carried out in the same way as that of quasi-static structures.

It should be noted that with *FE* idealization of large systems (order of magnitude 1000 degrees of freedom), the higher vibration modes are of no practical importance; often only the first two vibration modes are of interest. Nevertheless, all vibration modes must be determined in the modal analysis, to permit the transformations to be carried out with T or L, etc. (see, e.g. (6.41)). The procedure described here therefore cannot be considered as unreservedly effective for large systems, and the result generally is greatly simplified system idealizations. Usually the cross-spectra $S_{v_i v_j}$ of the generalized (modal) excitation forces are disregarded from the start in view of the greatly simplified system idealization, so for example in (6.43), only summing with respect to i is to be carried out. As in (6.44) and (6.46), we can then represent the participation factor for each degree of freedom i of the generalized deformation w_i in the form of the basic equation of the spectral analysis method

$$
\begin{aligned}
S_{w_i w_i} &= \hat{H}_{wv_i}(\omega) \cdot H_{wv_i}(\omega) \cdot \hat{H}_{v_i z}(\omega) \cdot H_{v_i z}(\omega) \cdot S_{zz}(\omega) \\
&= \hat{H}_{w_i z}(\omega) \cdot H_{w_i z}(\omega) \cdot S_{zz}(\omega) \\
&= |H_{w_i z}(\omega)|^2 \cdot S_{zz}(\omega).
\end{aligned}
\tag{6.49}
$$

Furthermore, in our considerations up to now we have tacitly assumed that the hydrodynamic masses and dampings are the same for all relevant frequencies. This assumption is best treated as an approximation, so in practice we must usually count on a constant mean value for these frequencies.

6.2 Classical Methods of Long-Term Evaluation

In Sect. 6.2.1 we first study the evaluation of the load-bearing capacity (strength) of marine structures under extreme environmental conditions. By extreme environmental conditions we mean natural events which are expected to occur once in very long periods of time of the order of 50 to 100 years. A related ultimate limit state (ULS) failure of structures under the resultant extreme loads would generally involve the loss of human life and property, as well as the possibility of great environmental damage, so the study of extreme environmental loads over long-term service is naturally very important. In this context the question of rational superposition of various environmental forces will be a focus of our study.

We must not, however, overlook the fact that marine structures are subject to almost continuously moderate, but essentially more frequent environmental loads during the time between the very rare extreme environmental loads. In fact, most partial damage observed in steel structures is due to fatigue. Apart from the fact that damage to parts of a structure always reduces the reliability of the whole structure under subsequent extreme loads, it can restrict or even halt normal operation. For this reason, evaluation of serviceability limit states (SLS) of marine structures under repeated load cycles at various load levels below extreme loads is just as relevant for safety and economical reasons as evaluation of their behaviour under extreme loads. In Sects. 6.2.2 and 6.2.3 we therefore discuss the most important principles and applications of some of the current methods of serviceability analysis of marine structures in long-term use at sea.

The methods outlined in Sects. 6.2.1 to 6.2.3 are used in modern design practice and rely on a database which has grown continuously in the last two to three decades. They are, however, generally the same as those methods, which were already known about three decades ago at the start of the development of marine structures. So in these sections we are dealing with essentially classical methods of long-term evaluation, even though its practical importance has to some extent only been recognized recently due to the spread of electronic data processing. We also familiarize the reader with some recent stochastic methods of analysing serviceability, so that he will be able to follow emerging modern methods of analysis in this field. We then give an introduction to analysis methods which can be called modern in the true sense in Sect. 6.3.

6.2.1 Design Values for Environmental Loads

We call responses of a marine structure to environmentally determined excitations simply environmental loads, which we generally abbreviate to the symbol y (for deterministic values) or Y (for stochastic values). This concept includes actual loads, internal and external, as well as all effects of environmentally determined excitations, which contribute directly or indirectly to loads, i.e. even elastic deformations or rigid-body motions, etc. are considered as loads in this section.

Representing design values for seaway loads entails the basically simple extension of, for instance, the methods already considered in detail in Sect. 5.2.2.2 (long-term statistics) for the design values of the seaway parameters themselves. This extension is particularly important here, as the same environmental load, e.g. a stress or deformation (deflection), can have various simultaneous causes (e.g. seaway or wind excitation), so we must deal with the combination of different stochastic processes which cause the same environmental load, as the simple totalling of separately calculated design values would generally be unnecessarily conservative. In presenting some principles of the combination of random processes we start with a sometimes practical comparative evaluation method and then examine the actual representation of design values of combined load processes.

6.2.1.1 Design Value of an Individual Wave Load

If, instead of the Gaussian seaway processes $\{Z(t)\}$ used in Sect. 5.2.2, we consider the Gaussian load processes $\{Y(t)\}$ of linear systems in the seaway, by using the characteristic parameters of narrowband seaway and load processes, we can derive the following relationship, analogous to (5.106), for determining specific values y of the seaway load against the associated up-crossing (exceedance) frequency $n_Y^+(y)$, in the same way as (5.106) itself. We find

$$n_Y^+(y) = T \cdot \sum_{i=1}^{I} \sum_{j=1}^{J} v_{Y_{ij}}^+(0) \cdot \exp\{-y^2/(2m_{Y0_{ij}})\} \cdot P_{ij} \tag{6.50}$$

where

$$v_{Y_{1j}}^{+}(0) = 1/(2\pi)\cdot\sqrt{m_{Y2_{1j}}/m_{Y0_{ij}}}, \tag{6.51}$$

$$m_{Yn_{1j}} = \int_{-\infty}^{\infty} \omega^{n}\cdot S_{YY_{1j}}(\omega)\mathrm{d}\omega, \tag{6.52}$$

$$S_{YY_{1j}}(\omega) = |H_{YZ}(\omega)|^{2}\cdot S_{ZZ}(\omega|(H_{Vi}, T_{Vj})) \tag{6.53}$$

(see (5.20), (5.27), (5.54)); we can calculate $S_{ZZ}(\omega|(H_{V_1}, T_{V_j}))$, for example, according to (5.90), (5.92) and (5.93) for the $I\cdot J$ seaway classes (H_{V_1}, T_{V_j}), which occur with probability P_{1j}. To the right of the symbol '|' is the 'condition' of the seaway class for which $S_{ZZ}(\omega)$ is calculated, and we use the somewhat more precise notation $n_Y^{+}(y)$ or $v_Y^{+}(0)$ instead of n_Y^{+} or v_{Y0}^{+} (see (5.20)) to represent some of the relationships to be discussed below. In using a long-term statistic $P_{1j} = P[H_{V_1}, T_{V_j}]$ in (6.50), the same principles should be observed as have already been discussed in detail in Sect. 5.2.2.2 for the example of the calculation of a design wave height. The (iterative) calculation of a design value $y = y_B$ is then carried out with the stipulation $n_{YB}^{+} = n_Y^{+}(y_B)$ for a design period $T = T_B$, as described in connection with (5.106), but for practical applications it is very useful to calculate in advance the associated up-crossing frequencies $n_Y^{+}(y)$ within the design period T_B for a series of values y, to give an up-crossing frequency distribution for y. We will illustrate this in the example below.

If necessary, we can of course extend (6.50) to broadband random processes and other parameters (seaway directions, as in Sect. 6.3.3.3 below, loading conditions, etc. – see [23]), but these are seldom applied details which do not give us any fundamentally new information.

6.2.1.2 Comparative Evaluation of Seaway Loads

To introduce the problems of the superposition of environmental loads to form one combined design value, we deal first with a comparative method, useful for practical applications, for establishing reference values for the same effects of the seaway on different marine structures [24–26].

By way of example, we consider two semisubmersible structures which display quite different characteristics in their motion behaviour in heavy seas. These are the semisubmersible design RS-35 (see Fig. 2.39 of Vol. I), the immersion behaviour of which we have already discussed in some detail with Fig. 3.76 of Vol. I, and the relatively large, semisubmersible platform SEAGAS, which was developed for the production and storage of LPG [27, 28] (see Fig. 3.89 of Vol. I). The two semisubmersibles are shown drawn to scale in Fig. 6.12.

For the two semisubmersibles we have calculated, using (6.50), up-crossing frequency distributions of the relative motion r_1 between the outermost lower edge of the platform and wave elevation ζ, calculating the transfer functions $|H_{RZ}(\omega)| = r_{ia}/\zeta_a$ by linear motion analysis in regular waves of amplitude ζ_a and frequency ω (index $i = 1$ for RS-35 and index $i = 2$ for SEAGAS). Figure 6.13 gives the up-crossing frequency distributions n^{+} of the wave elevation ζ in accordance with

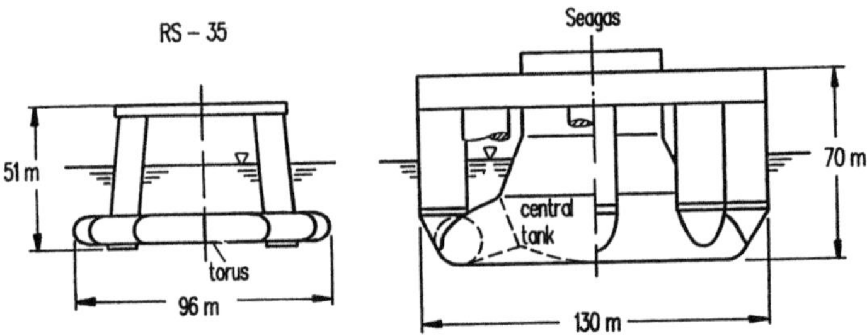

Fig. 6.12. RS-35 and SEAGAS semisubmersibles.

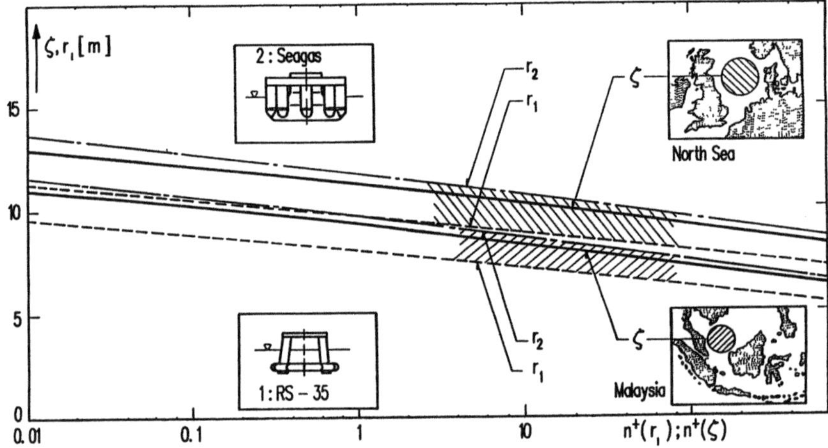

Fig. 6.13. Up-crossing frequencies n^+ for wave elevation ζ and relative motions r_i.

(5.106) and of the relative motions r_1 and r_2 in accordance with (6.50) for the two semisubmersibles, taking into account the seaway statistics of two service regions with quite different wave climate (long-term seaway statistics). We use the long-term statistics of the characteristic wave parameters in [29], AREA 3, for the North Sea, and in [30], AREA 31, for the sea area east of Malaysia.

We can see that the smaller (hydrodynamically transparent) RS-35 semisubmersible exhibits amplitudes of relative motion which are considerably smaller than the wave amplitudes, while the larger (hydrodynamically compact) SEAGAS semisubmersible has amplitudes of relative motion which are somewhat larger than the wave amplitudes: so RS-35 follows the wave contour better than SEAGAS in the long-term mean. If we now consider the provision of sufficient platform height (space between the underside of the platform and the still water surface), it is obvious that the larger SEAGAS semisubmersible requires greater platform height than the RS-35 semisubmersible for the same protection from excessive wave impact. The problem is to define the platform height in such a way as to match the different motion characteristics of the two semisubmersibles.

At the moment there is no binding regulation which would tell us explicitly for what mean up-crossing frequencies of relative motion the design values of the (minimum) platform heights above the still water level should be set to ensure adequate safety for the operation of the platforms during their service in the sea area under consideration. Here we can only refer to a requirement of the IACS (International Association of Classification Societies), according to which fixed platforms are regarded as sufficiently protected from wave impact if an air gap of 1.20 m above the crest of the 100-year design wave is provided. Our task is therefore to translate this safety requirement, defined for fixed platforms, to the two floating platforms under consideration here in such a way that their (minimum) platform heights can be defined in the assumed service conditions.

For this we will refer back to the considerations associated with (5.116): we know that the design value y_B, which occurs only once in the period of time T_n of n years, will be exceeded in another time period T_k of k years with the probability

$$P_{nk} = 1 - \exp\{-v_n \cdot T_k\} = 1 - \exp\{-T_k/T_n\} \tag{6.54}$$

where $v_n = 1/T_n$ is the rate of occurrence of the Poisson process of the (long-term) seaway load y_B which we are considering. Below we define the concept of the service time period $T_E = T_k$ (e.g. $k = 25$ years), in a similar manner to the concept of the (operational) design period $T_B = T_n$ (e.g. $n = 100$ years), and obtain

$$P_{BE} = 1 - \exp\{-T_E/T_B\}. \tag{6.55}$$

We can now, of course, declare each value y_S read off at any up-crossing frequency $n_{YS}^+ = n_Y^+(y_S)$ in an up-crossing frequency distribution (such as that in Fig. 6.13 with $y = r_i$) to be a safe design value, if we think that the design value read off at $n_{YB}^+ = n_Y^+(y_B)$ implies too high or too low a seaworthiness. In design regulations a safe design value is either subjected to a safety margin Δy ($y_S = y_B + \Delta y$) or formed with a safety factor γ ($y_S = \gamma \cdot y_B$). For the safe design value y_S obtained thus, designated here as a safety value, similarly to (6.55) with T_S as the (operational) safety period

$$P_{SE} = 1 - \exp\{-T_E/T_S\}. \tag{6.56}$$

With the plausible assumption, valid for stationary Poisson processes, that in the long term (several years) there is a linear relationship between the number of times the values y_E and y_s are exceeded and the reciprocal value of the associated waiting times T_E and T_S, when $n_{YE}^+ = n_Y^+(y_E)$

$$P_{SE} = 1 - \exp\{-n_{YS}^+/n_{YE}^+\} \tag{6.57}$$

is the probability that the safety value y_S read off at n_{YS}^+ will be exceeded during the service period T_E. We illustrate this result with the example of the RS-35 and SEAGAS semisubmersibles for three different service periods T_E (see Fig. 6.14).

The results shown have been developed on the basis of Fig. 6.13 and (6.57) such that the up-crossing probabilities calculated with (6.57) for $T_E = 15;\ 20;\ 25$ years (i.e. $n_{YE}^+ = 6.7;\ 5;\ 4$) are shown against the safety values $y_S = \zeta_S$ or $y_S = r_{iS}$ ($i = 1$ or 2 for the two semisubmersibles) read off at variable values n_{YS}^+ from Fig. 6.13. We

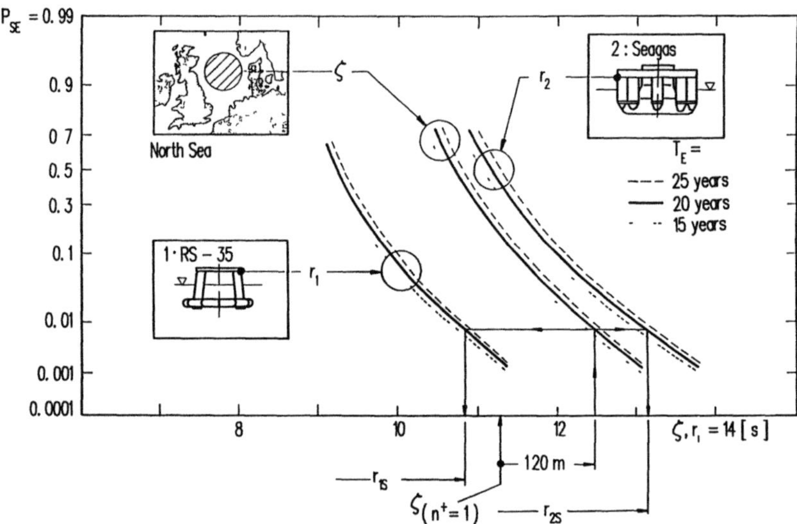

Fig. 6.14. Platform heights above still-water level at comparative risk values.

can see at once that the effect of service period T_E is not great. We will therefore concentrate on only one service period (20 years).

Figure 6.14 demonstrates graphically the concept pursued here of a comparative evaluation of seaworthiness according to the criteria of a minimum platform height above the still water level for adequate protection against wave impact, initially for the wave climate (long-term seaway statistics) of the central North Sea: we add to the 100-year design value of the wave amplitude the value 1.20 m (IACS requirement for fixed platforms, see above) and find the probability P_{SE} that this value will be exceeded within the service period (T_E) of 20 years. For the same probability (here $P_{SE} \simeq 0.009 = 0.9\%$) we read off the associated safety values of the relative motions r_{iS} from Fig. 6.14 for the two semisubmersibles, and obtain by comparative evaluation of the seaworthiness of two semisubmersibles with a fixed platform the required (minimum) platform heights for the semisubmersibles themselves. We have thus made the same risk P_{SE} of wave impact for all three platforms the criterion of comparable seaworthiness.

The same naturally applies to considering the effect of the wave climate of another area of the sea. This is demonstrated in Fig. 6.15 using the RS-35 as an example.

For the sea area east of Malaysia (i_2) we obtain a platform height 1.66 m lower than that for the area of the central North Sea (i_1). We can also see that the internationally agreed guideline of a 1.20 m air gap over the crest of the 100-year reference wave for fixed platforms does not define a uniform risk P_{SE} for different service areas. This is a general disadvantage of the use of safety margins, which are applied frequently today as a means of ensuring safety in structural engineering because they are easy to use. Seen in the light of a rational evaluation of the reliability of marine structures, they do not give an objective picture of the safety achieved: in the example given here of the use of the RS-35 platform in the sea

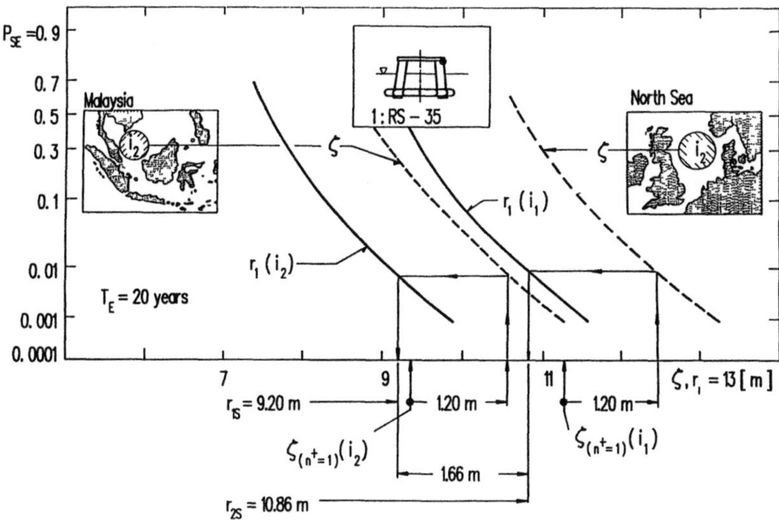

Fig. 6.15. Platform heights for different sea areas

area east of Malaysia, comparison with the application of the same platform in
the sea area of the central North Sea shows that on the basis of the same risk, a
somewhat smaller air gap is justified in the sea area east of Malaysia than in the
central North Sea.

6.2.1.3 Superposition of Load Processes

By considering the same seaway effect on two semisubmersibles we have indirectly
addressed a first fundamental principle of determining reference values which relate
to different, simultaneously acting, stochastic causes.

The joint risk of exceeding the design values of two simultaneously acting loads
should be equal to the risk of exceeding the design value of each individual load,
providing the other load is exactly zero. This principle, simple though it sounds,
does, however, contain some practical problems, which we will examine briefly
with the aid of a further example.

Here we regard the stress on a specific element of a marine structure as the
effect of two internal seaway (sectional) loads, bending moment and normal force,
which are indeed both due to the same seaway process, but whose reference values
can be regarded in a statisfactory approximation as stochastically independent of
one another, due to the differing transfer mechanisms between the external seaway
load distribution on the structure and the internal loads on the element of the
structure under consideration.

The fixed structure shown in Fig. 6.16 serves as an example, namely the North
Sea Research Platform operating to the northwest of Heligoland in the Bay of
Heligoland, with which we can use the unit load concept of [5] mentioned in
Sect. 6.1.1 for calculating internal loads. In this way we calculate transfer functions
of the bending stresses σ_B and the normal stresses σ_N at the denoted beam end B.

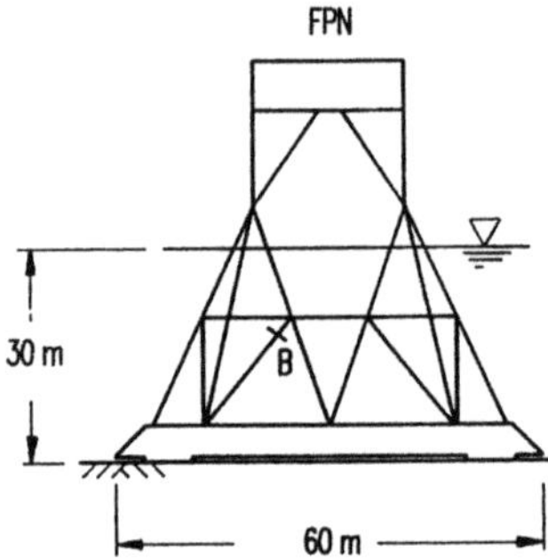

Fig. 6.16. North Sea Research Platform schematic.

With (6.50) to (6.53) and (5.106) we determine, formally as in the last example, frequency distributions for exceeding σ_B, σ_N and ζ for the wave climate in the southern North Sea [29]. These distributions are shown in Fig. 6.17a. Figure 6.17b arises from it on the basis of (6.57) for various assumed safety values $y_S = \sigma_{BS}$ or σ_{NS} and $n_{YE}^+ = 5$ ($T_E = 20$ years).

Assuming stochastic independence between bending and normal stress

$$P_{SE}[\sigma > \sigma_B + \sigma_N] = P_{SE}[\sigma_{BN}]$$
$$= P_{SE}[\sigma_B] \cdot P_{SE}[\sigma_N] = P_{SE}[\sigma > \sigma_B] \cdot P_{SE}[\sigma > \sigma_N], \tag{6.58}$$

from which we obtain Fig. 6.17c by evaluating Fig. 6.17b, various values of $P_{SE}[\sigma_{BN}]$ being varied as parameters.

We can see that there is an infinite number of combinations of σ_B and σ_N, which all have the same risk of exceedance $P_{SE}[\sigma_{BN}]$. However, our task is not indeterminate, because we can see from Fig. 6.17d that there is a single combination of parameters which represents a maximal load with a given joint risk of exceedance. That is of course the design value of the combined bending and normal stress to be used. We have used the factor K to indicate that a local stress concentration can be expected in the node.

The method described briefly here, and clarified with details in Fig. 6.17, is particularly suitable for load processes with different causes (e.g. wind and seaway). The linear load combination illustrated in the example does not mean a restriction of the procedure, which can also be applied to non-linear load combinations. Dealing with a problem in this way is indeed complex, but is appropriate in relation to the high engineering costs of a marine structure, for which the analysis process must in any case be carried out in many stages for other reasons.

As another alternative method of determining the design values related to different, simultaneously acting stochastic causes, we consider another rule using the example of linear load combination, but in this connection we also refer to more general principles and extension to the treatment of non-linear load combinations in [31] and [32] or [33].

The risk of exceeding the reference value of a stochastic process formed as the superposition of participating random load processes of the individual loads should be equal to the risk of exceeding the design value of each individual load, provided that all other loads are zero.

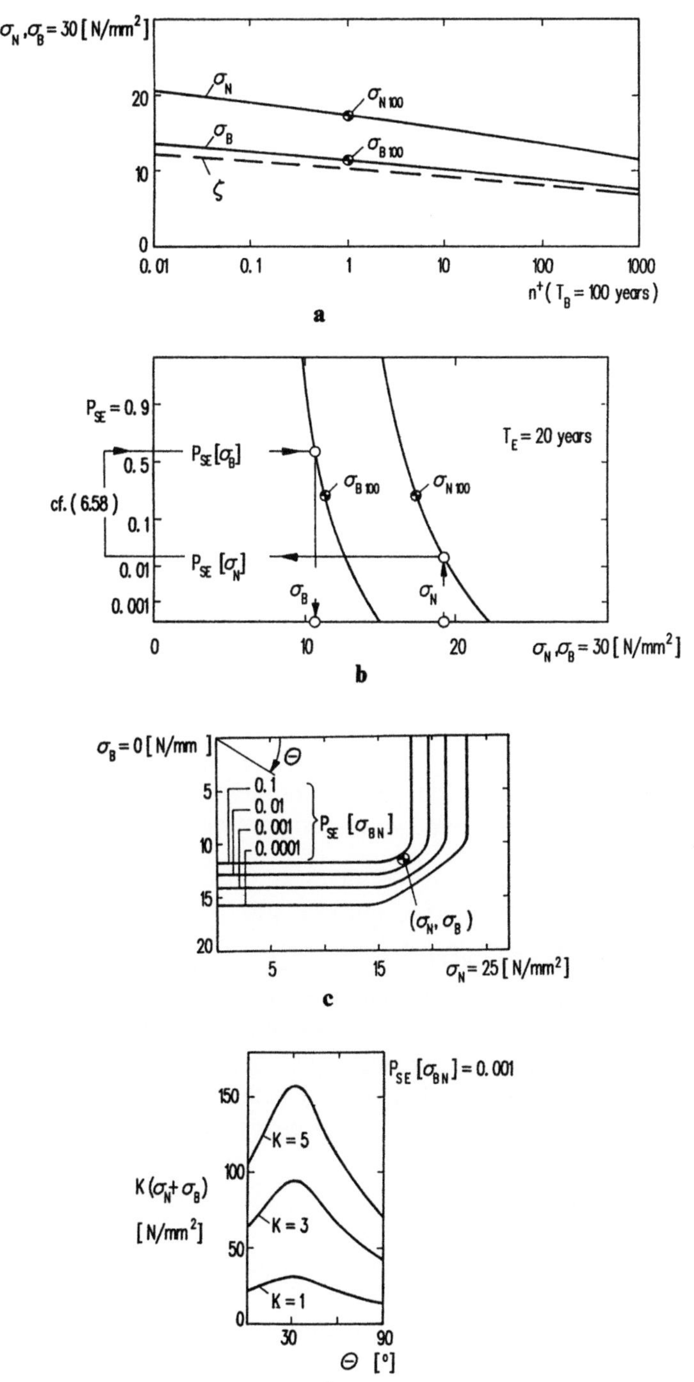

Fig. 6.17a–d. Examples of superposition of stochastically independent stresses.

As an illustration, we again consider only two stochastically independent load processes $\{Y_1(t)\}$ and $\{Y_2(t)\}$. The question of a design value y_B or safety value y_S for the combined loading process $\{Y(t)\} = \{Y_1(t) + Y_2(t)\}$ can generally be answered on the basis of the distribution function $F_{Y\max,T}$ of the maximum value of the combined load, as then y_B or y_s can be represented straight away as the maximum value $y_{\max}$ as a $p\%$ fractile on the basis of an accepted risk of exceedance $q = 1 - p$ by

$$y_{\max} = F^{-1}_{Y\max,T}(p). \tag{6.59}$$

(The exponent -1 symbolizes the inverse distribution function of F.) We define the random value $Y_{\max,T}$ as the maximal value of the sum (linear combinations) of the random values Y_1 and Y_2 during the time interval T

$$Y_{\max,T} = \underset{[0,T]}{\text{Max}}[Y_1 + Y_2]. \tag{6.60}$$

By applying the total probability theorem we obtain in accordance with (A1.19) firstly the distribution function $F_Y(y)$ of the combined load

$$F_Y(y) = P[Y_1 + Y_2 \leqslant y]$$

$$= \int_{-\infty}^{\infty} P[Y_2 \leqslant y - y_1 | Y_1 = y_1] \cdot f_{Y1}(y_1) dy_1$$

$$= \int_{-\infty}^{\infty} F_{Y2}(y - v) f_{Y1}(v) dv, \quad (v \equiv y_1). \tag{6.61}$$

Furthermore, similarly to (5.115), for N independent random samples of uniformly distributed random values Y_n, the extreme value distribution is

$$F_{Y\max}(y) = P[Y_1 \leqslant y, Y_2 \leqslant y, \ldots, Y_N \leqslant y]$$
$$= P[Y_1 \leqslant y] \cdot P[Y_2 \leqslant y] \cdot \ldots \cdot P[Y_N \leqslant y]$$
$$= (P[Y \leqslant y])^N = [F_Y(y)]^N. \tag{6.62}$$

In this connection, N is designated as the number of repetitions. We will now consider the two load process $\{Y_1(t)\}$ and $\{Y_2(t)\}$ simplified as random processes, which can be modelled by load stages (square wave processes), each load stage being constant for a time span θ_1 or θ_2 (see Fig. 6.18). Furthermore, $\theta_1 > \theta_2, T/\theta_2 = N, \theta_1/\theta_2 = M, N$ and M are integer numbers. Then instead of (6.60) we can write

$$Y_{\max,T} = \underset{[0,T]}{\text{Max}}\left[Y_1 + \underset{[t,t+\theta_1]}{\text{Max}}[Y_2] \right]. \tag{6.63}$$

With (6.61) and (6.62), therefore,

$$F_{Y\max,T}(y) = \left\{ \int_{-\infty}^{\infty} [F_{Y2}(y - v)]^M \cdot f_{Y1}(v) dv \right\}^N. \tag{6.64}$$

Calculations for three or more stochastically independent load processes follow the same pattern. However, for engineering applications such calculations are generally too time-consuming, and in many cases too complex, to be carried out

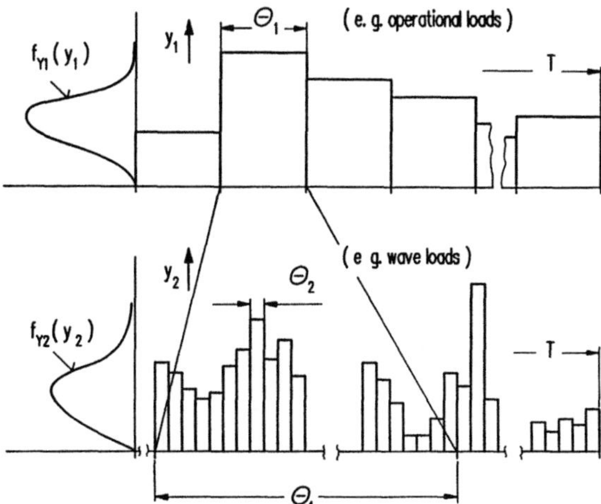

Fig. 6.18. Square wave processes of various loads.

precisely. For this reason, the following approximation is often used instead of (6.63):

$$Y_{\max,T} \simeq \mathrm{Max}\left[\mathop{\mathrm{Max}}_{[0,T]}[Y_1] + Y_2, \; Y_1 + \mathop{\mathrm{Max}}_{[0,T]}[Y_2] \right]. \qquad (6.65)$$

We use this approximation, called the Turkstra rule, firstly to determine an upper limit for the joint up-crossing rate $v_Y^+(y)$ of a value y of the random process $\{Y(t)\}$ of superposed loads, which interests us in connection with all practical applications based on (6.50). With $N_Y^+(y)$ as the random number of up-crossings of a value y of the random process $\{Y(t)\}$ in $[0, T]$, we estimate

$$P[Y_{\max,T} > y]$$

$$\leqslant P[Y_{\max,0} > y] + P[N_Y^+(y) \geqslant 1]$$

$$\leqslant P_0 + \sum_{n=1}^{\infty} P[N_Y^+(y) = n] \leqslant P_0 + \sum_{n=1}^{\infty} n \cdot P[N_Y^+(y) = n]$$

$$= P_0 + E[N_Y^+(y)] \simeq E[N_Y^+(y)] = v_Y^+(y) \cdot T, \; P_0 \simeq 0. \qquad (6.66)$$

This result is precise for the conditions of a Poisson process (see Sect. 5.1.3.1 (postulates 1 to 4)), and it is a very good estimate in the case where the probability of the occurrence of more than one up-crossing in T is small, i.e. for safety values $y = y_S$.

Furthermore, using the Turkstra rule in accordance with (6.65)

$$P[Y_{\max,T} > y] \leqslant P\left[\left(\mathop{\mathrm{Max}}_{[0,T]}[Y_1] + Y_2 \right) > y \right] + P\left[\left(Y_1 + \mathop{\mathrm{Max}}_{[0,T]}[Y_2] \right) > y \right]. \qquad (6.67)$$

With the total probability theorem as in (A1.19), using (6.66), we find, for example, the first term in (6.67) to be

$$P\left[\left(\underset{[0,T]}{\mathrm{Max}}[Y_1] + Y_2\right) > y\right]$$

$$= \int_{-\infty}^{\infty} P[Y_{1\,\mathrm{max,T}} > v \,|\, Y_2 = y - v] \cdot f_{Y2}(y - v) \mathrm{d}v$$

$$= T \cdot \int_{-\infty}^{\infty} v_{Y1}^{+}(v) \cdot f_{Y2}(y - v) \mathrm{d}v.$$

We obtain an analogous result for the second term in (6.67), and finally find

$$v_Y^{+}(y) \leqslant \int_{-\infty}^{\infty} v_{Y1}^{+}(v) \cdot f_{Y2}(y - v) \mathrm{d}v + \int_{-\infty}^{\infty} v_{Y2}^{+}(v) \cdot f_{Y1}(y - v) \mathrm{d}v \qquad (6.68)$$

for the up-crossing rate of a value y of the random process $\{Y(t)\} = \{Y_1(t) + Y_2(t)\}$.

This result, derived as an approximation on the basis of the Turkstra rule, can more generally be represented as a good upper limit [19], so we acknowledge the Turkstra rule in (6.65) to be a practically acceptable approximation under the given conditions. The limit in (6.68) can be extended to three or more independent processes $\{Y_1(t)\}$, which need not necessarily be square wave processes (see also [19], where there are further details of a possible expansion of the load combination concept indicated in (6.68), and of the associated (plentiful) literature on this subject).

When applying (6.68) to e.g. our example in Fig. 6.17, we require up-crossing rates $v_{Yk}^{+}(v)$, $y_1 = \sigma_B$ and $y_2 = \sigma_N$, which we can calculate for $v_{Yk}^{+}(v) = n_{Yk}^{+}(v)/T$, $k = 1$ and 2. We can then make $v_Y^{+}(y)$ equal to an up-crossing rate which is suitably small for safety values (e.g. 0.01) and solve (6.68).

For this we require the distribution densities $f_{Yk}(y - v)$ of the processes considered $\{Y_k(t)\}$, $k = 1$ and 2, at any point in time. We calculate these distribution densities similarly to (6.50) by means of

$$f_{Yk}(y) = \mathrm{d}F_{Yk}(y)/\mathrm{d}y.$$

Assuming a narrowband random process for Y_k the probability of exceeding a value y, i.e. $P[Y_k > y]$, is equal to $n^{+}(y)/n^{+}(y = 0)$, n^{+} being the average number of up-crossings of y or 0 in a (long-term) reference time interval, respectively. Division of numerator and denominator by that reference time interval gives $v^{+}(y)/v^{+}(y = 0) = v^{+}/v_0^{+}$. Hence, by definition

$$F_{Yk}(y) = 1 - P[Y_k > y] = 1 - v^{+}(y)/v^{+}(y = 0).$$

The relative (up-crossing) frequencies v^{+} and v_0^{+} can be calculated with (6.50) to give

$$v^{+}(y) = \sum_{i=1}^{I} \sum_{j=1}^{J} v_{ij}^{+}(y = 0) \cdot \exp\left\{-\frac{y^2}{2m_{Y0ij}}\right\} \cdot P_{ij},$$

$$v^{+}(y = 0) = \sum_{i=1}^{I} \sum_{j=1}^{J} v_{ij}^{+}(y = 0) \cdot P_{ij}.$$

Use of the approximation

$$v^+(y) \simeq v^+(y=0) \cdot \sum_{i=1}^{I} \sum_{j=1}^{J} \exp\left\{ -\frac{y^2}{2m_{Y0ij}} \right\} \cdot P_{ij},$$

which is an identity for $y = 0$, gives

$$F_{Yk}(y) = 1 - \sum_{i=1}^{I} \sum_{j=1}^{J} \exp\left\{ -\frac{y^2}{2m_{(Yk)0ij}} \right\} \cdot P_{ij}, \quad k = 1 \text{ and } 2.$$

6.2.2 Fatigue Strength Models

Fatigue of a material means a process characterized by a gradual reduction in the capacity of the material to withstand repeated loads. Damage means the reduction in strength after a certain number of repeated loadings. We generally distinguish three development stages in material damage, namely crack formation, crack propagation and failure. The first stage, crack formation, often occurs before the first load cycle, and is then due mainly to manufacturing factors, e.g. welding which we will not investigate in detail. Modelling the third stage, the final failure process, which usually happens relatively quickly in the form of a spontaneous fracture, would be of little practical benefit here, and so is not considered any further.

Thus, in Sect. 6.2.2.1, we describe the crack formation phase under repeated loading with a fatigue strength model, and in Sect. 6.2.2.2 we deal with a fracture model suitable exclusively for describing crack propagation. Finally, in Sect. 6.2.2.3 we enhance these models with some findings on stochastic fatigue strength evaluation.

6.2.2.1 Fatigue Strength Modelling

We examine the damage indicator D, henceforth called simply damage, the value of which we will consider to be zero before the first load cycle and one after the last cycle (failure). We define the increase in damage per load cycle for the n-th load cycle as $\Delta D_n = D_n - D_{n-1}$.

This increase in damage is influenced by the whole previous damage process $(D_1, D_2, \ldots, D_{n-1})$ and the stress range s_n in the n-th load cycle. Within the framework of an interaction-free damage accumulation theory, the increase in damage ΔD_n is considered for simplicity as independent of damage further back than D_{n-1}, i.e. $\Delta D = \varphi(D_{n-1}, s_n)$.

In other words, the sequence of damage accumulation before the $(n-1)$-th load cycle is to play no part in the damage accumulation in the n-th load cycle itself. If the damage accumulates relatively slowly and continuously with the load cycles n, ΔD_n can be approximated by the derivation dD/dn, and we obtain the kinetic law of interaction-free damage accumulation

$$dD/dn = \varphi(D, s). \tag{6.69}$$

Any dependence of the damage accumulation on a constant mean stress can often be disregarded for marine structures. If we can assume, furthermore, that the stress range s is the same in each load cycle (as in most fatigue tests today), we can consider the associated damage indicator D_s as dependent on the ratio $n(s)/N(s)$, i.e. the load cycles $n(s)$ related to the total load cycles $N(s)$, and on the stress range s

$$D_s = \gamma(n(s)/N(s), s). \tag{6.70}$$

If, in addition, we can disregard the explicit dependence of the damage indicator D_s on the stress range s, we obtain the stress-independent model of the damage

$$D_s = \gamma(n(s)/N(s)). \tag{6.71}$$

From this we find, in accordance with (6.69), the stress-independent, kinetic law of interaction-free damage accumulation

$$dD_s/dn(s) = 1/N(s)\cdot\gamma'(n(s)/N(s)), \tag{6.72}$$

or using the inverse function γ^{-1} in (6.71),

$$dD_s = 1/N(s)\cdot\gamma'(\gamma^{-1}(D_s))dn(s), \tag{6.73}$$

where $dn(s)$ is the number of load cycles between stress ranges of magnitude s and $s + ds$. Separation of the variables gives

$$dn(s)/N(s) = dD_s/\gamma'(\gamma^{-1}(D_s)), \quad 0 \leqslant D_s \leqslant 1.$$

As, due to the limits of D_s, the function values are $\gamma(0) = 0$ and $\gamma(1) = 1$, the substitution $D_s = \gamma(x)$ gives

$$\gamma'(\gamma^{-1}(D_s)) = dD_s/dx,$$
$$dn(s)/N(s) = dx, \quad 0 \leqslant x \leqslant 1,$$

and after integration over all stress ranges s, we find, with D_R as the total damage, the failure condition

$$D_R := \int_0^\infty dn(s)/N(s) = \int_0^1 dx = 1. \tag{6.74}$$

If we consider I separate areas with a mean stress variation $s_i, i = 1, 2, \ldots, I$, we obtain the failure condition

$$D_R := \sum_{i=1}^{I} n(s_i)/N(s_i) = \sum_{i=1}^{I} n_i/N_i = 1. \tag{6.75}$$

This failure condition for material fatigue under repeated loading, known as the Miner rule, allows us to link test results $N_i = N(s_i)$, obtained with constant stress ranges s_i, to stress range functions $s_i = s(n_i)$ or the inverse functions $n_i = n(s_i)$, $i = 1, 2, \ldots, I$, as occur with the real operation of marine structures and can be calculated in advance from up-crossing frequency distributions using the spectral analysis methods described in Sect. 6.2.1, to predict fatigue strength of structural elements. We go into this in more detail in Sect. 6.2.3, but we can see already how the service life of a structural element can be derived directly from (6.75): if we have specified $n_i, i = 1, 2, \ldots, I$, for a certain reference time period T_R, e.g. for the time period $T_R = T_B = 100$ years as in Sect. 6.2.1, the service life is

$$T_L = T_R/D_R. \tag{6.76}$$

Due to the simplifications introduced to derive the Miner Rule, for which we used a version given in [19], we cannot expect that this rule will give uniformly reliable fatigue predictions or service lives for every material, but it has rightly achieved particular practical importance as a criterion for the comparative evaluation of the relative fatigue strength of selected structural elements of a structure. We should point out that in our derivation we have so far not assumed a linear relationship between damage D_s and $n(s)/N(s)$ as, for example, in Miner's original paper [34], so we do not need the concept, since widely disseminated, of the linear damage accumulation hypothesis in conjunction with (6.74) and (6.75). If, instead of (6.71), we reckon on a linear relationship

$$D_s = n(s)/N(s) \tag{6.77}$$

then the damage accumulation per load cycle is

$$\Delta D_s = \Delta D(s) = dD_s/dn(s) = 1/N(s). \tag{6.78}$$

In this connection we can talk of a linear, stress-independent, kinetic law of interaction-free damage accumulation, to which (6.74) and (6.75) of course also apply.

A realistic representation of the load cycles $N_s = N(s)$ which can be endured by the material at stress ranges s is

$$\begin{aligned}
N_s &= k \cdot s^{-m}, & s &> s_0, & m &> 0, \\
&= \infty, & s &\leqslant s_0,
\end{aligned} \tag{6.79}$$

where k, m and s_0 can be determined more precisely by tests with a constant stress range s. Assuming that the sample in the test has indefinite fatigue strength when $s \leqslant s(N_0) = s_0$, we can give the first alternative in the above equation in the following form (see Fig. 6.19)

$$N_s/N_0 = (s/s_0)^{-m}, \quad s > s_0, \quad k = s_0^m \cdot N_0. \tag{6.80}$$

The representations (6.79) and (6.80) are usually referred to as S–N curves. We should point out that with experimentally determined S–N curves, as s becomes lower or N_s becomes higher, higher scatter occurs in the parameters (N_0, s_0) or (k, m). This means in practice that, for safety, it is assumed there is no unlimited fatigue strength, $s_0 = 0$, if the number of tests (samples) required for a reliable (statistical) prediction is too high for cost reasons, and then an S–N curve linearly corrected at (N_0, s_0) for $s < s_0$ without an unlimited fatigue strength component can be used (see Fig. 6.19).

Furthermore, for marine applications it is important to know the type of failure mode described by the S–N curve: N_s describes the number of stress cycles which lead either to crack initiation at the stress range level s or to a through-thickness crack in the test piece. For pipe joints the service life T_L calculated with the crack initiation curve is about an order of magnitude shorter than that calculated with the through-thickness crack curve!

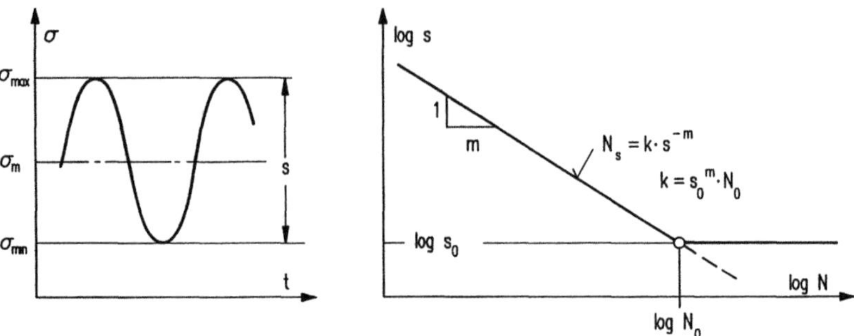

Fig. 6.19. Principle of $S-N$ curve.

6.2.2.2 Crack Propagation Modelling

In many structures the question of crack propagation is particularly important because it must be assumed that a large proportion of load-bearing structural elements will have relatively small initial cracks or fine notches at the outset (e.g. all welded steel structures) which are very important for marine engineering. Crack propagation begins generally only after a certain initial period of scarcely noticeable crack growth, which is overvalued by the crack propagation model considered here. Then, however, a crack of length a begins to grow with each load cycle, and the propagation rate $\Delta a_n = da/dn$ per load cycle grows in accordance with the Paris–Erdogan Rule, derived from observations of crack propagation [35]

$$\Delta a_n = da/dn = C(\Delta K)^M, \quad \Delta K > 0. \tag{6.81}$$

We have illustrated this relationship in Fig. 6.20. In (6.81) C is a constant determined by material properties (elasticity, yield stress and fracture strength), M is a coefficient of model influence between 2 and 4, and ΔK is the fluctuation range of a stress concentration factor K which is dependent on the size and type of the crack, and is intended to take account of the redistribution of a calculated (e.g. in a linearly elastic manner) stress σ, which could be encountered at the location of the crack in an intact body, at the crack point. On the basis of a purely mechanical

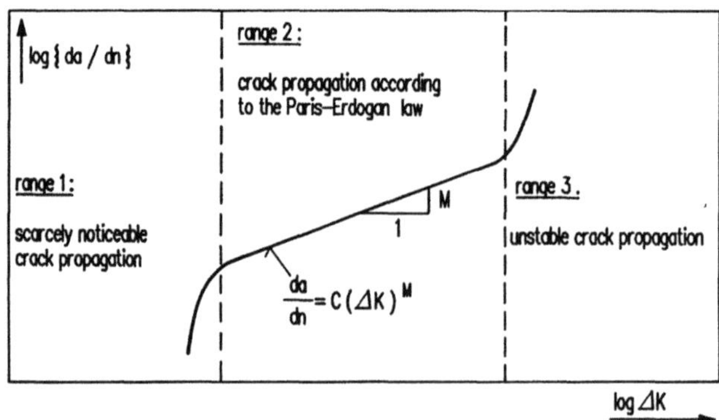

Fig. 6.20. Principle of Paris–Erdogan rule

consideration of the problem (e.g. in [36]), the K factor can be given as

$$K = Y(a) \cdot \sqrt{\pi a} \cdot \sigma, \tag{6.82}$$

$$\Delta K = Y(a) \cdot \sqrt{\pi a} \cdot s, \tag{6.83}$$

s again being the stress range per load cycle, and $Y(a)$ a geometry correction factor which depends on the crack shape, the crack size (crack length a), and the shape of the component. As a first approximation we can make $Y(a) = 1.1$ [37]. Then according to (6.81) and (6.83)

$$da/dn = C(1.1\sqrt{\pi s} \cdot s)^M. \tag{6.84}$$

6.2.2.3 Stochastic Evaluation of Fatigue Strength

In the form (6.84) the Paris–Erdogan rule can only be applied to constant stress ranges s. When applying it to marine structures we must also consider the natural (not harmonic) variations of the stress variations due to loadings induced by seaway.

If we separate the variables in (6.84) and integrate (index i and f for start or end), we find, assuming a constant rate $v = n/t = N_f/T$ for the number n or N_f of stress range cycles per time interval t or T,

$$\int_{a_i}^{a_f} a^{-M/2} da = C(1.1\sqrt{\pi})^M \cdot \int_0^{N_f} s^M(n) dn = C(1.1\sqrt{\pi})^M \cdot v \cdot \int_0^T s^M(t) dt$$

$$= C(1.1\sqrt{\pi})^M \cdot N_f \cdot (1/T) \cdot \int_0^T s^M(t) dt$$

$$= C(1.1\sqrt{\pi})^M \cdot N_f \cdot \langle s(t)^M \rangle = C(1.1\sqrt{\pi})^M \cdot N_f \cdot E[S^M].$$

The last equation applies when assuming ergodicity of the random process of the stress ranges $\{S(t)\}$ having a random sample $s(t)$ (see note to (5.8)). This evaluation of the Paris–Erdogan rule in the form of (6.84) differs from an analogous evaluation with a constant stress range s only in the use of $E[S^M]$ instead of s^M; we consider the crack propagation from a_i to a_f as independent of the sequence of the total of N_f stress variations (interaction-free model of crack propagation). Therefore, for ergodic random processes $\{S(t)\}$ we define an equivalent stress range $\bar{s}$ as

$$\bar{s}^M := E[S^M] = \sum_{i=1}^{I} E[S^M | S_i] \cdot P[S_i]. \tag{6.85}$$

(For the last term in this equation see (5.102); here we are considering a stochastic stress range S, which assumes values s from domains S_i.) To evaluate (6.85) we estimate

$$P[S_i] = n_i/N,$$

where $n_i = n(s_i)$ is the number of stress ranges in the class s_i, which represents the domain S_i, and N is the total number of stress ranges in all I stress range classes

s_i in a steady-state seaway. Similar to the definition given by (6.85), we make

$$E[S^M|S_i] := s_i^M.$$

From (6.85) we thus obtain the equivalent stress range required to apply the Paris–Erdogan rule to the natural ergodic seaway

$$\bar{s} \simeq \left(\sum_{i=1}^{I} n_i \cdot s_i^M / N \right)^{1/M}, \quad N = \sum_{i=1}^{I} n_i. \tag{6.86}$$

A generalized analytical representation of $E[S^M]$ will be given in Sect. 6.3.3.1 in the derivation of (6.180).

We find the total number of load cycles N_f for crack propagation from a_i to a_f from integration of the Paris–Erdogan rule in the form of (6.84), taking account of (6.86). After separation of the variables we find

$$\int_{a_i}^{a_f} a^{-M/2} da = C(1.1\sqrt{\pi} \cdot \bar{s})^M \cdot \int_0^{N_f} dn,$$

$$N_f = 2[a_i^{(2-M)/2} - a_f^{(2-M)/2}]/[(M-2) \cdot C(1.1\sqrt{\pi} \cdot \bar{s})^M], \tag{6.87}$$

where $M > 2$. We note the effect of the initial crack length a_i, for which we generally know only statistical estimates.

We can also develop a stochastic statement, corresponding to the Miner rule, on the expected damage: for this we consider the damage $D(s) = D_s$ after I load cycles within a reference time T_R at the stress range level s as the sum of the individual damage accumulations $\Delta D(s)$, i.e. $D_s = \Delta D_1(s) + \Delta D_2(s) + \cdots + \Delta D_I(s)$. On the plausible assumption that the same stress ranges s do not follow one another directly in time in the natural seaway process, and the damage accumulations $\Delta D_i(s)$ are therefore stochastically independent of one another

$$E[D_s] = E[\Delta D_1(S)|S = s] + E[\Delta D_2(S)|S = s]$$
$$+ \cdots + E[\Delta D_I(S)|S = s] = I \cdot E[\Delta D(S)|S = s].$$

For the total damage after s stress range cycles within T_R, we obtain, similarly to (5.102)

$$E[D_R] = n \cdot \int_0^{\infty} E[\Delta D(S)|S = s] \cdot f_S(s) ds, \tag{6.88}$$

taking account of all possible stress ranges $S = s$ with an identical distribution density $f_S(s)$ due to $I = n \cdot f_S(s) ds$, $f_S(s) ds$ being the probability $P[S]$ in the interpretation in (A1.10). According to (6.78) the damage expected at the stress range level s is $E[\Delta D(S)|S = s] \simeq 1/N(s)$, $N(s)$ being the total number of load cycles with the stress range s which leads to failure. Thus

$$E[D_R] \simeq n \cdot \int_0^{\infty} 1/N(s) \cdot f_s(s) ds. \tag{6.89}$$

If we consider I separate classes with average stress range $s_i, i = 1, 2, \ldots, I$, we find

$$E[D_R] \simeq n \cdot \sum_{i=1}^{I} 1/N(s_i) \cdot P[S_i] = n \cdot \sum_{i=1}^{I} 1/N_i \cdot P_i. \tag{6.90}$$

The number of damage-relevant stress range cycles s in a random sample of a stress process $\{\sigma(t)\}$ depends on the counting method used to allocate amplitudes of the stress σ to the stress range s. At this point we will initially allocate a half stress range to each of these amplitudes, i.e. $\sigma = s/2$, which is realistic for a narrow-band process. As we can assume the Rayleigh distribution for the distribution density of the peak values with this assumption, we find from (6.89) with (6.79)

$$E[D_R] = (n/k) \cdot 1/2 \int_0^\infty (2\sigma)^m \cdot \sigma/\mathrm{Var}[\Sigma] \cdot \exp\{-\sigma^2/(2\,\mathrm{Var}[\Sigma])\}\,d\sigma$$

$$= (2\sqrt{2})^m \cdot n/k \cdot \sqrt{\mathrm{Var}[\Sigma]}^m \cdot \Gamma\{1 + m/2\}. \tag{6.91}$$

Here the Γ function is defined as

$$\Gamma\{(a+1)/b\} = b \cdot \int_0^\infty u^a \exp\{-u^b\}\,du, a = m+1, b = 2, u = \sigma/\sqrt{2\,\mathrm{Var}[\Sigma]}. \tag{6.92}$$

Determining the variance of D_R, $\mathrm{Var}[D_R]$, is somewhat more complex, but can also be given in an analytically closed form [38].

In Sect. 6.3.3 we will present further developments of the stochastic evaluation of fatigue strength on the basis of fracture mechanics in conjunction with the reliability analysis of mechanical components of load-bearing structures, which we will introduce in Sect. 6.3.1, but we will give only a few results and introductory explanations within the limited framework of this book, as such developments are not complete and have not become established practice in marine structures. The interested reader will find more information on this in [39, 40].

As a preliminary conclusion, however, we would like to describe a probabilistic model of fatigue strength, in which the (uncertain) damage present before the first load cycle can be explicitly taken into account. We will start here with the simplest version of this model, the extension and applications of which are dealt with in detail in [41].

We consider duty cycles of equal intensity, duty cycles being defined as fairly long, regularly occurring periods of operation in which the structural element is subjected to an increase in damage. By the term equal intensity we mean the assumption that everything that happens in one duty cycle also happens in every other duty cycle. The increase in damage in a duty cycle depends only on the damage at the start of the duty cycle, and by dividing the damage into different classes $d_i, i = 1, 2, \ldots, I$, the damage only progresses into the next highest damage class when it increases, i.e. after a duty cycle it cannot have jumped a class. The damage d_I is one, i.e. reaching the I-th class indicates failure.

Under these assumptions the damage accumulation can be described as a Markov chain, which we have already met in Sect. 5.1.3.2. In this case the matrix of the transition probabilities P (transition matrix) has elements only on the main diagonal (persistance probabilities p_{ii} in a class d_i) and on the upper subsidiary diagonal (transition probabilities $p_{ij} = p_{i,i+1} = 1 - p_{ii}$ into the next higher class d_{i+1})

$$P = \begin{bmatrix} p_{11} & p_{12} & & & & \\ & p_{22} & p_{23} & & & \\ & & \cdots & & \cdots & \\ & & & p_{ii} & p_{i,i+1} & \\ & & & & \cdots & \cdots \\ & & & & & p_{II} \end{bmatrix}. \tag{6.93}$$

We describe the probability of initial damage before the first duty cycle as a histogram with the vector

$$\pi_0^T = (\pi(1), \pi(2), \ldots, \pi(i-1), 0). \tag{6.94}$$

If we indicate reaching damage d_i at times $t = 1, 2, \ldots, x$, etc., by D_x, i.e. if we define

$$P[D_x = d_i] = \pi_x(i)$$

as the probability of damage d_i at time $t = x$, we find the histogram of the damage probabilities of the different classes at time $t = x$ as a result of the Markov chain as in (5.32) to be

$$\pi_x^T = \pi_0^T \cdot P^x, \quad x = 1, 2, \ldots, \tag{6.95}$$

where the vector π_x is defined as

$$\pi_x^T = (\pi_x(1), \pi_x(2), \ldots, \pi_x(I)). \tag{6.96}$$

The probability of damage-free operation up to $t = x$ is therefore

$$P[D_x < d_I] = \sum_{i=1}^{I-1} \pi_x(i) = 1 - \pi_x(I).$$

This describes the probabilistic fatigue model, designated in [41] as a B-model of cumulative damage, in its simplest terms. We can only indicate the range of practical extensions here: it is quite possible to take account of unequal intensity of duty cycles, necessary for applications in marine technology when the stress is analysed spectrally. The B-model is beautifully simple in concept and easily used in practice, if realistic information on the initial distribution π_0 and the transition matrix P is available. The transition matrix can generally only be indicated with statistically reliable data for the transition probabilities, after a considerable amount of testing, and it should be pointed out that (6.93) can easily be extended to a general upper triangular matrix. Only the future will show to what extent marine technology can make use of the possibilities of the B-model.

6.2.3 Fatigue Strength under Seaway Loads

We would like to emphasize the particular importance of fatigue strength for marine structures, especially fixed platforms, by mentioning some cost factors in the production and use of tubular joints:

1. Quality of joint material.
2. Quality of joint manufacture (i.e. welding, heat treatment, seam treatment, etc.).
3. Frequency of joint repairs, (i.e. length of service interruptions, inspection requirements, etc.).

Due to such factors, the expense, already accepted at the design stage, of fatigue strength evaluation, which proceeds methodically at the maximum level of what is analytically and numerically possible, is generally very high (see [42, 43]).

Common to all evaluations of serviceability is the necessity of translating results of a global stress analysis of the structure to the special situation of the structural element under consideration, for there is no fatigue strength of the whole structure in the real sense, only the fatigue strength of its elements. If one or more elements fail in service, the residual load-bearing capacity of the partially damaged structure is naturally open to question, but this is a special subject in the field of load-bearing capacity in extreme environmental conditions (Sect. 6.2.1), and not of load-bearing capacity in service conditions, which we are dealing with here.

To evaluate the fatigue strength of highly-loaded joints in marine structures (e.g. in jacket platforms) we first carry out a global strength analysis of the whole structure to determine, on this basis, local stress ranges at the intersections of the structural elements (beams or bars) at the joints: by using stress concentration factors K, which take account of the influence of local geometrical factors, as well as microscopic defects, etc., the local stress ranges at each relevant point of an intersection can be calculated from the results of the global stress analysis, so by using the fatigue strength models presented in Sect. 6.2.2, the fatigue strength of any joint can be evaluated.

Examining all the joints of a structure is generally so expensive that only a few joints which are regarded as particularly critical for fatigue are evaluated. Various methods are used depending on the particular characteristics of the structures and their operation, and on the different stages of detail specification in the design process. Here we will show the most important details of four methods variously suited to different structures and purposes, first three methods of fatigue strength evaluation arranged in rising order of complexity (Sect. 6.2.3.1 to 6.2.3.3), and finally a method which is particularly suitable for the first design for fatigue strength (Sect. 6.2.3.4).

6.2.3.1 Deterministic Method of Analysis

In this method, realistic modelling of the natural seaway is dispensed with from the outset. Rather 'blocks' of periodic single waves of specified height H_i and period T_j are used, $i = 1, 2, \ldots, I, j = 1, 2, \ldots, J$. The number of juxtaposed single waves in each block depends on the relative frequency $P_{ij} = P[H_{Vi}, T_{Vj}]$ of the seaway class (H_{Vi}, T_{Vj}) in the reference time period T_R, of which the individual wave block under consideration is representative. In this very simplified method, often only the marginal histogram $P_i = P[H_{Vi}]$ of the wave height H_{Vi} is used, so the single wave period T_i chosen must be representative of the associated seaway class (H_{Vi}). It is quite possible and, within the framework of the deterministic analysis method, more precise, firstly to estimate a cumulative long-term distribution of the individual wave heights H_1, and from this to determine the occurrence frequencies $n_1 = n(H_i)$ – as described later in Sect. 6.2.3.2 in more detail for any seaway effect y.

In the simplest case which we are considering first, the i-th wave block has about $n_i = T_{\mathrm{R}} \cdot P_i / T_i$ single waves of height $H_i = H_{\mathrm{Vi}}$. We will now designate such a wave block in accordance with our previous nomenclature as an elementary wave of finite length. For each of these I elementary waves of finite length, a statical structural analysis is carried out, the wave forces being determined using the methods described in Chap. 3. As a result of the subsequent strength analysis carried out by the methods described in Chap. 4, we obtain I stress ranges $s = s(H_i)$ at the considered structural elements for each wave height H_i, $i = 1, 2, \ldots, I$, which then have to be multiplied by the applicable stress concentration factors K (see Fig. 6.21).

As we know $n_i = n(H_i)$ and $s_i = s(H_i)$, we directly obtain the number of stress ranges at level i, that is $n_i = n(s_i)$, $i = 1, 2, \ldots, I$, necessary for applying the Miner rule as in Sect. 6.2.2.1, and by using the S–N curve $N_i = N(s_i)$, $i = 1, 2, \ldots, I$, applicable to the structural element under consideration, we obtain with (6.76) a nominal damage D_{R} per reference time T_{R} or with (6.76) the nominal service life T_{L}.

Figure 6.21 also indicates that it is possible to consider dynamic effects indirectly from previous calculations with the aid of dynamic amplification factors D, if the period T_i lies in the domain of a natural period of the structure. These dynamic amplification factors can be represented as a ratio of the dynamic stress range to the quasi-static stress range calculated here.

Of course, we can calculate the ratio n_i / N_{f_i} for each wave block, e.g. with (6.87) for $\bar{s} = s_i$, and hence with

$$D_{\mathrm{R}} = \sum_{i=1}^{I} n_i / N_{f_i}$$

we can calculate a nominal damage with the Paris–Erdogan rule: $D_{\mathrm{R}} < 1$ means $a < a_f$ (see (6.87)).

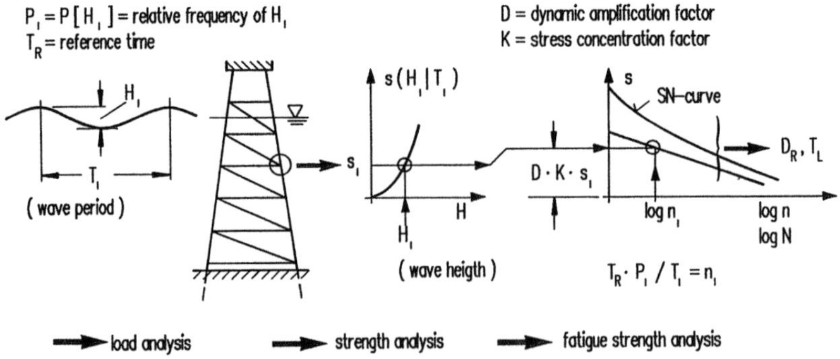

Fig. 6.21. Fatigue strength evaluation with elementary waves of finite length.

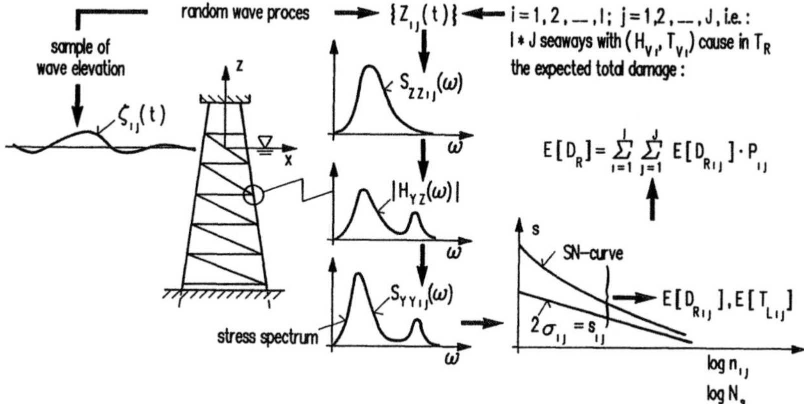

Fig. 6.22. Fatigue strength evaluation using spectral analysis method.

6.2.3.2 Spectral Analysis Method

We proceed basically in the same way as when giving up-crossing frequencies $n_Y^+(y)$ of the value y of any seaway effect Y in a fairly long measurement period T (see (6.50) to (6.53)): let the partial damage expected as a seaway effect in the class (H_{Vi}, T_{Vj}) during a reference period $T = T_R$ be $E[D_{Rij}]$. Then the expected total damage $E[D_R]$ can be shown with (6.90) similarly to (5.103) as follows (see also Fig. 6.22):

$$E[D_R] = \sum_{i=1}^{I} \sum_{j=1}^{J} E[D_{Rij}] \cdot P_{ij} = \sum_{i=1}^{I} \sum_{j=1}^{J} n_{ij} \left(\sum_{l=1}^{L} 1/N_l \cdot P_{lij} \right) \cdot P_{ij}. \tag{6.97}$$

Here $P_{ij} = P[H_{Vi}, T_{Vj}]$ are the probabilities of the occurrence of the $I \cdot J$ seaway classes (H_{Vi}, T_{Vj}) considered, $P_{lij} = P[S_l|(H_{Vi}, T_{Vj})]$ is the probability of the occurrence of a value s of the stochastic stress range S from the domain S_l in the relevant seaway class, N_l is the number of load cycles which leads to failure of the structural element under consideration with stress ranges from the domain S_l, and n_{ij} is the average number of possible waves in the seaway class under consideration during the reference time T_R.

In the case of a quasi-static structure with a narrowband response process (spectrum) we can use, for example, the Rayleigh distribution of $y = \sigma$ to determine P_{lij}, i.e.

$$P_{lij} = \exp\left\{ -(y_l - \Delta y_l/2)^2/(2m_{Y0ij}) \right\} - \exp\left\{ -(y_l + \Delta y_l/2)^2/(2m_{Y0ij}) \right\} \tag{6.98}$$

with Δy_l is the width of the stress class $y_l = \sigma_l$, which is represented by the index l, and

$$n_{ij} = T_R/T_{Y0ij} = T_R/(2\pi) \cdot \sqrt{m_{Y2ij}/m_{Y0ij}}, \tag{6.99}$$

the moments of the stress spectrum m_{Y0ij} and m_{Y2ij} being determined with (6.52). We therefore consider the stress process $\{Y(t)\} = \{\sigma(t)\}$, i.e. each value y in (6.97) to (6.99) is to be interpreted as a peak value (amplitude) of the effective stress σ (including local stress concentrations) (see Fig. 6.19). For narrowband spectra the stress range s is equal to twice the amplitude value of the stress, $s = 2\sigma$; we can generally disregard the effect of any mean stress σ_m in the case of marine structures. For narrowband processes we should also consider the analytically closed representation of $E[D_{R_{1j}}]$ in (6.91) instead of (6.90), for which we must also evaluate (6.99). See Sect. 6.2.3.3 for a correction for wideband processes.

The evaluation of (6.97) to (6.99) for static structures with the spectral method of analysis is therefore based on determining all response spectra $S_{YYij}(\omega)$ for the various seaway classes (H_{Vi}, T_{Vj}), e.g. according to the unit load concept described in Sect. 6.1.1, as well as on the reference time T_R which is arbitrarily definable in principle (one year is normal), on the $S-N$ curve $N_l = N(s_l)$ of the structural element under consideration and on the wave climate (i.e. the long-term seaway statistics P_{ij}) of the operating area of the structure).

For quasi-static structures we can generally accept a further simplification of the spectral method of analysing the fatigue strength problem developed so far: it is quite customary to use the long-term up-crossing frequencies $n_Y^+(y = \sigma)$ which can be determined with (6.50) to (6.53) for $T = T_R$ directly to evaluate the Miner rule with (6.75), the loading function $n_1 = n(\bar{y}_1)$ being defined by

$$n_i = n_Y^+(y_{iu}) - n_Y^+(y_{i0}),$$

for a series of representative values $\bar{y}_1$, where y_{i0} and y_{iu} are the upper and lower limits of the interval with the representative value $\bar{y}_1$. With the narrowband response process, a representative stress range $\bar{s}_i = 2\bar{y}_1$ belongs to each value n_i, $i = 1, 2, \ldots, I$, thus defined, so we know the stress range function $\bar{s}_1 = \bar{s}(n_1)$ for the effect of all seaways and must apply the Miner rule, as in Fig. 6.21, only once.

With $E[D_{R_{1j}}] = n_{1j}/N_{f_{1j}}$ for n_{1j} in (6.99) and $N_{f_{1j}}$ in (6.87) and each seaway class (H_{Vi}, T_{Vj}), we can carry out an analysis analogous to (6.97) with the Paris–Erdogan rule (see also notes at the end of Sect. 6.2.3.1).

6.2.3.3 Simulation

For various reasons we are forced at times, especially when dealing with non-linear and/or dynamic systems, to dispense partly or wholly with the spectral method of analysis and to evaluate fatigue strength by simulating random samples of the relevant random stress process. Three important stages in this simulation are shown in Fig. 6.23.

We have already dealt in detail with the first step of such a simulation, namely the representation of a random sample $\zeta_{1j}(t)$ of class (H_{Vi}, T_{Vj}) seaway, in Sect. 5.2.1, and we demonstrated an application of the simulation technique in Sect. 6.1.2. In the second stage of the fatigue strength evaluation we are interested in the variation of the stress $\sigma_{1j}(t)$ applied to the structural element by a random seaway sample, which we can determine for each time interval with the loading calculations in Chap. 3 and the strength calculations in Chap. 4. In connection with the fatigue strength evaluation, we must mention here the third step, i.e. we

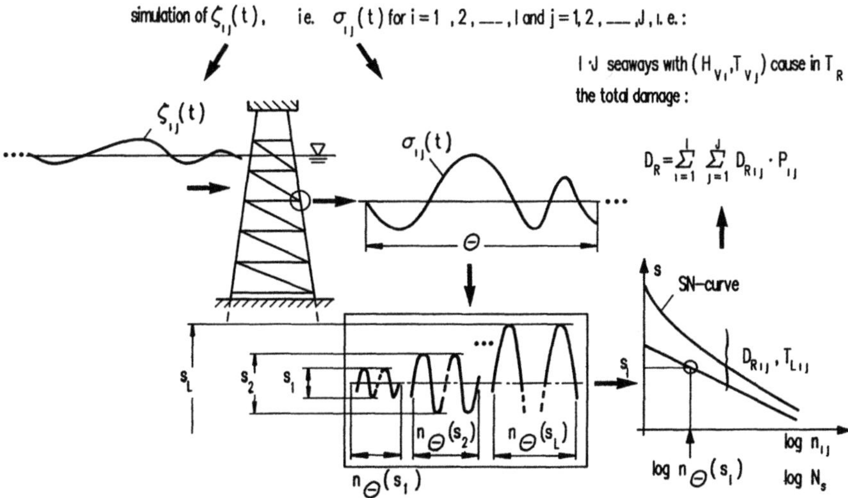

Fig. 6.23. Fatigue strength evaluation with simulation in the time domain

must describe proven methods of determining the relative frequency $n_\theta(s_l)$ of the stress range class s_l in a random sample of length θ, in particular for random samples which do not belong to narrowband processes, as these are especially problematical for counting the stress variations.

Below are three important procedures for counting the stress ranges of a stress range class s_l from a random sample $\sigma(t)$ of length θ:

1. Counting the stress maxima: with the procedure all stress maxima above zero are counted and combined with the value of a fictitious minimum of the same size to form a stress range s. The maximum and minimum must not follow one another, so local maxima below zero are disregarded, and successive extremes above zero are rated with excessively high stress range numbers. We are generally on the safe side with counting the stress maxima, because high stress variations are counted more frequently than with the other two procedures.
2. Counting stress ranges: in this procedure the occurrence of successive extremes is counted as a half stress cycles. With this the large stress ranges of longer periods are not taken into consideration.
3. Counting by the rain-flow method: in this procedure we imagine the random sample $\sigma(t)$ with the time axis going vertically down (Fig. 6.24). Then we regard the random sample as a sequence of imaginary roof slopes subjected to rain, from which the rain flows according to the following rules [44]:

   — the flow starts on the upper side of a roof slope at every extreme.
   — a flow which begins to the left of the (vertical) t axis (see Fig. 6.24) stops when the next extreme is further to the left than the starting point. The opposite is true when the start is to the right of the t axis.
   — the flow always stops when it meets a flow which started earlier.

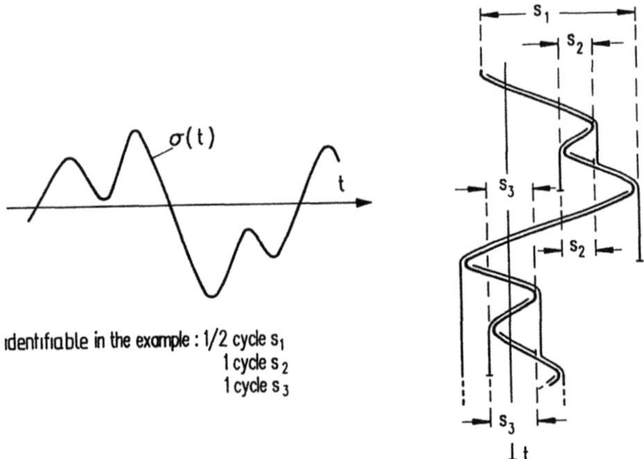

Fig. 6.24. Principle of rain-flow method.

Under these conditions, a half cycle is always allocated to the stress range s_l projected onto the (horizontal) stress axis between the start and end of a flow path. If a stress range class is counted in this way N times, then $n_\theta(s_l) = N/2$.

The rain-flow method takes into account short- and long-period load cycles, and therefore is the most realistic of the three counting procedures for wideband random processes. Generally, the ralationship $\sigma = s/2$ used, for example, in the derivation of (6.91) is not precise, so a rain-flow correction $\lambda(\alpha, m)$ obtained with the third counting procedure is usually used, α being defined in (5.26) and m in (6.79) (see Appendix II in [45]).

We can mention [37] as an example of a detailed evaluation of the fatigue strength of a marine structure by the simulation technique with alternative application of the Miner and the Paris–Erdogan rules, as well as a whole series of different parameters for the mechanical model of the structure (including sea floor modelling). The example used there shows how complex this procedure is overall, because we have to deal with all the seaway classes one after the other; of course, we have to weight the counting result of each random sample of length θ with $T_R \cdot P_{1j}/\theta$ (T_R = reference time, $P_{ij} = P[H_{Vi}, T_{Vj}]$), to obtain the estimated value $n_{ij}(s_l)$ for the stress range class s_l in a seaway of class (H_{Vi}, T_{Vj}). For reasons of limited statistical confidence, θ itself should not be too short. Thus a hybrid technique is often used in dynamic systems, combining the spectral analysis method with simulation in the time domain [46].

We can avoid the counting processes described by using stress functions $\sigma(t)$ obtained by simulation to give the related stress spectra by means of Fourier transformation of their autocorrelation functions, and to obtain the value of the transfer function by division of the stress spectra by the seaway spectrum (see (6.1)). If we have specified some of these transfer functions (e.g. for light, medium and heavy seaway conditions), we can obtain the transfer functions for the intermediate seaway conditions as an approximation by interpolation, and derive the missing (pseudo) response spectra with (5.54) by multiplication with the related seaway spectra.

Then we are in a position to proceed with the concept of the spectral analysis method, as in Sect. 6.2.3.2, again.

6.2.3.4 Design for Fatigue Strength

In design it is important to make realistic estimates of the fatigue strength in an effective manner. The procedure given below for developing design stress ranges [47] is suitable for this purpose: Prerequisites are

1. Applicability of the Miner rule (see (6.75)).
2. Validity of the Weibull distribution for wave heights (see (5.109)) for the special case of the significant wave height $H_{1/3}$.
3. The ability to represent the relationship between stress range s and wave height h in the form

$$s = \gamma \cdot h^g \tag{6.100}$$

(see the representation $s(H_i | T_i)$ in Fig. 6.21.

From this we obtain, with (6.79),

$$1/N_s = s^m/k = \gamma^m \cdot h^{gm}/k$$

and, because of (6.100), (6.74) gives

$$D_R = \int_{h=0}^{\infty} dn(h)/N(h) = \gamma^m/k \cdot \int_{h=0}^{\infty} h^{gm} dn(h), \tag{6.101}$$

$dn(h)$ being interpreted as the number of waves with heights between h and $h + dh$. With the probability density $f_H(h)$, the wave height H regarded as a stochastic value and the number n_R of all waves in the reference time period T_R

$$dn(h) = n_R \cdot f_H(h) dh,$$

i.e. with (6.101) and (A1.15)

$$D_R = n_R \cdot \gamma^m/k \cdot \int_0^{\infty} h^{gm} dF_H(h).$$

For the assumed Weibull distribution we find with the form parameters b and d similarly to (5.109)

$$dF_H(h) = bu^{b-1} \exp\left\{ -u^b \right\} du, \quad u = h/d,$$

so, by using the Γ function defined in (6.92) with $a = gm + b - 1$ we obtain

$$D_R = n_R \cdot \gamma^m/k \cdot d^{gm} \cdot b \cdot \int_0^{\infty} u^a \exp\left\{ -u^b \right\} du = n_R \cdot \gamma^m/k \cdot d^{gm} \cdot \Gamma\{1 + gm/b\}. \tag{6.102}$$

If we apply this result, which is valid for arbitrary wave heights h under the conditions given, to the particular case of a steady-state, narrowband Gaussian seaway process, the Weibull distribution becomes a Rayleigh distribution, i.e.

$b = 2, d = H_{1/3}/\sqrt{2}$ (see (5.24) for $a = h/2$ as well as (5.77) and (5.80)). We obtain

$$D_{RK} = n_R \cdot \gamma^m \cdot \Gamma\{1 + gm/2\} \cdot (H_{1/3})^{gm}/(k\sqrt{2^{gm}}) \qquad (6.103)$$

(index K indicates short-term). Extending this equation to a long-term analysis gives, with the total probability theorem as in (A1.19),

$$D_{RL} = n_R \cdot \gamma^m \cdot \Gamma\{1 + gm/2\}/(k\sqrt{2^{gm}}) \cdot \sum_{i=1}^{I} (H_{1/3i})^{gm} \cdot P[H_{1/3i}].$$

The Weibull distribution can be used as a long-term distribution, so an analysis identical to the above derivation gives finally [47]

$$D_{RL} = n_R \cdot \gamma^m \cdot d^{gm} \cdot \Gamma\{1 + gm/2\} \cdot \Gamma\{1 + gm/b\}/(k\sqrt{2^{gm}}) \qquad (6.104)$$

(index L indicates long-term). We determine the parameters b and d, as demonstrated with (5.110) in Fig. 5.13, from statistical data, for n_R we use, e.g., the number of times the values $y = 0$ is exceeded as in (6.50) to (6.53); we can, however, estimate n_R quite accurately with a little experience by assuming a mean wave period (see (5.108)). The parameters k and m characterize the S–N curve of the structural element under consideration, and the parameters γ and g can be determined, for example, in the way illustrated in Fig. 6.21, if necessary with the additional use of a regression analysis, if $s = s(h)$ is undulating.

The special feature of the result (6.104) is of course the analytically closed representation of the long-term damage D_{RL}. We now have a practical starting point for application to the design, which we examine briefly.

We consider a stress range s_R, which we assume will be exceeded precisely once in the reference time T_R. Then according to (6.79) with (6.100)

$$k = \gamma^m \cdot N_R \cdot (h_R)^{gm}$$

with N_R as the number of stress cycles to failure with stress range s_R, and h_R as the wave height which is exceeded once in all n_R waves which occur in T_R. After substitution in (6.102) we find

$$D_R = n_R/N_R \cdot (d/h_R)^{gm} \cdot \Gamma\{1 + gm/b\}$$

as the total damage from all seaway classes, related to a design case (n_R, h_R, $N_R = N(s_R)$). For the assumed Weibull distribution

$$F_U(u) = 1 - \exp\{-u^b\}, u = h_R/d,$$
$$\ln\{1/(1 - F_H(h_R))\} = (h_R/d)^b.$$

Because $1 - F_H(h_R)$ is the relative up-crossing frequency of h_R, which we have assumed once in n_R waves, $1/(1 - F_H(h_R))$ is equal to the number n_R of waves in T_R. Therefore

$$D_R = n_R/N_R \cdot \Gamma\{1 + gm/b\}/(\ln\{n_R\})^{gm/b}.$$

According to (6.79)

$$s_R = (k/N_R)^{1/m},$$

so for the reliable operation of a structural element according to $D_R < 1$, the greatest permissible stress range s_{max} must be

$$s_{max} \leqslant s_R = [(k/n_R)/\Gamma\{1 + gm/b\}]^{1/m} \cdot [\ln\{n_R\}]^{g/b}. \tag{6.105}$$

s_R defines the stress range which must not be exceeded if h_R occurs, if fatigue strength is to be guaranteed throughout the whole reference time with n_R load cycles. We speak more generally of the reference value of the stress range or of the design stress range it $T_R = T_B$. The parameters (k, m) are defined by the S–N curve of the structural element under consideration, the parameter b is determined by the wave climate with the prescribed Weibull distribution, and the parameter g describes in accordance with (6.100) a characteristic strength parameter of the structure under wave loading.

As an illustration we use

— S–N curve parameter: $m = 4.13$; $k = 4.2 \cdot 10^{14}$; Fig. 6.25, parameter Q
— wave climate parameter: $b = 1$
— strength parameter: $g = 1.3$, see Sect. 7.3.2
— number of waves in T_R: $n_R = 10^8$ ($T_R \simeq 20$ years).

Thus, in accordance with (6.105)

$$s_{max} \leqslant [4.2 \cdot 10^6/\Gamma\{6.37\}]^{1/4\,13} \cdot \ln^{1\,3}\{10^8\} = 476\ \text{N/mm}^2.$$

The stress range must not be greater than $476\ \text{N/mm}^2$ for wave height h_R, to guarantee fatigue strength for all stress ranges expected in the long-term.

We should point out that the wave climate can also be represented by two combined Weibull distributions to take account of the special characteristics of certain areas of operation. In the example, we chose the strength parameter g according to recommendations in [48], and obviously the reference time T_R can be defined as e.g. the design time period T_B (e.g. 100 years) for safety, or, conversely,

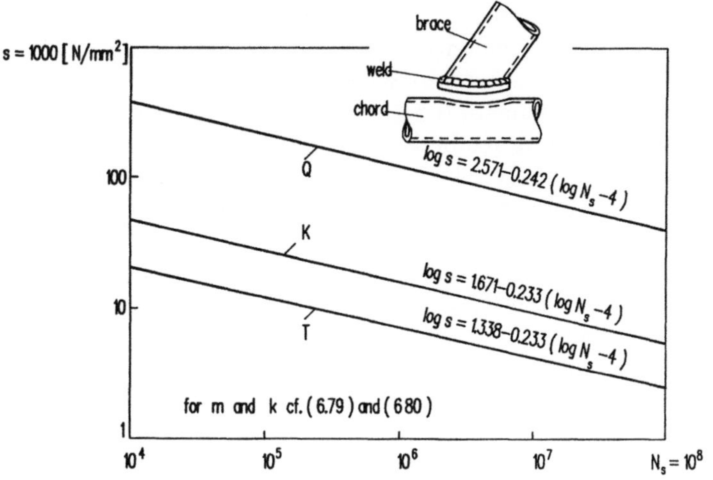

Fig. 6.25. S–N curves for simple joints [89].

the failure criterion $D_R < 1$ could be used instead of $D_R = 1$, which makes it possible to take influences such as seawater corrosion into account as well. These and similar variations occur in practice. In this connection we must remind ourselves again that we may have developed rational methods for evaluating the load-bearing capacity of marine structures in service conditions which are eminently suitable for obtaining relative fatigue strength predictions, but that we are not in a position to make absolute statements on fatigue strength.

During the design of a marine structure it becomes possible gradually to estimate the required parameters more precisely, so (6.105) is a valuable criterion for the preliminary examination of highly loaded structural elements with regard to their fatigue strength. This design criterion can be compared with regulatory criteria in which values such as acceptable tensile, buckling and warping stresses, etc. are prescribed. The parameters m, k and g can naturally be influenced during the course of design decision-making.

6.3 Modern Methods of Reliability Analysis

In the discussion of the fatigue strength problem we have seen that the capacity of a marine structure to resist the loads is influenced by various strength or resistance parameters, and that these parameters, like the loading parameters, can be stochastic values, which are partly determined by random processes. So, in principle, strength should be regarded as a random process, which we should compare with the random process of the combined stress, if we are looking for a statement of the mechanical reliability of the structure. In the following we first consider the relevant principles of reliability analysis for a certain instant of time during the course of these random processes of loading and strength. Then we will extend our approach to include the influence of time.

In Sect. 6.3.1 we find a collective summary of the first order methods of determining the mechanical reliability of load-bearing structural elements, and in Sect. 6.3.2 we will extend these fundamentals to the current state of reliability analysis of marine structures. In both sections we restrict ourselves to that part of modern reliability technique which opens up new, more extensive possibilities for rational decision-making for the design engineer working in marine technology as an extension of existing methods of analysis. For more comprehensive information on the subject of the reliability of load-bearing structures we would refer the interested reader to a selection of suitable standard works [19, 49–55].

We point out right from the start that the influence of time, which we will exemplify in Sect. 6.3.3, can be dealt with formally on the same basis in reliability analysis if appropriate transformations have been performed first (e.g. [56]), so the point in time principles represent a generally applicable basis for more extensive observations in the time domain.

We call reliability techniques modern because they are still being developed. In marine technology, in particular, the development of a practically applicable reliability theory which takes the influence of time into account offers the prospect of advances in the joint consideration of advanced design and inspection or maintenance strategies.

6.3.1 Reliability of Load-Bearing Structural Elements

We restrict our observations initially to two parameters, namely the stress on a structural element on the one hand, and the available resistance of this structural element on the other. By stress we mean both an external load and the associated internal loading on the structural element such as, for instance, a sectional load in a simple beam or bar or the resultant stress in the material. The same applies to the resistance, where the value of the resistance which is equal in magnitude to the load is always considered, e.g. the bearable external load, the sectional load or the strain, that the structure is capable of withstanding.

If the stress can be clearly represented by a value a and the resistance by a value w, a structural element is in a safe state if $a < w$. However, as a rule it is not possible to make such concrete statements on stress and resistance. We can only indicate from time to time what values we expect for the greatest stress or the resistance actually existing, and to what extent these values can deviate from our expectation in the end. From the point of view of the method, it is not important whether we base this information on general assumptions (*a priori*) or use statistical data or theoretical considerations. In the following we are concerned with rationally comprehending the uncertainty about stress and resistance of the structural element expressed by our expectation and the possible deviation from this expectation: subjective evaluation is to be replaced with a mathematical response to the question of to what extent it is probable that reality will produce a pair of values (a, w) which have the characteristic $a < w$.

For this purpose, we use a probability statement P, and define reliability R alternatively in the form of a 'difference statement',

$$R = P[Z > 0], \quad Z = W - A, \tag{6.106}$$

a 'quotient statement'

$$R = P[\Gamma > 1], \quad \Gamma = W/A, \tag{6.107}$$

or a 'comparison statement'

$$R = P[W > A]. \tag{6.108}$$

Reliability is thus defined as a value between 0 and 1, as this is the domain of the probability P. We also get the complementary statement for R

$$Q = 1 - R, \tag{6.109}$$

where we designated Q as the risk. We express this more generally below.

If we regard safety, as we generally do, as the certainty that a structural element will not fail, then reliability is defined here as the probability of safety: safety is 100% reliability. We should reiterate at this point that we indicate random values by capitalized symbols and their realization with lower case symbols, e.g. random value A with realization a, etc.

To develop a rational (mathematical) formalization with which the reliability of a structure can be evaluated, we need to discuss the definition equation (6.106) briefly. For this it is useful to consider a different analytical form known from probability theory (see also Appendix 1 (A1.14))

$$R = 1 - F_Z(z=0) = \int_0^\infty f_Z(z)\,dz. \tag{6.110}$$

F_Z is the probability distribution function of Z, $f_Z(z)$ the associated probability density. If we consider the case where both the random stress parameter A and the random resistance parameter W follow the Gaussian normal distribution (see (5.18)); the difference parameter $Z = W - A$ also follows this distribution, if W and A are stochastically independent of one another (see [53]). The first two stochastic moments of the distribution density of Z are then, for the mean value,

$$m_Z = m_W - m_A, \quad m_Z = E[Z] = \int_{-\infty}^\infty z \cdot f_Z(z)\,dz, \tag{6.111}$$

and for the standard deviation

$$\sigma_Z = \sqrt{\sigma_W^2 + \sigma_A^2}, \quad \sigma_Z^2 = \mathrm{Var}[Z] = \int_{-\infty}^\infty (z - m_Z)^2 f_Z(z)\,dz. \tag{6.112}$$

We define the standardized random parameter U with the transformation equation

$$U = (Z - m_Z)/\sigma_Z, \tag{6.113}$$

and make $Z = z = 0$ to determine the realiability as in (6.110) (see Fig. 6.26). Thus, in accordance with (6.111) to (6.113), we obtain a realiability index

$$\beta = U(Z=0) = -m_Z/\sigma_Z = -(m_W - m_A)/\sqrt{\sigma_W^2 + \sigma_A^2}. \tag{6.114}$$

With (6.110), using the conventional abbreviation Φ for F and φ for f when we are dealing with standardized Gaussian normal probability functions (see Fig. 6.26)

$$R = 1 - \Phi(\beta), \quad \beta = \Phi^{-1}(1 - R). \tag{6.115}$$

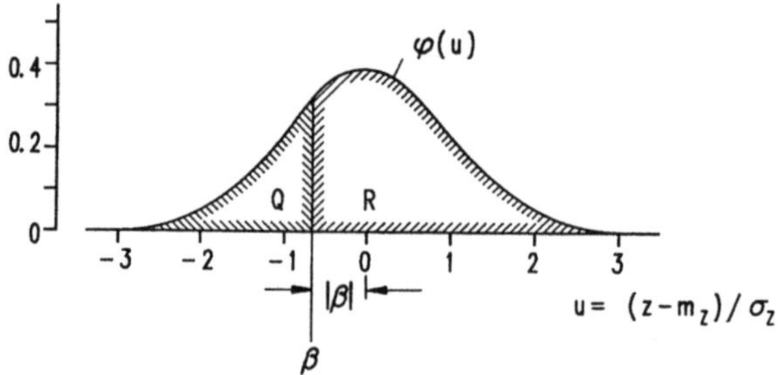

Fig. 6.26. Reliability index and standardized Gaussian normal distribution.

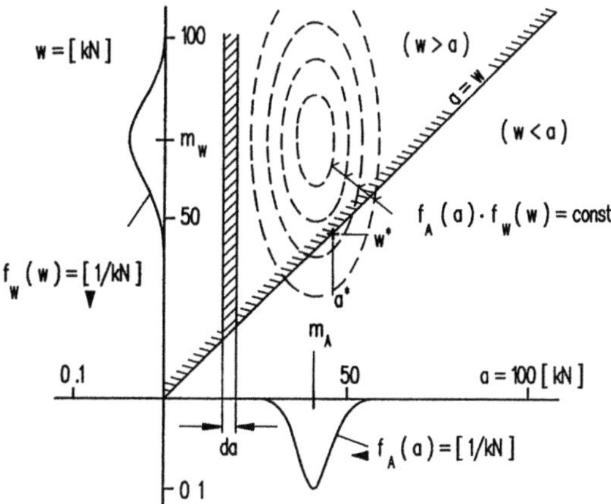

Fig. 6.27. Probability densities of A and W and failure boundary $a = w$.

Φ^{-1} symbolizes the inverse function of Φ and with $\varphi(u)$ in accordance with (5.19)

$$\Phi(\beta) = \int_{-\infty}^{\beta} \varphi(u)\,du, \quad \varphi(u) = 1/\sqrt{2\pi}\cdot\exp\{-u^2/2\}. \tag{6.116}$$

We consider the definition equation (6.108) of reliability in somewhat more detail, on the basis of the assumption that stress A and resistance W are random parameters which are stochastically independent of one another. We can then represent a joint density function $f_{AW}(a, w)$ with the product of the densities $f_A(a)\cdot f_W(w)$ as a set of probability contours over the domain of P defined by orthogonal coordinates a and w (see Fig. 6.27).

To simplify our further considerations we will only deal with the first quadrant of the domain of P, that is we will only permit positive values of A and W at first. The definition equation (6.108) now requires that we determine the partial volume of the joint probability density which lies above the straight line $w = a$, for in this area $w > a$. We thus find

$$R = \int_{a=0}^{\infty} \int_{w=a}^{\infty} f_A(a)f_W(w)\,da\,dw = \int_{a=0}^{\infty} f_A(a)(1 - F_W(a))\,da \tag{6.117}$$

We can carry out the same procedure to find R with (6.109) from a representation of Q, calculating Q as the partial volume of the probability contour below the straight line $w = a$. Then we find

$$R = \int_{w=0}^{\infty} F_A(w)f_W(w)\,dw. \tag{6.118}$$

(6.117) and (6.118) also apply to an extension to any negative values a and w, if we make the lower limit $-\infty$ instead of 0. We see this from the fact that, for example (6.118) can be regarded formally as the total probability theorem for all 'conditions' w (see (A1.19)).

We now want to convert Fig. 6.27 to a somewhat different form by standardizing the random values of stress A and resistance W as in (6.113)

$$U_A = (A - m_A)/\sigma_A, \quad U_W = (W - m_W)/\sigma_W. \tag{6.119}$$

Assuming that the marginal distributions $f_A(a)$ and $f_W(w)$ follow the Gaussian density function, the transformation given by (6.119) maps the joint density function $f_{AW}(a, w)$ on the two-dimensional Gaussian standard normal distribution $\varphi(u_A, u_W)$ which is symmetrical to the origin of the coordinates (see Fig. 6.28).

The failure boundary, given as a limit state function that we define as the geometrical location of all points which separate the domains $a < w$ and $a > w$, has the equation $a = w$ here, i.e.

$$u_W = \sigma_A/\sigma_W \cdot u_A + (m_A - m_W)/\sigma_W. \tag{6.120}$$

The coordinates (u_A^*, u_W^*) of the intersection of the failure boundary with a straight line through $(0,0)$, which is vertical to the failure boundary, has the equation $u_W = -\sigma_W/\sigma_A \cdot u_A$. Therefore

$$u_A^* = \sigma_A(m_W - m_A)/(\sigma_W^2 + \sigma_A^2),$$
$$u_W^* = \sigma_W(m_A - m_W)/(\sigma_W^2 + \sigma_A^2). \tag{6.121}$$

The distance between the points $(0,0)$ and (u_A^*, u_W^*) is found from

$$\sqrt{u_A^{*2} + u_W^{*2}} = (m_W - m_A)/\sqrt{\sigma_W^2 + \sigma_A^2} = |\beta|, \tag{6.122}$$

and on comparison with (6.114) we recognize in (6.122) a geometrical interpretation of the reliability index: the absolute value of the reliability index is equal to the minimum distance between the point $(0, 0)$ and the failure boundary in standardized coordinates. Often the reliability index is defined as the value of the second term in (6.122), but this definition is not suitable for distinguishing faithfully between risk and reliability in all cases. Therefore, in this text we use exclusively the definition given by (6.114), or compatible versions.

Furthermore, standardized coordinates are unwieldy for practical calculations. We use the transformations in (6.119) and obtain from (6.121) the coordinates (a^*, w^*)

$$a^* = m_A + \beta \cdot \alpha_A \cdot \sigma_A,$$
$$w^* = m_W + \beta \cdot \alpha_W \cdot \sigma_W. \tag{6.123}$$

Here α_A and α_W are designated as sensitivity factors

$$\alpha_A = -\sigma_A/\sqrt{\sigma_W^2 + \sigma_A^2}, \quad \alpha_W = \sigma_W/\sqrt{\sigma_W^2 + \sigma_A^2} \tag{6.124}$$

which we define more generally later. If we divide (6.121) by β, we recognize that the sensitivity factors can also be interpreted as direction cosines of the β vector in Fig. 6.28. As a^* and w^* lie on the (45°) failure boundary, as shown in Fig. 6.27, $a^* = w^*$, and we find, according to (6.123),

$$a^* = w^* = (m_A\sigma_W^2 + m_W\sigma_A^2)/(\sigma_W^2 + \sigma_A^2). \tag{6.125}$$

Thus we can calculate the point $a^* = w^*$ in non-standardized coordinates (i.e. in

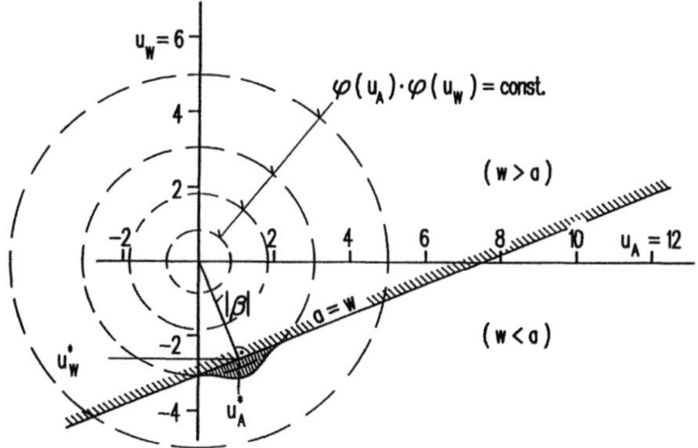

Fig. 6.28. Two-dimensional probability density of A and W and failure boundary $a = w$ in standardized coordinates.

the physical domain of basic parameters), without needing to know the reliability index or the reliability. We know, however, that the point (a^*, w^*) is of great importance for reliability, as can be seen, for example, by looking at the standardized form in Fig. 6.28: as the contour lines of equal probability density appear as circles, a circle whose tangent is formed by the failure boundary in (u_A^*, u_W^*) must have the relatively highest probability density of all points along the failure boundary (it lies closest to the coordinate origin). This obviously means that the probability of failure is greatest close to (u_A^*, u_W^*) or (a^*, w^*). Consequently, this point is generally called the design point because it seems appropriate to specify nominal values of the physical basic parameters – e.g. (a_N, w_N) – in design codes or similar regulations in such a way that they have a suitable relationship to the design values of these parameters – here (a^*, w^*). We return to these considerations, which are important for design practice in Sect. 7.4.1.

Stress and resistance are themselves influenced by a whole range of further physical parameters which can be of a stochastic nature. In the case of steel, for example, the resistance parameters yield stress, ultimate stress, modulus of elasticity, etc., and it is therefore of great practical importance to extend our observations to more than two physical basic parameters, which may also be correlated, and to formulate our reliability model which has been developed only for A and W so far, more generally in mathematical terms. Therefore, instead of the basic parameters A and W, we consider the physical basic parameters $X_1, X_2, \ldots, X_K$, which we combine in the vector X. We attempt to standardize these basic parameters in a similar manner to (6.113), i.e. to transform the X domain of the basic parameters formally with

$$U = C_X^{-1/2}(X - m_X), \quad X = m_X + C_X^{1/2} U \qquad (6.126)$$

into the U domain of the standardized basic parameters $U_1, U_2, \ldots, U_K$, which are combined in the vector U. C_X is the covariance matrix already defined in (5.17), and m_X symbolizes the vector $(m_{X1}, m_{X2}, \ldots, m_{XK})^T$. We now show that the random values U_k obtained from the generally correlated basic parameters X_k with the transformation (6.126) are standardized, i.e. that all have the mean value zero and the standard deviation one, and furthermore they can be considered stochastically independent under certain circumstances.

We can see immediately that (6.126) gives the results of (6.119), if $X_1 = A$, $X_2 = W$ and $K = 2$, as with stochastic independence the covariance matrix is equal to the diagonal matrix, i.e.

$$C_X^{-1/2} = \lceil 1/\sigma_A \; 1/\sigma_W \rfloor,$$

The symbol $\lceil \cdots \rfloor$ summarizes the content of the main diagonal, all other elements being zero. For the general case of K basic parameters X, which are correlated, the matrix $C_X^{-1/2}$ can be calculated as

$$C_X^{-1/2} = WL^{-1/2}W^T, \quad W = [w_1 w_2 \cdots w_K], \tag{6.127}$$

the matrix W summarizing the normalized eigenvectors w_K which are obtained from the eigenvectors ω_k as a normalized solution of the K eigenvalue problems

$$(C_X - I\lambda_k)\omega_k = O, \quad k = 1, 2, \ldots, K.$$

The eigenvalues λ_k are summarized in the diagonal matrix

$$L = \lceil \lambda_1 \lambda_2 \cdots \lambda_K \rfloor, \quad L^{1/2} = \lceil \sqrt{\lambda_1} \sqrt{\lambda_2} \cdots \sqrt{\lambda_K} \rfloor$$

and $I = \lceil 11 \cdots 1 \rfloor$ is the unit (diagonal) matrix. Under these conditions $C_X^{1/2}C_X^{1/2} = C_X$ (see, e.g. [49] or [53 part II]), i.e. in (6.126)

$$E[U] = C_X^{-1/2}(E[X] - m_X) = O, \tag{6.128}$$

$$C_U = C_X^{-1/2}C_X C_X^{-1/2} = I, \tag{6.129}$$

as $C_Z = AC_X A^T$ applies in general to linear transformations $Z = AX$ [49], and according to (6.127)

$$(C_X^{-1/2})^T = (WL^{-1/2}W^T)^T = WL^{-1/2}W^T = C_X^{-1/2}.$$

Because of the characteristics of the transformation with (6.126) expressed by (6.128) and (6.129), this can be regarded as a way of transforming the generally correlated physical basic parameters X into stochastically independent, standardized random parameters U. Another characteristic of this transformation is to transform physical basic random parameters X, which follow Gaussian normal distribution, in such a way that the standardized random parameters U created follow the standardized Gaussian normal distribution, as with the generally applicable transformation condition

$$f_U(u) = f_X(x(u))|C_X|^{1/2}$$

($|C_X|^{1/2}$ corresponds to the Jakobi determinant, applied here to (6.126)) we obtain

$$f_U(u) = 1/\sqrt{(2\pi)^K} \cdot \exp\{-u^T u/2\} = \prod_{k=1}^{K} \varphi(u_k). \tag{6.130}$$

(For the definition of $\varphi(u_k)$ see (6.116) with index k.) Therefore, we can designate the U domain as the domain of stochastically independent random values with standard normal distribution, if it arose by means of (6.126) as the transformation of the correlated normally distributed physical basic random parameters of the X domain. As the transformation (6.126) is linear, it maps a linear function $g_L(X)$ on a function $h_L(U)$ in the U area, which is also linear. For any transformations, even non-linear, we find with (6.126)

$$g(X) = h(U) = h(C_X^{-1/2}(U - m_X)). \tag{6.131}$$

In the following we use $g(X)$ to represent the mechanical model of the physical basic parameters; by the mechanical model of a structural element or a load-bearing structure we mean the analytical description of the behaviour of the element or structure under the influence of loadings. We define $g(X)$ such that

$$g(X) \quad \begin{cases} <0 & \text{indicates failure, failure state,} \\ =0 & \text{indicates indifference, limit state,} \\ >0 & \text{indicates load-bearing capacity, safe state.} \end{cases} \tag{6.132}$$

In this connection $g(X)$ is also designated a structural performance or state function, abbreviated structure function. With X as a vector of all basic random parameters, $G = g(X)$ defines a random parameter. The definition $Z = g(A, W) = W - A$ used in (6.106), also called the safety margin, and the definition $\Gamma = g(A, W) = W/A$, known as the safety factor, are simple examples of such random parameters. The failure boundary (limit state function) $g(X) = 0$ is a hypersurface of dimension $K - 1$, K being the number of physical basic random parameters considered. A practical example for $g(X)$ will be discussed in Sect. 6.3.3.

In the U domain we identified the value of the reliability index as the shortest distance between $(0, 0)$ and the failure boundary $h(U_A, U_W) = 0$ (see (6.120), (6.122) and Fig. 6.28). With the vector of the unit normals e_U of the failure boundary directed into the failure domain $h(U) < 0$, we find more generally in accordance with Fig. 6.29

$$u^* = -\beta \cdot e_U^*, \quad h(u^*) = 0. \tag{6.133}$$

These are $K + 1$ equations for the K components of the design point u^* and the

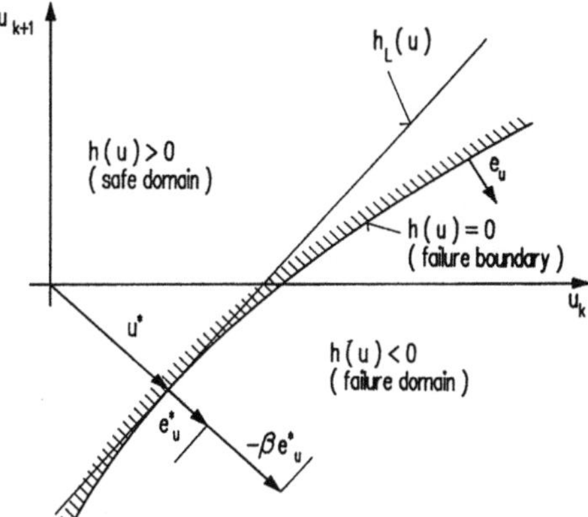

Fig. 6.29. Design point and reliability index in standardized coordinates.

reliability index β. We will not go into further detail here on the range of familiar possibilities for algorithmic treatment of these equations and their solution (see, for example [57, 58]), but Fig. 6.29 shows clearly that a solution can exist.

From (6.133) we find, by multiplication with e_U^{*T}

$$\beta = -e_U^{*T} u^* = -u^{*T} e_U^* \tag{6.134}$$

and similarly to (6.133) we obtain

$$\sqrt{u^{*T} u^*} = \sqrt{\beta^2} = |\beta|.$$

We now transform u^* in (6.133) and (6.134) by means of (6.126) into the X domain of the physical basic random parameters. For this we summarize the partial derivatives according to the K coordinate directions in the Nabla vector ∇

$$\nabla_U^T = (\partial/\partial u_1 \, \partial/\partial u_2 \cdots \partial/\partial u_K),$$
$$\nabla_X^T = (\partial/\partial x_1 \, \partial/\partial x_2 \cdots \partial/\partial x_K),$$

and by applying the chain rule of differential calculus to the transformation equation (6.126) we find

$$\nabla_U^T = \nabla_X^T \cdot [\partial x_i/\partial u_k] = \nabla_X^T C_X^{1/2}.$$

Here $[\partial x_i/\partial u_k]$ is a quadratic matrix of the dimension K, i is the row index, k the column index. We apply this result to the structure function $h(U)$ at the design point u^* and obtain the partial derivatives

$$a_U^T = (\nabla_U^T h(U) | U = u^*)^T$$
$$\quad = (\nabla_X^T g(X) | X = x^*)^T C_X^{1/2} = a_X^T C_X^{1/2},$$
$$a_X^T = (\nabla_X^T g(X) | X = x^*)^T \tag{6.135}$$

with $g(X)$ as in (6.131) and a_X^T as a gradient vector of the structure function at the design point. Now the vector of the unit normals at the design points can be represented as

$$e_U^* = -a_U/\sqrt{a_U^T a_U} = -a_X^T C_X^{1/2}/\sqrt{a_X^T C_X a_X} = -\alpha, \tag{6.136}$$

and with this definition equation for the sensitivity factors α we recognize the important characteristic

$$e_U^{*T} e_U^* = 1 = \alpha^T \alpha = \sum_{k=1}^{K} \alpha_k^2. \tag{6.137}$$

The equations (6.133) for the design point in the X domain of the correlated physical basic random parameters are thus

$$x^* = m_X + \beta \cdot C_X^{1/2} \alpha, \quad g(X) = 0. \tag{6.138}$$

Similarly to (6.134) we obtain the reliability index with (6.126) and (6.136)

$$\beta = \alpha^T C_X^{-1/2}(x^* - m_X),$$
$$\beta = a_X^T(x^* - m_X)/\sqrt{a_X^T C_X a_X}. \tag{6.139}$$

This is the general definition of the reliability index. With $X = (A W)^T$, $m_X = (m_A m_W)^T$, $C_X = \lceil \sigma_A^2 \sigma_W^2 \rfloor$, $g(X) = W - A$, i.e. $a_X^T = -(-1 \; 1)$, we of course obtain from this (6.114) again

$$\beta = (-a^* + m_A + w^* - m_W)/\sqrt{(-1 \; 1)\lceil \sigma_A^2 \sigma_W^2 \rfloor(-1 \; 1)^T},$$

$$\beta = -(m_W - m_A)/\sqrt{\sigma_W^2 + \sigma_A^2},$$

as $a^* = w^*$ (see (6.125) or Fig. 6.27). Furthermore, in accordance with (6.136)

$$(\alpha_A, \alpha_W) = (-\sigma_A/\sqrt{\sigma_W^2 + \sigma_A^2}, \; \sigma_W/\sqrt{\sigma_W^2 + \sigma_A^2}), \tag{6,140}$$

(see (6.124)). Apart from this simple case of a linear structure function, used to illustrate the general definition of the reliability index and the associated equations, the reliability index, as already mentioned, can be determined iteratively either in the U domain by means of (6.133), or in the X domain by means of (6.138), together with the design point coordinates u^* and x^*. To estimate the risk we can approximate a non-linear failure boundary linearly at the design point

$$Q = P[g(X) \leqslant 0] \simeq P[g_L(X) \leqslant 0] = P[h_L(U) \leqslant 0]. \tag{6.141}$$

With the Taylor series development in u^*

$$h_L(U) \doteq h(u^*) + a_U^T(U - u^*)$$

(a_U^T as in (6.135), the symbol $\doteq$ indicates termination after first-order terms), we find, taking (6.133), (6.136) and (6.134) into account,

$$Q \simeq P[\alpha^T U \leqslant \beta] = \Phi(\beta), \tag{6.142}$$

as the sum $\alpha^T U$ is again a standard normal distributed random parameter because $E[\alpha^T U] = \alpha^T E[U] = 0$ and $\mathrm{var}[\alpha^T U] = \alpha^T \alpha = 1$. The use of (6.115) therefore represents a reliability statement for the generalized case of K correlated, normally distributed, physical basic random parameters, which can be regarded as a linear failure boundary approximation at the design point. In this connection we speak of a first-order reliability method, which of course gives results if we take no acount of the distribution function $F_X(x)$ associated with the first and second moments m_X and σ_X^2, or simply assume that X follows the Gaussian normal distribution. This is usually called the First-Order Second Moment (FOSM) method.

As for correlated physical basic random parameters X, which follow the Gaussian normal distribution, the standardized, stochastically-independent random values in (6.126) also follow Gaussian (standard) normal distribution, the design point x^* or u^* is a point of the relatively largest probability density along the failure boundary (see $h(U)$ in Fig. 6.28). For this reason, the linear failure boundary approximation in the design point in the evaluation in (6.142) gives, in most practical cases, a good approximation of the risk of failure Q or the reliability $R = 1 - Q$. We can achieve comparatively good approximations for basic parameters X which do not follow Gaussian normal distribution, if we succeed in replacing both distribution function and density function with the respective Gaussian normal distributions such that the original and the fictitious Gaussian dummy distributions coincide at the design point x^*. Therefore, for stochastically independent initial values X we should

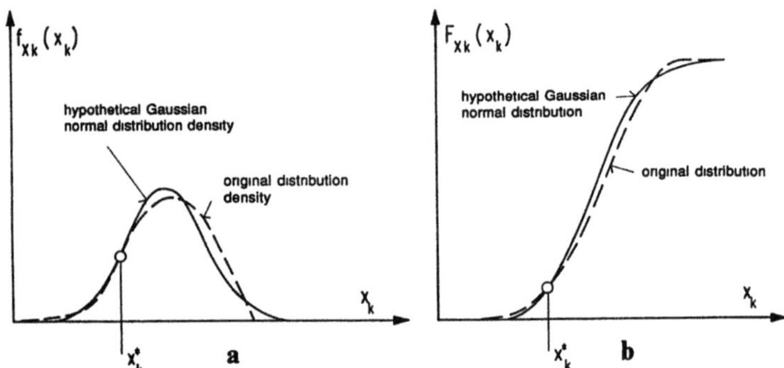

Fig. 6.30a, b. Fictitious Gaussian normal and original distributions.

obtain (see Fig. 6.30)

$$\Phi((x_k^* - m_{Xk})/\sigma_{Xk}) = F_{Xk}(x_k^*), \quad k = 1, 2, \ldots, K,$$
$$\varphi((x_k^* - m_{Xk})/\sigma_{Xk}) = f_{Xk}(x_k^*) \cdot \sigma_{Xk}. \tag{6.143}$$

The conditions (6.143) are fulfilled if we calculate the parameters of the fictitious Gaussian normal distributions as

$$\sigma_{Xk} = \varphi(\Phi^{-1}(F_{Xk}(x_k^*)))/f_{Xk}(x_k^*),$$
$$m_{Xk} = x_k^* - \sigma_{Xk} \cdot \Phi^{-1}(F_{Xk}(x_k^*)).$$

This procedure, called a 'Normal Tail Approximation', must be repeated after each step in the iterative search for the design point x^* or u^*. In this connection we also speak of the Rackwitz–Fiessler algorithm, which was applied for the first time in [32].

To simplify complex calculations, which we will deal with in more detail in Sect. 6.3.2, the iterative procedure is sometimes dispensed with, and the design point is simply estimated with multiplication factors ψ_k [55]. Similarly to (6.133) and (6.138), we write as an approximation

$$x^* = m_X + \lceil \psi_1, \psi_2, \ldots, \psi_K \rfloor \cdot \sigma_X, \tag{6.144}$$

estimating the multiplication factors by means of $\psi_k = \beta \cdot \alpha_k$ with suitable values for β and α_k (see Sect. 7.4.1).

We can, of course, also regard the conditions (6.143) as one of many transformation equations possible in addition to (6.126) between all points of the X and U domain

$$\Phi(u) = F_X(x), \quad u = \Phi^{-1}(F_X(x)), \tag{6.145}$$

as, for stochastically independent, normally distributed basic random parameters, this transformation would be identical with the transformation in (6.126) (see derivation of (6.130)). We now consider how we can extend this transformation to correlated basic random parameters of X which are not normally distributed, in such a way that the random parameters U again have standard normal distribution and are stochastically independent. We first consider two correlated physical basic parameters X_1 and X_2. So then with stochastic independence of the associated standardized random parameters U_1 and U_2 we have, similar to (A1.18) or (A1.21),

$$\Phi(u_1)\Phi(u_2) = F_{X1}(x_1)F_{X2}(x_2|x_1).$$

With stepwise transformation of the two basic random parameters we get

$$u_1 = \Phi^{-1}(F_{X1}(x_1))$$
$$u_2 = \Phi^{-1}(F_{X2}(x_2|x_1)).$$

For K random values we simply continue this process, and thus obtain the transformation rule suggested by Hohenbichler and Rackwitz [59], generally known as the Rosenblatt transformation

$$u_3 = \Phi^{-1}(F_{X3}(x_3|x_1x_2))$$
$$\vdots$$
$$u_k = \Phi^{-1}(F_{Xk}(x_k|x_1x_2\ldots x_{k-1})), \quad k = 1, 2, \ldots, K. \tag{6.146}$$

In the iterative search for the design point u^*, we only need the transformation at certain points, i.e. for certain prescribed values $X_1 = x_1$, $X_2 = x_2$, $X_3 = x_3$, etc. For correlated basic random parameters $Y_k = (X_k - m_k)\cdot\sigma_{Xk}$, the linear transformation $Y = Au$ corresponds to the inverse Rosenblatt transformation

$$y_k = F_{Xk}^{-1}(\Phi(u_k)|y_1, y_2, \ldots, y_{k-1})$$

with values a_{ij} of the matrix A all being zero for $j > i$ according to the above equation (there is only one lower triangular matrix). This is the way which is particularly suitable for practical calculations with computer programs. For further information see [60].

An application of the Rosenblatt transformation which is important for evaluating the reliability of marine structures within the scope of the first-order method described here is given in [19]. We will not repeat the details here, but will point out that the example relates to the representation of long-term up-crossing probabilities (cumulative distribution functions) of wave elevation ζ in a specific sea area. With the last formula in Sect. 6.2.1.3 we indicated how distribution functions $F_{Yk}(y)$ of any seaway effects Y_k, and with $Y_k = Z$, of the stochastic wave elevation Z itself, can be calculated, but a FOSM procedure with the Rosenblatt transformation (as described in [19]) gives additional information on the effect of statistical uncertainty, which derives from the data of the local wave climate (long-term seaway statistics) in the example. These sample calculations show that such statistical uncertainties can be meaningful in rarely occurring reference values of wave elevation which are of practical interest (Fig. 6.31 from [19]): with an up-crossing risk of between 10^{-5} and 10^{-4} the uncertainty reaches the order of magnitude of the air gap of $1.20\,\text{m}$ above the crest of the design wave (see Sect. 6.2.1.2).

Eventually, we should realize that the vector of distribution parameters, e.g. the vector of the mean values of random parameters $X_1, X_2, \ldots, X_I$ i.e. $\mu^T = (m_{X_1}, m_{X_2}, \ldots, m_{X_I})$, may be statistically uncertain. If the basic random parameters μ have density

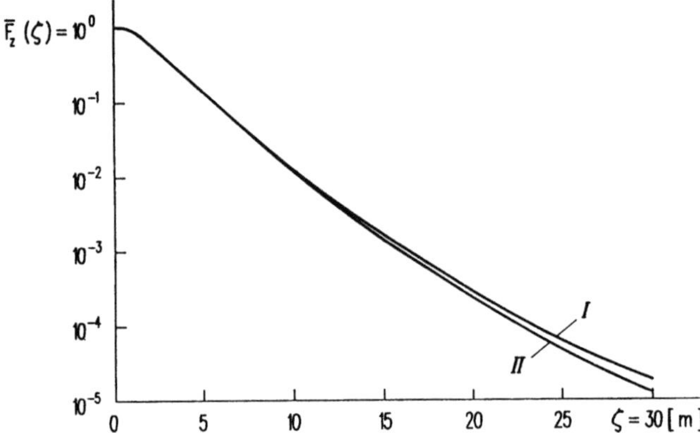

Fig. 6.31. Effect of statistical uncertainties on risks of exceedance of wave elevation [19]. I: with; II without statistical uncertainty

functions $f_\mu(\mu)$ then the failure probability is

$$Q = \int_M \int_F f_X(x|\mu) f_\mu(\mu)\, dx\, d\mu$$

with F the failure domain in the physical random parameter space and M the domain of the uncertain parameter vector μ.

6.3.2 Reliability of Load-Bearing Structural Systems

With all load-bearing structures the correlation between elements of the structure is particularly important for calculating failure probability, so we will clarify some aspects of this in advance.

From Fig. 6.28 in Sect. 6.3.1, we obtain the equation of the linear failure boundary $h(u_A, u_W)$ in the U domain of the stochastically independent random parameters U_A and U_W from the condition: stress equal to resistance $(a = w)$. Taking account of (6.120) as well as (6.114) and (6.124), we find

$$h_i(u) = \alpha_{W_i} \cdot u_W + \alpha_{A_i} \cdot u_A - \beta_i = 0 \tag{6.147}$$

for any element i. For random parameters U_A and U_W with standard normal probability distribution, (6.147) defines a normally distributed random value H_i with standard deviation equal to 1. Because (6.147) also applies to any other element l, the correlation coefficients between two elements i and l are

$$\varrho_{il} = \text{Cov}\,[H_i, H_l] = E[H_i \cdot H_l] - E[H_i] \cdot E[H_l]$$

$$= \alpha_{W_i} \cdot \alpha_{Wl} + \alpha_{Ai} \cdot \alpha_{Al} \tag{6.148}$$

(see (A1.28)). From Fig. 6.32, in which the failure boundaries of the two elements in the U domain of the standardized random parameters are shown, (6.124) gives

$$\cos \psi = u_A^*/|\beta| = \alpha_A, \quad \sin \psi = u_W^*/|\beta| = -\alpha_W. \tag{6.149}$$

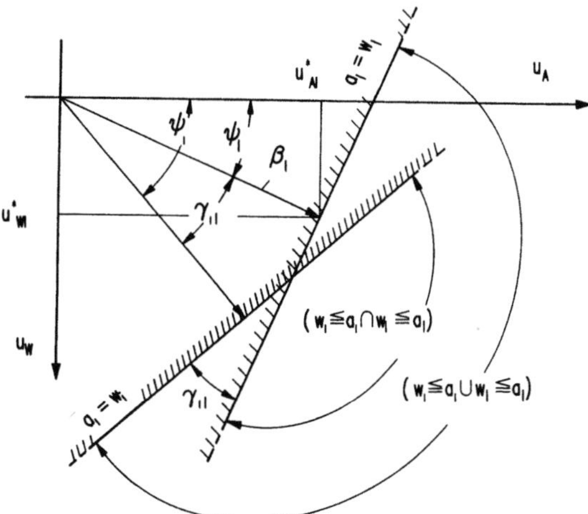

Fig. 6.32. Failure boundaries of elements i and l.

Thus (6.148) gives a geometrical interpretation of the correlation coefficient between elements i and l of the structure, which is important for our further considerations

$$\varrho_{il} = \cos\{\psi_i - \psi_l\} = \cos\gamma_{il}, \tag{6.150}$$

i.e. the correlation coefficient of two load-bearing structural elements corresponds to the cosine of the angle between the two design-point vectors or the linearized failure boundaries in the U domain of the stochastically independent standardized random parameters. To calculate it we must first determine the sensitivity factors, and by generalizing (6.148) we then form the correlation matrix of the elements with a linear or linearized structure function

$$P = A \cdot A^T, \tag{6.151}$$

$$A = \begin{bmatrix} \alpha_{11} & \alpha_{12} & \cdots & \alpha_{1K} \\ \alpha_{21} & \alpha_{22} & \cdots & \alpha_{2K} \\ \vdots & \vdots & & \vdots \\ \alpha_{I1} & \alpha_{I2} & \cdots & \alpha_{IK} \end{bmatrix},$$

K being the number of basic parameters considered, and I the number of structural elements considered. Matrix P has dimension $I \cdot I$, and is filled with units in the main diagonal (see (6.137)).

Having made these preliminary comments, we can now deal with the reliability of load-bearing structures in more detail.

6.3.2.1 Structures Behaving Like Series or Parallel Systems

Definition: a structure is regarded as a logical series system if the failure of any individual load-bearing element causes the failure of the structure. (Series systems only function if all elements function.)

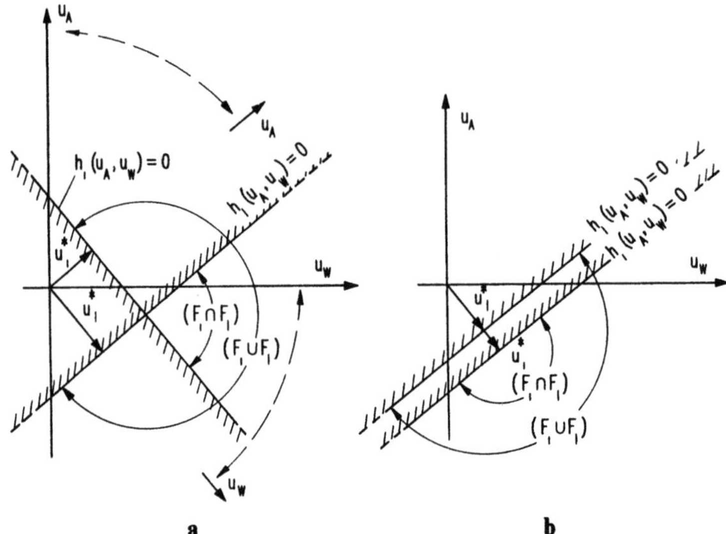

Fig. 6.33. Failure boundaries of elements i and l. **a** with stochastic independence; **b** with full correlation.

Figure 6.33 shows the linearized failure boundaries $h(u_A, u_W) = 0$ of two elements i and l in the U domain of the stochastically independent, standardized random parameters U_A and U_W of load and resistance, and the sub-domains in which the random values H_i and H_l of the structural elements i and l obtained with (6.147) become either simultaneously ($\cap$) or separately ($\cup$) smaller than 0 (failure domain), which is an indicator of the failure of the structural elements (see (6.141)): F is a failure indicator as an abbreviation for $h(U_A, U_W) \leqslant 0$.

Thus we can read the failure risk Q_S for a structure with two elements in a logical series system directly from Fig. 6.33. With stochastic independence of the elements i and l in the failure domain $F_i \cup F_l$ we have

$$Q_S = \int_{|u_i^*|}^{\infty} \varphi(u_A) du_A \int_{-\infty}^{|u_l^*|} \varphi(u_W) du_W + \int_{-\infty}^{\infty} \varphi(u_A) du_A \int_{|u_i^*|}^{\infty} \varphi(u_W) du_W.$$

(6.152)

Taking into account $|u^*| = |\beta|$ (see (6.134)), and the identity

$$\Phi(-\beta) = 1 - \Phi(\beta)$$

(6.153)

(6.152) gives us

$$Q_S = 1 - (1 - \Phi(\beta_i))(1 - \Phi(\beta_l)), \quad \varrho_{il} = 0$$

(6.154)

for stochastic independence of the elements i and l. We can read the corresponding expression for full correlation directly from Fig. 6.33b, again for $F_i \cup F_l$

$$Q_S = \text{Max}[\Phi(\beta_i), \Phi(\beta_l)], \quad \varrho_{il} = 1,$$

(6.155)

and by generalizing (6.154) and (6.155) to I elements we obtain from this the trivial bounds for structures behaving like logical series systems

$$\overset{I}{\underset{i=1}{\text{Max}}}\,\Phi(\beta_\mathrm{i}) \leqslant Q_\mathrm{S} \leqslant 1 - \prod_{i=1}^{I}(1 - \Phi(\beta_\mathrm{i})), \quad \varrho_\mathrm{il} \geqslant 0. \tag{6.156}$$

These bounds can be tightened considerably, but such details are of minor importance for our observations. The interested reader is, however, referred to a summary which deals with the subject of reliability bounds in more detail [61], as well as to a more recent illustration of the numerical consequences [62]. Here we are primarily concerned with clarifying the importance of the correlation between structural elements for the failure risk of logical series systems: the lower bound applies to full correlation, the upper bound to stochastic independence of the elements of the system.

Definition: a structure is regarded a logical parallel system if it fails if all elements fail. (Parallel systems function if at least one element functions.)

Looking at the failure domains $F_\mathrm{i} \cap F_\mathrm{l}$ in Fig. 6.33a, we can see that for two stochastically independent structural elements of a parallel system we must have

$$Q_\mathrm{P} = \int_{|u^*_\mathrm{i}|}^{\infty} \varphi(u_\mathrm{A})\mathrm{d}u_\mathrm{A} \int_{|u^*_\mathrm{i}|}^{\infty} \varphi(u_\mathrm{W})\mathrm{d}u_\mathrm{W} = \Phi(\beta_\mathrm{i})\Phi(\beta_\mathrm{l}), \quad \varrho_\mathrm{il} = 0, \tag{6.157}$$

and from Fig. 6.33b we can read off, also for $F_\mathrm{i} \cap F_\mathrm{l}$ with fully correlated elements

$$Q_\mathrm{P} = \text{Min}[\Phi(\beta_\mathrm{i}), \Phi(\beta_\mathrm{l})], \quad \varrho_\mathrm{il} = 1. \tag{6.158}$$

So, generalizing to more than two elements then gives the trivial bounds for parallel systems

$$\prod_{i=1}^{I}\Phi(\beta_\mathrm{i}) \leqslant Q_\mathrm{P} \leqslant \overset{I}{\underset{i=1}{\text{Min}}}\,\Phi(\beta_\mathrm{i}), \quad \varrho_\mathrm{il} \geqslant 0. \tag{6.159}$$

Comparison with the bounds in (6.156) and (6.159) shows that positive correlation has the opposite effect with parallel systems to that with series systems: in this case (Q_P), the failure risk of the parallel system, is increased by positive correlation (upper bound); in the other (Q_S), the failure risk of the series system, positive correlation reduces the risk (lower bound).

The following guide rule applies to load-bearing structures behaving like series systems (chains, statically-determinate truss works, etc.): the fewer components there are and the higher their positive correlation is, the lower is the loss of reliability of the system compared to a single component. For structures behaving like parallel systems the more components there are and the lower their positive correlation is, the greater is the gain in reliability compared to a single component.

The trivial bound in (6.159) can be considerably tightened like the trivial bounds in (6.156) of the series system. In the case of load-bearing structures behaving like parallel systems, however, there is another special factor which influences the failure risk of the structure; namely, the effect of the material law, which we must discuss briefly.

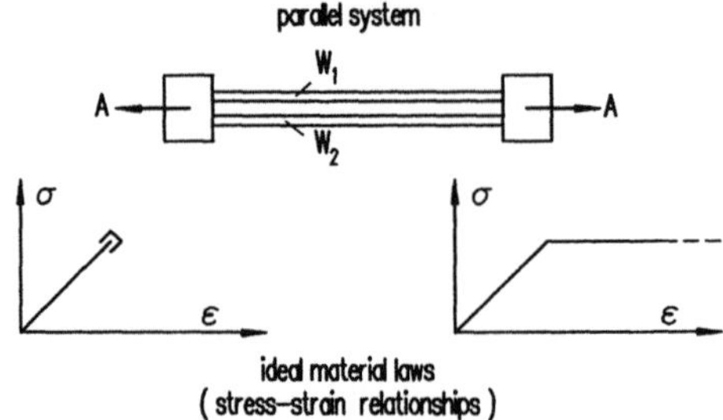

Fig. 6.34a,b. Structure with logical parallel system and two material laws.

Using the example of a structure with two parallel tension members, we consider two idealized material laws; namely, perfectly elastic/brittle material (Fig. 6.34a) and perfectly elastic/plastic material (Fig. 6.34b).

We can imagine the failure mechanism such that either element 1 fails first followed by element 2 or, conversely, first element 2 then 1. Thus we get a series system as a logical representation of this parallel structure, consisting of two sub-systems (modules), each containing two elements forming a parallel system. Now we represent the failure domains of these two sub-systems as a Venn diagram, taking into consideration that in the case of the material law in Fig. 6.34b perfectly plastic strength remains after failure of an element, while this cannot be assumed with the material law in Fig. 6.34a (see Fig. 6.35). The difference in size of the two failure domains shows us clearly the effect of the material law on the failure risk of the parallel structure

$$Q_P(b) \leqslant Q_P(a). \tag{6.160}$$

The construction material used for many marine structures (jackets, semisubmersibles, etc.) is high-strength steel, which can be idealized better by Fig. 6.34b than Fig. 6.34a under tensile or bending stress, so in general we can consider the case of the material law in Fig. 6.34b as the representative one for the interests of

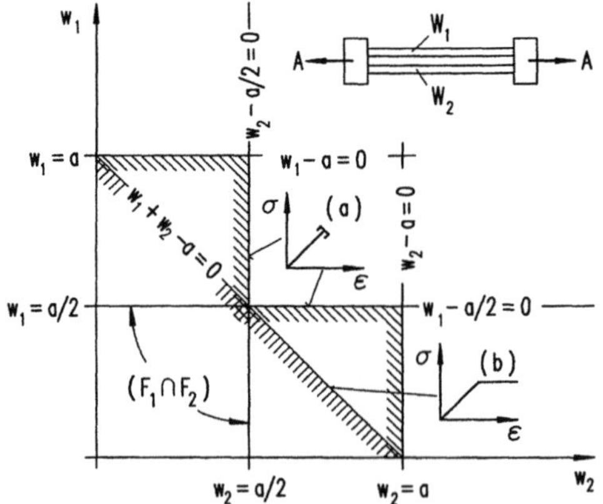

Fig. 6.35. Venn diagram of a structure with logical parallel system and two material laws.

steel structure design in marine technology. With (6.160), however, we then always obtain only a lower bound of failure probability! Irrespective of this, in practice (e.g. when considering a tension-leg platform as a real example of a parallel system consisting of tension girders), if there is a failure of one leg similar to a brittle fracture, we have to expect a subsequent shock loading of the other legs, so the actual risks of system failure are much higher [63].

6.3.2.2 Structures Behaving Like Redundant Systems

Definition: a structure is regarded as a redundant system if more than one element must fail for the system to fail.

According to this definition, all structures behaving like parallel systems are to be considered as special cases of redundant systems. Furthermore, all statically indeterminate structures are redundant, as we will show with the example of a portal frame of perfectly elastic/plastic material as in Fig. 6.36.

This frame is obviously statically indeterminate, and if we want to remove this indeterminacy by inserting pliable joints (hinges) at the most highly loaded points in the frame, we must use at least two hinges; a third could be enough to make the structure unstable. If we assume the perfectly elastic/plastic material law (Fig. 6.34b), we can call them plastic hinges from the outset. In this case it is easy to demonstrate [53 part II] that the occurrence of the following successive plastic hinge combinations (independent of one another and independent of the sequence) would lead to the failure of the frame: 1–3–4, 1–3–5, 2–3–4, 2–3–5.

A generally applicable algorithm for finding all independent plastic hinge combinations, which can also be designated as fundamental failure mechanisms, was developed in [64]. The structure functions of the failure mechanisms (see (6.132)) can be calculated from the deformation work in the simple example considered here (see Sect. 4.4.2). With the abbreviation W as a random value for

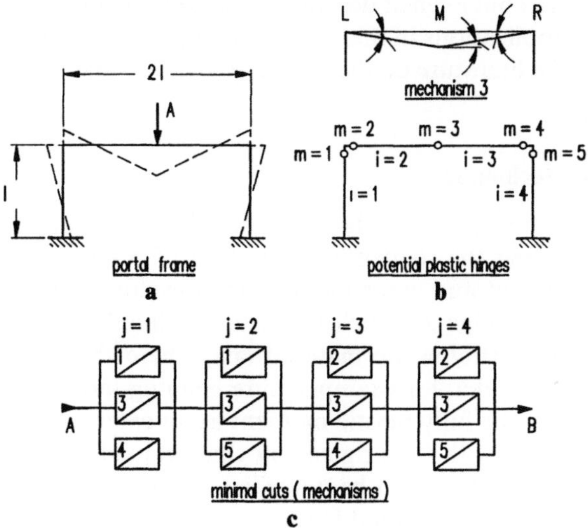

Fig. 6.36a–c. Portal frame with potential plastic hinges and failure mechanisms.

the fully plastic moment, we find at failure (see Fig. 6.36b)

$$W_L \vartheta + W_M \vartheta + W_M \vartheta + W_R \vartheta = Al\vartheta, \tag{6.161}$$

and we obtain four stochastic structural parameters $G_j = g_j(W, A)$ for the failure mechanisms

$$G_1 = W_1 + 2W_3 + W_4 - Al, \quad G_2 = W_1 + 2W_3 + W_5 - Al,$$
$$G_3 = W_2 + 2W_3 + W_4 - Al, \quad G_4 = W_2 + 2W_3 + W_5 - Al. \tag{6.162}$$

The failure risk of the redundant portal frame is given by the condition that at least one mechanism occurs (one structural parameter G_j becomes smaller than zero). Therefore

$$Q_R = P[G_1 \leqslant 0 \cup G_2 \leqslant 0 \cup G_3 \leqslant 0 \cup G_4 \leqslant 0], \tag{6.163}$$

i.e. the redundant structure of the portal frame also behaves like a series system, the 'elements' of which are sub-systems (modules) which themselves represent pure parallel systems, each consisting of three elements, as three elements must always fail for a failure mechanism to develop (Fig. 6.36c). We have discussed the two basic logical systems, series and parallel, in Sect. 6.3.2.1. At this point we would like to make clear that the failure mechanisms (parallel systems) must be determined first in a reliability analysis of redundant structures, and they are then used as 'elements' of a series system to form the logical system of the redundant structure. We determine the correlation between these 'elements' with (6.151) (see [53, part II]).

In practice, this is very expensive for most marine steel structures with a high degree of redundancy, because these structures generally have many fundamental failure mechanisms and combinations thereof. For this reason, procedures have been developed which make it possible to represent, with an accuracy which is acceptable for most practical applications, only the mechanisms which are probably the most significant (dominant), without looking for all possible failure mechanisms. Here we will give a brief but explicit description of the two most familiar procedures with indications of applications in marine technology; we must leave the details to the extensive specialist literature on the subject (see the detailed and clear summary in [55]).

'Branch-and bound' technique

1. Branching: to develop a search algorithm suitable for computer programming we start by representing the sequential formation of plastic hinge series, i.e. we trace, with a type of trial and error method, the development of possible failure paths up to the plastic hinge at which the structure finally fails.

With the perfectly elastic/plastic material law, where there is a failure at the potential plastic hinge under, for example, pure bending moment stress, we insert a real hinge and apply the fully plastic moment as an external moment and carry out a further calculation with the original (stochastic) loads. (Other failure modes such as, for example, buckling, can be modelled in a similar way, including cases with interaction of different sectional loads.)

After each step $p = 1, 2, \ldots, p_K$ we can find, for example, that plastic hinge, out of all plastic hinges $M(p)$ possible for the next step, which would appear to be the most probable next one in the given circumstances. This immediately indicates a criterion by which this plastic hinge is chosen as having the greatest failure probability, i.e.

$$P[F_{\mathrm{m}}^{(p)}] := \underset{m \in M(p)}{\mathrm{Max}} \; P[F_{\mathrm{m}}(p)], \tag{6.164}$$

where $F_{\mathrm{m}}(p)$ serves as a stochastic failure indicator for the element under consideration. In (6.164) (p) means that we are at the p-th step of this branching algorithm, in which there are $M(p)$ potential plastic hinges to choose from; (p) should therefore be interpreted as an indicator of the partial damage state of the structure to be taken into consideration at the p-th step. With $(p) = (1)$, the structure is intact, i.e. $M(1)$ is the number of all possible plastic hinges or of all plastic hinges considered in the model. The relatively greatest failure probability found for an element m with the partial damage state of the structure indicated by (p) is therefore $P[F_{\mathrm{m}}^{(p)}]$.

A somewhat more complex, but generally very appropriate branching criterion is suggested in [65]. Here m is chosen at the pth step of the branching algorithm such that

$$P[F_{\mathrm{n}}^{(1)} \cap F_{\mathrm{m}}^{(p)}] := \underset{m \in M(p)}{\mathrm{Max}} \; [F_{\mathrm{n}}^{(1)} \cap F_{\mathrm{m}}(p)], \tag{6.165}$$

where $F_{\mathrm{n}}^{(1)} \equiv F_{\mathrm{m}}^{(1)}$ is determined as in (6.164) with $(p) = (1)$ for $n = m$.

The criteria (6.164) and/or (6.165) define the indices m of elements belonging to a probable plastic hinge series which develops sequentially.

2. Bounding: with the probabilities defined in (6.165), the probability of occurrence of a partial (incomplete) plastic hinge series $q(p)$ with p plastic hinges with different indices m can be estimated immediately. Figure 6.37, which shows the case $M(p) = 3$, $m = 1, 3$ and 4, in the form of a Venn diagram, gives us

$$\begin{aligned} P[F_{\mathrm{q}}^{(p)}] &= P[F_{\mathrm{m}}^{(1)} \cap F_{\mathrm{m}}^{(2)} \cap \cdots \cap F_{\mathrm{m}}^{(p)}] \\ &= P[(F_{\mathrm{m}}^{(1)} \cap F_{\mathrm{m}}^{(2)}) \cap (F_{\mathrm{m}}^{(1)} \cap F_{\mathrm{m}}^{(3)}) \cap \cdots \cap (F_{\mathrm{m}}^{(1)} \cap F_{\mathrm{m}}^{(p)})] \\ &\leqslant \underset{k=2}{\overset{p}{\mathrm{Min}}} \; P[F_{\mathrm{m}}^{(1)} \cap F_{\mathrm{m}}^{(k)}], \end{aligned} \tag{6.166}$$

and by using this upper limit we (clearly) get

$$P[F_{\mathrm{q}}^{(1)}] \geqslant P[F_{\mathrm{q}}^{(2)}] \geqslant \cdots \geqslant P[F_{\mathrm{q}}^{(p)}]. \tag{6.167}$$

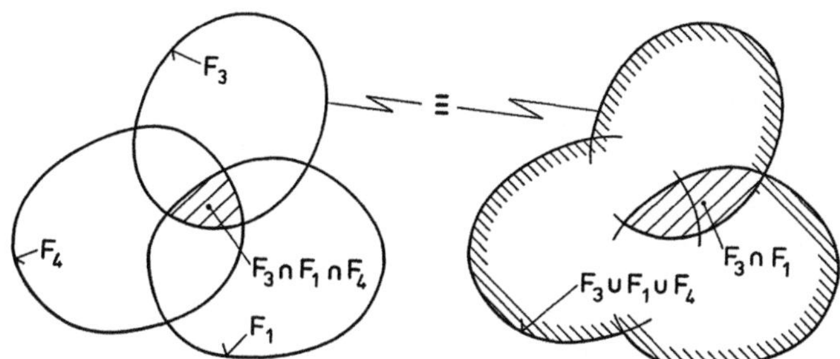

Fig. 6.37. Venn diagram for domains of three random parameters.

The characteristic (6.167) permits us to decide at the p-th step which potential plastic hinges can be discarded, and which must still be considered for the formation of dominant plastic hinge series: if we propose a limit at which a complete plastic hinge series, consisting of p_K plastic hinges, can be discarded as not dominant, e.g. at

$$P[F_\mathrm{q}^\mathrm{K}] \leqslant 10^{-n}, \quad (\text{e.g. } n=6), \tag{6.168}$$

such a criterion is fulfilled at $p = K$ because of (6.167) if it is already fulfilled at any step $p \leqslant \mathrm{K}$. The practical benefit of this 'branch-and-bound' algorithm thus lies in the ability to discard non-dominant plastic hinge series, before they have been completely defined. For the perfectly elastic/plastic material law, complete plastic hinge series, with any sequence of plastic hinges, always form one of the plastic hinge combinations sought, whose probability of occurrence can be further narrowed down by (6.166) and another non-trivial lower bound [55].

'Beta-unzipping' technique
In this procedure the logical system of the structure is at first modelled relatively crudely. The development of the reliability statement is observed in a subsequent stepwise, systematic creation of increasingly realistic logical system models, and this process stops when changes in the result are practically no longer relevant [55].

We will begin, for example, with the reliability of the structure at the lowest (crudest) level 0, where we simply make

$$Q_\mathrm{R} = \overset{M}{\underset{m=1}{\mathrm{Max}}}\, Q_\mathrm{m} = \overset{M}{\underset{m=1}{\mathrm{Min}}}\, |\beta_\mathrm{m}|, \quad \beta_\mathrm{m} \leqslant 0, \tag{6.169}$$

Q_m being the failure risk of the element (potential plastic hinge) m.

At the next level 1, we model the structure as a series system in which the elements have various reliability indices β_m and can be variously correlated ($\rho_\mathrm{mn} \neq \rho_\mathrm{mk} \neq \rho_\mathrm{nk}$), but we consider only elements with values β_m which exceed a prescribed minimum value, i.e. which fail with a minimum probability Q_m. Of course, for the series systems we use bounds which are considerably narrower than the trivial bounds given with (6.156) (see [66]).

At the next level 2, we then use a series system consisting of sub-systems which themselves have just two parallel elements; here it is necessary to find only critical element pairs. This is done first by inserting a plastic hinge in the element m whose failure probability is greatest (see (6.164)), and by adding an external load (moment) of the magnitude of the element resistance (fully plastic moment) at this point. Now this element m is combined with all the elements to form parallel systems consisting of two elements, which give a joint failure probability with the otherwise unchanged (stochastic) loads, which exceeds a prescribed level such as, for instance, in (6.168).

By continuing this algorithm for parallel systems, consisting of 3, 4, 5, etc., elements which then together form a simplified series system of the redundant structure, the level of the reliability analysis is raised to levels 3, 4, 5, etc.

We can see that the branch-and-bound and the beta-unzipping methods become increasingly similar from level 2 upwards, and results obtained so far confirm the

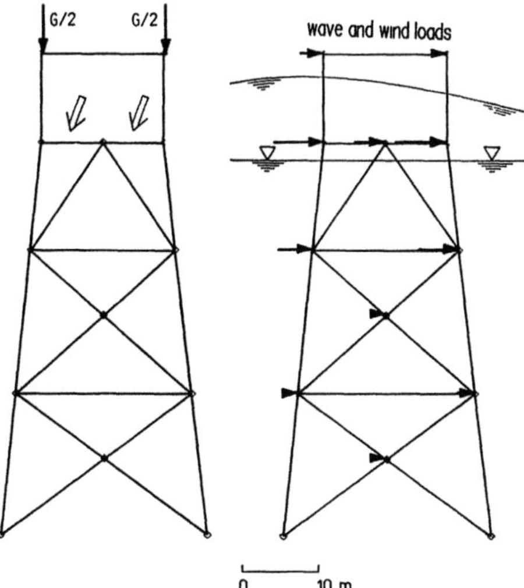

Fig. 6.38. Two-dimensional jacket module, idealization and loads.

practical usefulness of both methods for reliability analysis of load-bearing structures made of bars and beams. In marine technology these are essentially steel jackets and related structures, for which we will give some representative results below.

First, we consider two jacket frames under wind and wave loading (Figs. 6.38 and 6.39 from [67]). To compare these different structure types, we assume the same initial dimensions for tube diameter and wall thickness, and define the loadings with the same wave in the same phase. In addition, there is the wind loading on the same above-water part of the structure, and also the same platform weight loading G. For the purposes of our stochastic study, these values are to be

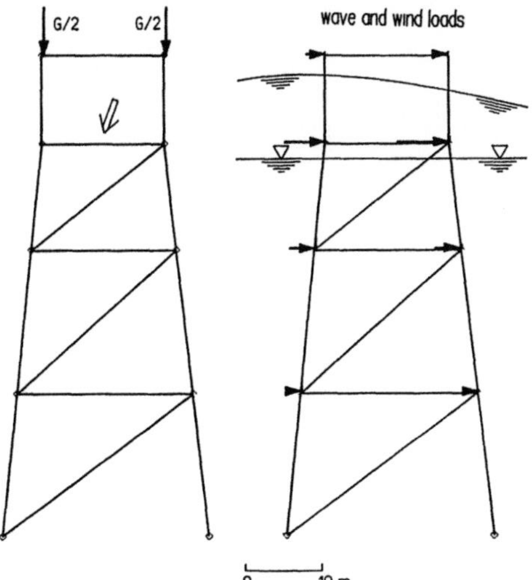

Fig. 6.39. Two-dimensional jacket module, idealization and loads.

considered as mean extremal loads, and the actual values can be lower or higher. The measurement which defines the possible deviation from the mean is the variation coefficient of variation, $V = \sigma/m =$ standard deviation/mean value, which we will set at 20% for all design parameters in this example (which is purely for illustration). All tube elements with the same diameter and the same wall thickness are regarded as being 90% correlated, and in the case of the same diameter but a different wall thickness we assume a lower correlation of 20%; otherwise, all design parameters are uncorrelated. The wave loads indicated in Figs. 6.38 and 6.39 at the joints of the jacket frame are, of course, linearly dependent on one another with a regular wave.

An impression of the degree of redundancy in both cases can be gained from the results of two sample calculations compared in Fig. 6.40, in which the complete failure mechanism with the greatest probability of occurrence (dominant mechanism) of each structure was determined with a branch-and-bound technique, which took into account the stochastic characteristics of the load and strength parameters. We cannot go into further detail here on this procedure, which is a stochastic finite element technique, but see [68] for detailed information.

It can be seen from Fig. 6.40 that with properly proportioned dimensions, the load-bearing capacity or the ultimate limit state of a jacket frame can scarcely be reached or exceeded: in principle, too much has to happen for anything to happen, and that is the essence of redundant structures. We modify this statement somewhat in a moment, but first we must say that when designing marine steel structures, reliability analysis allows us to see where particularly critical structural elements are to be found (in the example in Figs. 6.38 and 6.39 at the points indicated by double arrows). On this basis, we can take sensible constructional safety decisions or design in suitable inspection strategies for subsequent operation. This is the real point of the evaluation method for marine structures developed here: during an analysis cycle for a particular configuration of the structure and its dimensions, we can determine how close we have come to the target of a design which takes

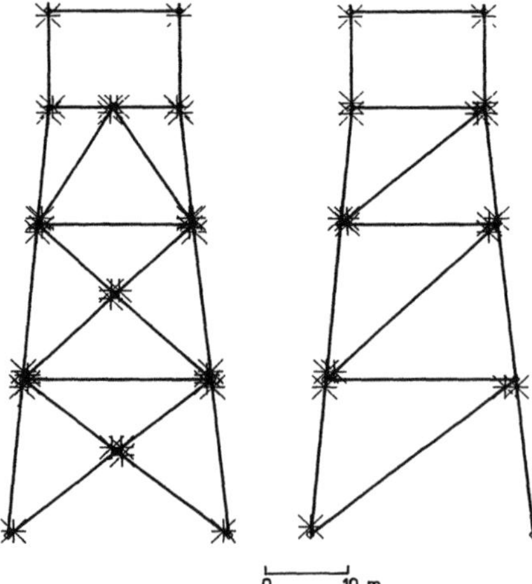

Fig. 6.40. Dominant failure mechanisms, modules as in Figs. 6.38 and 6.39.

proper account of safety. The more difficult it becomes, during the course of different analysis cycles to identify dominant failure mechanisms with the algorithms described or similar ones, i.e. the more plastic hinges each mechanism has, the closer we are to an optimally reliable structure.

This finding is largely confirmed in a very extensive study [69]. We will quote from this another result which is of general interest for design practice (see Fig. 6.41). This is the question of the effective use of bracing for jacket frames, X and K braces being compared for different ratios of vertical loads F_v to horizontal loads F_H.

The reliability of the frame configurations considered increases with negative values of β (see (6.115) based on our definition of β in, e.g. (6.114)). This value is shown in Fig. 6.41a for perfectly elastic/brittle material, and in Fig. 6.41b for perfectly elastic/plastic material. With relatively high horizontal loads, such as we would expect, for example in the 100-year storm, the K brace is just as reliable as the X brace when the material in Fig. 6.41a is used. This finding, which is perhaps initially surprising, indicates that structural redundancy loses its meaning as far as reliability is concerned in the case of brittle material: e.g. there would be no safety effect in constructing the leg of a tension leg structure from many parallel load-bearing tendons highly correlated to the same load, if they are made of a material which has an overwhelmingly brittle-fracture performance. This finding is verified more precisely in [69].

Of course, the examples given are idealized with respect to a series of subsidiary conditions, which we cannot deal with fully here, but they show nevertheless that evaluating marine structures for reliability has developed into a practice-related extension of the classical evaluation methods described in Sect. 6.1 and 6.2 (see [70]). The effect of this method of analysis on modern construction regulations

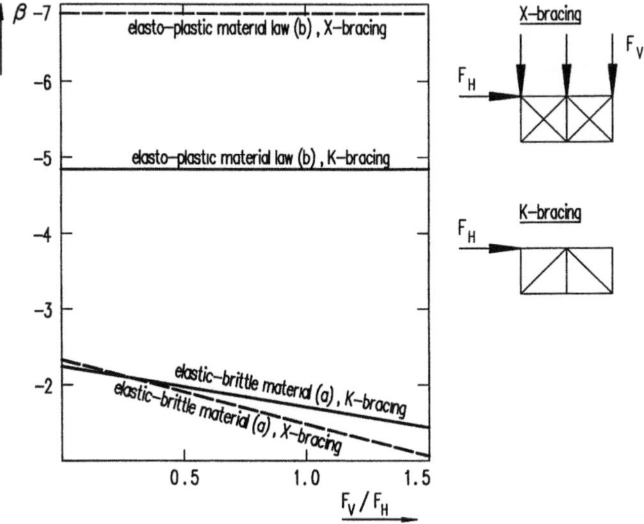

Fig. 6.41. β comparison of frames with X and K braces [69].

should no longer be neglected; we will return to this in Sect. 7.4. We view the concept presented here as part of the whole evaluation concept for marine steel structures, including the development of modern in-service inspection strategies, which we will outline in the following section.

6.3.3 Reliability and Risk as Functions of Time

We have already introduced the concept of time-variant probability when dealing with different load processes in Sect. 6.2.1.3 (see $P(T)$ in (6.67)). If we define T as a 'random time to failure' with distribution function $F_T(t)$, then for any time t

$$R(t) = P_s(t) = P[T > t] = 1 - F_T(t) = 1 - Q(t) \qquad (6.170)$$

is the probability of success, and

$$Q(t) = P_f(t) = P[T \leqslant t] = F_T(t) = 1 - R(t) \qquad (6.171)$$

is the risk of failure at time t. We will be using notations $P_s(t)$ for $R(t)$ and $P_f(t)$ for $Q(t)$ with index 's' for success and index 'f' for failure only in this final section of Chap. 6.

A simple application of (6.171) is the fatigue failure probability esimation based on the damage model discussed in Sect. 6.2.2.1. With (6.91), extended to I stress range classes with average stress range s_i and probability of occurrence P_i as in (6.90), the expected damage at a reference time t is

$$E[D(t)] = \frac{t}{k} \cdot \Omega, \ \ \Omega = (2\sqrt{2})^m \cdot \Gamma(1 + m/2) \cdot \sum_{i=1}^{I} P_i \cdot \frac{n_i(t)}{t} \sqrt{\mathrm{Var}(\Sigma_i)}^m,$$

where P_i has the interpretation of the probability $P(H_{v_i})$ if we consider stationary random stress processes $\{\Sigma_i(t)\}$ caused in a structural element by stationary random wave proceses $\{Z_i t)\}$, characterized by significant wave heights $H_{v_i} = 4 \cdot \sqrt{\mathrm{Var}(Z_i)}$, (cf. (5.80) and (5.92)). The term $\mathrm{Var}(\Sigma_i)$ and the relative number of stress ranges $n_i(t)/t = v_i$ can be calculated from stress spectra $S_{\Sigma_i \Sigma_i}(\omega)$ using (5.14) and (5.20), the latter with rain-flow corrections according to [45] if necessary (see Sect. 6.2.3.3). We recall that response spectra like $S_{\Sigma_i \Sigma_i}(\omega)$ are normally obtained from standard wave spectra $S_{Z_i Z_i}(\omega)$ like (5.88) with (5.89) via transfer functions $H_{\Sigma_i Z_i}(\omega)$ as in (6.53), where $H_{\Sigma_i Z_i}(\omega)$ must be calculated on the bases of hydrodynamic and stress analysis methods described in Chap. 3 Vol. I and Chap. 4 for a number of elementary wave frequencies ω, as discussed in Sect. 6.1.1 with (6.1).

Failure occurs if $E[D(t)] > E[\Delta]$, where $E[\Delta] = 1$ is used in (6.75). Hence, Δ is an uncertain basic parameter like D if we consider stochastic S–N-model parameters M and K with realizations m and k. Letting $D = \Delta$, the time T_n to fatigue failure of a structural element n and the related structure function $g_n(X)$ of the reliability analysis problem are

$$T_n = \frac{\Delta \cdot K}{\Omega(T_n)}, \ \ g_n(X(t)) = T_n - t, \ \ X^T(t) = (\Delta, K, M, t),$$

where t is a deterministic basic parameter, but Δ, K and M are random parameters for which stochastic models (distribution functions and/or moments) must be defined. From (6.171) we thus have

$$P_{n_f}(t) = P[T_n \leqslant t] = P[g_n(X(t)) \leqslant 0].$$

This is exactly the reliability analysis problem discussed in (6.141). Details of a solution for P_{nf} and the alternative usage of the fracture mechanics model of Sect. 6.2.2.2, applied in a similar form, can be found in [71].

The problem of the influence of time on the probability of failure is generalized introducing a so-called risk function

$$v(t) = f_T(t)/(1 - F_T(t)).$$

Then with (A1.15)

$$v(t)\mathrm{d}t = \mathrm{d}F_T(t)/(1 - F_T(t))$$

and integrating from 0 to t we have

$$P_f(t) = 1 - \exp\left\{ - \int_0^t v(\tau)\mathrm{d}\tau \right\}. \tag{6.172}$$

If $v(t)$ is constant in time it equals the event rate v of the Poisson process (cf. (5.30), (5.29)). Thus, without going into further details, we can imagine that the conditions behind the probability of failure function (6.172) are similar to those of a Poisson process (see Sect. 5.1.3.1).

The difficulties in determining $P_f(t)$ arise from the fact that $F_T(t)$ is generally not known, nor can it be determined statistically from observations with sufficient accuracy. Thus it must be derived from statistics and/or the physical context. This can be done using auxiliary (Poissonian) random-point processes $\{X(t)\}$ of excursions (exits) of the structure function $g(x(t))$ into the failure domain $V(t) = \{g(x(t); t) \leqslant 0\}$ (see Fig. 6.33 for examples of 2-dimensional failure domains). Denoting the number of excursions by $N_X^+(t)$, for small failure probabilities an asymptotic approximation for (6.172) can be derived [72]

$$P_f(t) \approx 1 - \exp\{ - E[N_X^+]\} \tag{6.173}$$

where $P_f(0) \approx 0$ is assumed for simplicity.

Special non-stationary random processes may be modelled as juxtapositions of piecewise stationary random processes being determined by parameters which are varying in time, e.g. class parameters (H_V, T_V) of the stationary wave elevation processes discussed in Chap. 5, Sect. 5.2.2.2. In such cases, (6.173) no longer holds. It is, however, possible to take account of these phenomena if the random state vector $X(t)$ of basic parameters is suitably separated into three components:

$$X(t)^T = [\mathscr{R}^T, \mathscr{Q}(t)^T, Y(t)^T]$$

— $\mathscr{R}$ = a vector of time-invariant random variables,
— $\mathscr{Q}(t)$ = a slowly varying ergodic random vector sequence of parameters defining a non-stationary random vector process as juxtapositions of piecewise stationary random vector processes $\{Y(t)\}$
— $\{Y(t)\}$ = a random vector process representing $\{X(t)\}$ as a stationary process

within small sub-intervals of short length (order of hours), thus $\{Y(t)\}$ is determined with a vector of parameter values $\mathcal{Q}(t) = q$.

Then, it has been shown [73] that the failure probability related to $\{X(t)\}$ can be approximated by

$$P_f(t) \sim 1 - E_{\mathcal{R}}[\exp\{-E_{\mathcal{Q}}[E[N_X^+(t; \mathcal{R}, \mathcal{Q})]\}], \tag{6.174}$$

and, as the point process of exits out of the safe domain is a regular process, i.e. Poisson conditions of Sect. 5.1.3.1 are satisfied, we have

$$P_f(t) \sim 1 - E_{\mathcal{R}}\left(\exp\left\{-E_{\mathcal{Q}}\left[\int_0^t v^+(V(\tau|\mathcal{R}, \mathcal{Q}))d\tau\right]\right\}\right) \tag{6.175}$$

where $E_{\mathcal{R}/\mathcal{Q}}[\cdots]$ denote expectations with respect to the random vectors $\mathcal{R}$ or $\mathcal{Q}$ respectively. It should be observed that the expectation operation with respect to $\mathcal{Q}$ is performed inside the exponent operator, whereas the expectation operation with respect to $\mathcal{R}$ is performed outside the exponent operator. In general, the dimensions of $\mathcal{R}$ and $\mathcal{Q}$ are far too large to perform the necessary integrations to obtain the expectations by a direct numerical integration scheme. However, it is possible to apply an advanced version of reliability analysis algorithms of the FOSM method, which we discussed in Sect. 6.3.1 in some detail, to find the expectations with respect to parameters $\mathcal{R}$ and $\mathcal{Q}$ [73–75]. For in sufficiently rare outcrossings, or if the random processes show pronounced narrowband character and thus exhibit pronounced clustering of the outcrossings, the Poissonian assumption is no longer valid, but usually errs on the conservative side.

In Sect. 6.3.3.1 to 6.3.3.6 we apply the above ideas to fatigue failure and the related inspection planning problem of a hatch corner of a container ship structure as one of the most challenging type of marine structures [76, 77], but the ideas may also be applied to offshore structures in a similar form [75].

6.3.3.1 Fatigue Failure Probability Modelling with Stress and Strength being Functions of Time

We shall restrict our considerations to the crack propagation phase only (see Fig. 6.20), but mention in passing that similar considerations can also be made for the endurance limit or initiation phase of crack development. Crack instability is taken as the governing failure mode. One of the simplest brittle failure (crack instability) criteria gives the failure domain $V(t)$ as

$$V(t) = \{1 - K_r(t) \leqslant 0\}, \tag{6.176a}$$

which may be extended to the interaction of plastic collapse and brittle failure to give

$$V(t) = \left\{\frac{\Sigma_r}{\left(\left(\frac{8}{\pi^2}\ln\left[\frac{1}{\cos(\pi\Sigma_r/2)}\right]\right)\right)^{1/2}} - K_r(t) \leqslant 0\right\}, \tag{6.176b}$$

where Σ_r and $K_r(t)$ are non-dimensional plastic collapse and fracture parameters, respectively. These quantities are defined by

$$\Sigma_r \approx 2 \frac{\Sigma_p}{(\Sigma_y + \Sigma_u)}, \tag{6.177}$$

$$K_r(t) = \frac{K_I(t)}{K_{Ic}} = \frac{Y(A(t))\sqrt{A(t)\pi}\Sigma_p}{K_{Ic}}, \tag{6.178}$$

with Σ_p the peak stress at the net cross section of the structural element. Σ_y and Σ_u are the yield strength and ultimate strength, respectively. (As before, we use capital letters to indicate stochastic parameters: Σ are random stresses with realizations σ.) K_{Ic} is the fracture toughness, $A(t)$ the crack length at time t, and $Y(A(t))$ is a geometry factor as introduced in (6.82). For a crack at the edge of a panel a constant geometry factor can be assumed: $Y_c = 1.12$. Other sometimes more realistic, experimentally based failure criteria can be found in [78].

For crack propagation the simple one-dimensional crack growth with time relationship is assumed as in (6.81) and (6.84)

$$\frac{dA(t)}{dN} = CY_c\sqrt{A(t)\pi}S(t))^M \tag{6.179}$$

with $S(t)$ the stress range at time t, C and M material parameters as in Sect. 6.2.2.2. Equation (6.179) has to be integrated in order to determine the crack length $A(t)$ at time t under the initial condition $A(0) = A_0$. For random loading an appropriate counting of the damage relevant stress cycles has to be performed. For simplicity of presentation, it is assumed that the stress process is a narrowband zero mean Gaussian process with variation $\mathrm{Var}[\Sigma] = m_0$ (cf. (5.26)), m_0 being defined in (5.27). The usually conservative peak counting method is applied for an effective stress range $s = 2\sigma$, with σ the local peak value of a random sample $\sigma(t)$. Thus as in (6.87) with $a_f \rightarrow A(t)$, $a_i \rightarrow A_0$ we have

$$\Psi(A(t)) = \frac{2}{2-M} \cdot \frac{1}{Y_c^M C\pi^{M/2}} (A(t))^{(2-M)/2},$$

$$\Psi(A_0) = \frac{2}{2-M} \cdot \frac{1}{Y_c^M C\pi^{M/2}} (A_0)^{(2-M)/2},$$

hence

$$\Psi(A(t)) - \Psi(A_0) = E[S^M]vt.$$

where $vt = \sqrt{m_4/m_2}$ is the number of local peak values of $\{\Sigma(t)\}$ and

$$E[S^m] = \int_0^\infty s^M \cdot f_S(s)ds$$

can be evaluated analogously to (6.91) giving

$$\Psi(A(t)) - \Psi(A_0) = \frac{t}{2\pi}\sqrt{\frac{m_4}{m_2}}(2\sqrt{2m_0})^M\Gamma\left(\frac{M+2}{2}\right). \tag{6.180a}$$

The Γ-function is given in (6.92) for $m \to M$, so $A(t)$, needed in (6.179), can be evaluated at any time t if the spectral moments of stress m_i are given. For an extension of (6.180a) to wideband processes (see [79])

$$\Psi(A(t)) - \Psi(A_0) = \frac{t}{2\pi} \sqrt{\frac{m_4}{m_2}} (2\sqrt{2m_0})^M \left\{ \frac{\alpha^{M+1}}{2\sqrt{\pi}} \Gamma\left(\frac{M+1}{2}\right) \right.$$
$$\left. + \varepsilon\Gamma\left(\frac{M+2}{2}\right) T_{M+2}\left(\frac{\varepsilon}{\alpha}\sqrt{M+2}\right) \right\}$$

with α as in (5.26), i.e. $\varepsilon^2 = m_2^2/(m_0 m_4) = 1 - \alpha^2$, and T_{M+2} is the Student's t-distribution with degree of freedom $M + 2$. This finally yields

$$A(t) = \left[A_0^{(2-M)/2} + C \cdot \frac{2-M}{4\pi} \cdot \sqrt{\pi^M} \cdot \sqrt{\frac{m_4}{m_2}} \cdot (2\sqrt{2m_0})^M \cdot t \right.$$
$$\left. \cdot \left\{ \frac{\alpha^{M+1}}{2\sqrt{\pi}} \Gamma\left(\frac{M+1}{2}\right) + \varepsilon\Gamma\left(\frac{M+2}{2}\right) T_{M+2}\left(\frac{\varepsilon}{\alpha}\sqrt{M+2}\right) \right\} \right]^{2/(2-M)}$$

$$(6.180b)$$

The outcrossing rate of the stress process $\{\Sigma(t)\}$ out of a safe domain into the failure domain $V(t)$ as specified with (6.176a/b) can be rewritten in terms of a time variant threshold value of strength $\Sigma_E(t)$, and the mean stress level $\bar{\Sigma}$ as

$$v^+(V(\tau|\mathcal{R}, \mathcal{Q})) = \omega_0 \cdot \varphi(s) \cdot [\varphi(\dot{s}/\omega_0) - (\dot{s}/\omega_0)\Phi(\dot{s}/\omega_0) \qquad (6.181)$$

(cf. (6.19)) and $s = (\Sigma_E - \bar{\Sigma})/\sqrt{m_0}$. The equivalent strength threshold $\Sigma_E(t)$ for the brittle failure mode is thus obtained from

$$V(t) = \{\Sigma_p \geqslant \Sigma_E(t)\}$$
with
$$\Sigma_E(t) = K_{Ic}/(Y_c\sqrt{A(t)\pi}). \qquad (6.182a)$$

A similar though more complicated derivation for the plastic-collapse/brittle-failure mode leads to

$$\Sigma_E(t) = \frac{(\Sigma_y + \Sigma_u)}{\pi} \text{acos}\left[\left\{ \exp -\frac{1}{8}\left(2\pi \frac{K_{Ic}}{Y_c\sqrt{A(t)\pi}(\Sigma_y + \Sigma_u)} \right)^2 \right\} \right] \qquad (6.182b)$$

which will be used in the container ship example below. The fatigue failure probability can now be obtained by taking the expectations with respect to $\mathcal{R}$ and $\mathcal{Q}$ as already indicated with (6.175) [73], where $\mathcal{R}^T = (A_0, C, M, K_{Ic}, \Sigma_y, \Sigma_u, Y_c, \bar{\Sigma})$ and $\mathcal{Q}^T = (m_0, m_2, m_4)$.

The spectral moments of m_0, m_2, and m_4 of piecewise stationary stress processes can be given in a quasi-analytical form using response surfaces as explained in [76]. This is an essential part of the necessary numerical evaluations, and we outline briefly how the response surfaces are defined and represented for reliability analysis. With this we present more advanced applications of the methods discussed in Chaps. 3 (Vol. I), 4 and 5.

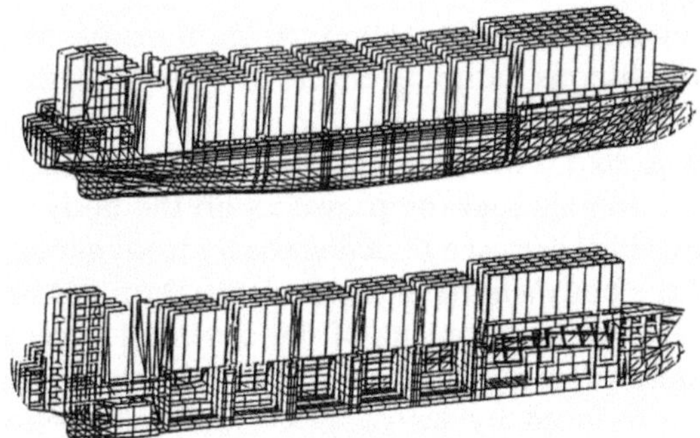

Fig. 6.42. Finite element mesh of a modern container ship

6.3.3.2 Hydrodynamic and Structural Analysis

For a realistic analysis of wave load induced stresses in a ship structure a computer program for hydrodynamic analysis must be coupled with a structural finite-element (FE) computer program. The finite-element idealization of a marine structure is rather complex, especially for ship structures as is illustrated with the FE mesh of a container ship's body in Fig. 6.42, where about 18 000 elements have been used to model the global structure.

The basis of our hydrodynamic approach is a three-dimensional linear potential theory of moving bodies in harmonic, gravity waves [80], which is an extended version of the diffraction theory presented in Chap. 3 of Vol. I. As we know from this the undisturbed (incident), the scattered (diffraction), and the body-motion-induced (radiation) wave potentials must satisfy the Laplace equation in the fluid domain subject to the relevant (linearized) boundary conditions:

1. The fluid particles must not penetrate the free surface.
2. The pressure is constant everywhere at the free surface.
3. Vertical velocity components vanish at the sea bottom (3a) and at the underwater surface of the floating body (ship or offshore structure) under consideration (3b).
4. Radiation waves propagate away from the moving body.

The underwater body of the marine structure (a ship in our example) is modelled by small surface elements (patches) represented by sources or sinks that satisfy boundary conditions 1 to 3a and 4 (Green's functions). Seven integral equations of the Fredholm type are solved for the unknown source or sink strengths that satisfy boundary condition 3b as well, one integral equation for incident waves with the body fixed and six equations for unit motions of the body's centre of gravity in six degrees of freedom. The solutions of each of the integral equations are numerically realized using a system of n linear equations, with n being the number of surface patches on the underwater surface of the body. The solutions for the incident waves yield the excitation forces, the solutions for the unit body motions yield the coefficients of hydrodynamic masses and mass moments of inertia, as well as the (potential) damping coefficients needed in the set of the six coupled linear differential equations which describe the equilibrium of forces in terms of the unknown motions. To obtain the motions of the floating body from this set of equations, viscous damping effects are also linearized (most details are given in Chap. 3, Vol I).

With the motions given it is then straightforward to calculate the resulting velocity potential in the fluid domain and the pressures on the body's underwater surface (Bernoulli equation) which are used to define the hydrodynamic nodal forces on the underwater part of the body's finite element idealization per unit wave amplitude. From the six acceleration components of the centre of gravity, the accelerations and, thereby, the inertia forces acting on the nodal points are obtained. Further, hydrostatic pressure induced by heave, pitch and roll motions of the body in waves is analytically described by a linear expression, and added to the nodal forces in the same manner. The nodal mass forces due to gravity have directions relative to the structure's coordinate system which change with pitch and roll motions.

Based upon the sum of these forces, stresses in structural elements of the body are obtained due to its motions in regular waves of unit wave amplitudes. Thus stress amplitude operators or stress transfer functions $H_{\Sigma Z}(\omega)$ of stresses in any element of the structure can be determined. They will be used for further evaluations explained below.

Some special features of the interface between force and stress calculation based on hydrodynamic (surface patch) modelling of the considered structure on one hand and its FE strength modelling on the other, are worth mentioning. Since the development of a realistic FE model of an offshore structure or ship is always expensive (time-consuming), the main principle observed when developing this interface in [76] was as follows: the FE model of the structure, once developed for different purposes of strength analyses – including, for instance, vibration analysis – should not be changed for the purpose of the hydrodynamic analysis. Normally, the number of surface elements of the underwater body of a floating structure is unmanageable in a hydrodynamic analysis as outlined above. The number N of equations to represent the Fredholm type integral equations was limited to about 600 for different reasons, the evaluation of Green's functions being the most important one. Furthermore, the waterline that determines the upper boundary of the underwater body is defined by the considered loading condition, which may change due to different design cases (operations, etc.), and thus cannot be expected to be identical with any boundary of the surface elements already determined for strength analysis. Therefore, an interface was developed as part of the hydrodynamic computer program with two provisions:

1. Definition of surface patches from a number of adjacent finite surface elements by means of a patch list being part of the input data.
2. Cut off of upper parts of finite surface elements which contain the waterline, and inclusion of the 'wet' part of these elements in the adjacent surface patches.

For each of the surface patches so defined the centre of gravity was determined to allow hydrodynamic calculations. The resulting hydrodynamic pressures were then used to define a constant pressure distribution over the area of each surface patch, and nodal forces were calculated applying the pressure directly to the finite surface elements. The applied finite-element method is well described in a number of text books, e.g. [81].

6.3.3.3 Multi-Dimensional Response Surfaces for Spectral Moments

As already mentioned, it is not practical to combine directly a computer code for the hydrodynamic and structural analysis with a computer code for the reliability analysis of fatigue-sensitive structural elements. It is, however, possible to generate response surfaces with a simple algebraic structure from the hydrodynamic and strength analyses. These response surfaces can then be used rather effectively in the reliability analysis. Response surfaces should fulfil certain requirements. First, the response quantity should be an explicit function of the input parameters. The surface should ideally at least be capable of double differentiation, otherwise application of the reliability methods discussed in Sect. 6.3.1.2 would be very difficult, if not prohibitive. A response surface can only be an approximate representation of the true input–output relationship. It is, however, sufficient that this approximation is of high quality in the domain of those basic parameters which are most significant for accurate reliability estimations.

The basic parameters essentially influencing the result of hydrodynamic and structural analyses are elementary wave period T, elementary wave direction ϕ with respect to the structure, speed of advance V, and loading condition l. The load–load effect relationship is assumed to be linear, so the influence of an arbitrary wave amplitude ζ_a can be determined by simply multiplying the transfer function $H_{SZ}(\omega, \phi, V, l, ...)$ by the value of ζ_a. Here, $\omega = 2\pi/T$ is the angular frequency of elementary waves. In the example, the transfer function was determined for a set of values of parameters ω and ϕ, but only for one value of V being 2/3 of the maximum ship speed, and for one average loading condition l.

We consider an ergodic narrowband Gaussian seaway process, which – according to the basic ideas developed in Chap. 5 – can be thought of as the superposition of an infinite number of elementary waves of different amplitudes ζ_a and frequencies ω at (random) phase shifts. The standard procedure to define spectral moments m_i of a linear system is based on the Pierson–Moskowitz seaway spectrum $S_{ZZ}(H_V, T_V, \omega)$ given in, e.g. (5.88), which can be extended to include broadband parameters like energy spreading over elementary waves from different directions to give the stress response spectrum as

$$S_{\Sigma\Sigma}(\omega, \vartheta, H_V, T_V) = S_{ZZ}(\omega, H_V, T_V) \int_{-\pi/2}^{\pi/2} H_{\Sigma Z}(\omega, \vartheta + \chi) G(\chi) d\chi. \qquad (6.183)$$

Here, a spreading function of the form $G(\chi) = A \cos^B(\chi)$ is considered where A and B are empirical constants; χ is the angle between the direction of the elementary wave relative to the main direction of the seaway and ϑ is the angle between the main direction of the seaway and the orientation of the structure, determined, for example, as the direction of the speed V of a ship. The constants of the spreading function must be detemined such that the above integral equals unity when setting $H_{\Sigma Z}(\omega, \phi)$ equal to 1, e.g. $A = 2/\pi$ and $B = 2$ describe the observed spreading sufficiently well for moderate sea states. The transfer functions $H_{\Sigma Z}(\omega, \phi)$ between stress Σ and elementary wave elevation Z are already determined for angles $\phi = \vartheta + \chi$ between elementary waves and the orientation of the structure. The i-th spectral moments m_i can thus be determined from (6.183) and (5.27) to give

$$m_i(\vartheta, H_V, T_V) = \int_0^\infty \omega^i S_{ZZ}(\omega, H_V, T_V) \int_{-\pi/2}^{\pi/2} H_{\Sigma Z}^2(\omega, \vartheta + \chi) A \cos^B(\chi)\, d\chi\, d\omega.$$

$$(6.184)$$

To allow the use of m_i as part of a quasi-analytic limit state function in a reliability analysis, we define response surfaces for m_i, $i = 0, 2, 4$, on the basis of the discrete parameter set as node-points, which are corner points of an n-dimensional rectangular net; n is the number of parameters under consideration.

First, in all calculated node-points of the response surface the directional derivatives with respect to any parameter are determined using one-dimensional standard spline functions. These derivatives are stored for later use. Second, the n-dimensional rectangular net is searched for the interval that contains the interpolation values, i.e. the values which are required for the reliability calculations. The n-dimensional response surface in the interval thus defined is then built up by products of elementary Hermite polynomials. All function values and all derivatives of these polynomials are zero in the corners except one function value or one directional derivative which is unity. Thus, the n parameter values in the interpolation point between the corners are obtained as the sum of all Hermite function values in that interpolation point, weighted with the actual corner point function values and derivatives. The number of Hermite polynomials used depends on the number n of parameters and is equal to $(n + 1)\cdot 2^n$, e.g. 12 in case $n = 2$ and 32 in case $n = 3$, etc. An example for a representation of a two-dimensional surface is given in Fig. 6.43. The response surface is continuous and has continuous derivatives everywhere in the parameter space, since interpolation values depend on continuous functions and derivatives at the boundaries of and within the intervals considered. The use of spline functions to define directional derivatives in the corner points yields a relatively smooth surface over the whole response surface within the boundaries of the calculated node-points. It should be recognized that the type of response surface chosen is capable to fit very well even complicated input–output relationships. It requires a relatively large number of node-point evaluations and is, therefore, suitable only for problems where the parameter vector is not too large.

6.3.3.4 Stochastic Models of Basic Parameters and Data

We have seen that the spectral moments m_i depend on stochastic parameters H_V, T_V and ϑ. From Sect. 6.3.1.2 we recall that for reliability estimations stochastic models and possible correlations between these parameters must be established. Suitable stochastic models for H_V and T_V can be inferred from long-term seaway statistics which are available in the form of scatter diagrams representing relative frequencies $P(H_V, T_V)$ of observed values of H_V and T_V (cf. (5.109) and (5.110)). Based on data from [82], which, in our example, we have averaged over four of the original observation areas 15/16/24/25 to include the significant part of the

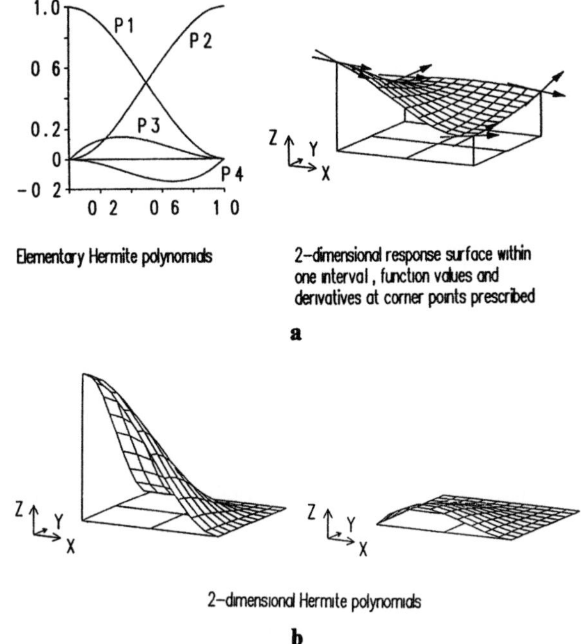

Elementary Hermite polynomials

2-dimensional response surface within one interval, function values and derivatives at corner points prescribed

2-dimensional Hermite polynomials

Fig. 6.43a, b. Example of a two-dimensional response surface represented by Hermite polynomials.

main East–West shipping routes of the North Atlantic, the parameters of the distribution of H_V and T_V can be estimated. From these data the correlation coefficient between H_V and T_V was estimated as $\rho_{HT} = 0.5$. The marginal probability distribution of H_V could well be approximated by a least-square Weibull fit

$$F(H_V) = 1 - \exp\left[-\left(\frac{H_V}{2.71}\right)^{1.54} \right]. \tag{6.185}$$

Here, H_V has the dimension of metre. It is interesting to observe that the marginal distribution of T_V could fit the Weibull distribution rather well (see Fig. 6.44)

$$F(T_V) = 1 - \exp\left[-\left(\frac{T_V - 2.0}{6.1}\right)^{4.4} \right]. \tag{6.186}$$

Here, I_V has the dimension of second. Figure 6.44 demonstrates that Weibull fits are acceptable representations of the North Atlantic seaway. The joint distribution of H_V and T_V is given as

$$F(H_V, T_V) = F(H_V) \cdot F(T_V | H_V).$$

Following [83], the conditional distribution $F(T_V | H_V)$ can be approximated as

$$F(T_V | H_V) = 1 - \exp\left[-\left(\frac{T_V - \kappa}{u(H_V)}\right)^{k(H_V)} \right] \tag{6.187}$$

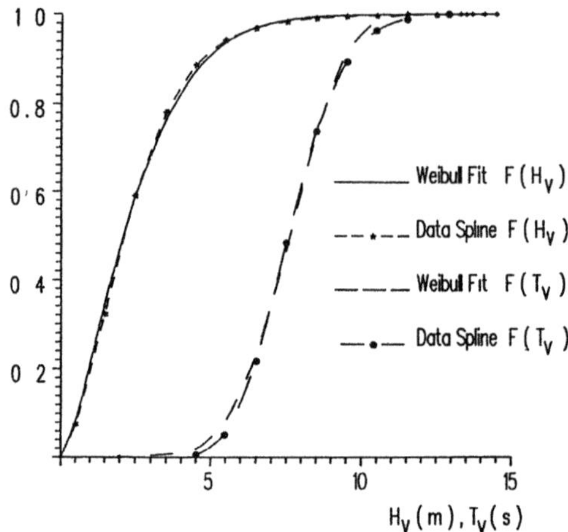

Fig. 6.44. Original data and Weibull fit for the marginal probability distribution function for H_V (significant wave height) and T_V (characteristic wave period).

where, for simplicity, the parameter κ is taken stochastically independent of H_V and

$$u(H_V) = a \exp [b\, H_V]. \tag{6.188}$$

$$k(H_V) = c \exp [d\, H_V]. \tag{6.189}$$

A maximum likelihood estimation of the parameters in (6.188) and (6.189) for the North Atlantic seaway under consideration verified the parameters in (6.185) and, further, gave $a = 5.7, b = 0.05, c = 4.6, d = 0.02$ and $\kappa = 2$.

A stochastic model of the angle ϑ between the main direction of the seaway and the orientation of the structure is obtained from

$$p(\vartheta) = \sum_{k=1}^{K} p(\gamma_k) p(\gamma_k + \vartheta) \tag{6.190}$$

with K discrete angles $\gamma_k, p(\gamma_k)$ being the probability of the main direction γ_k of the seaway with respect to North, and $p(\gamma_k + \vartheta)$ is the probability of the orientation of the structure with respect to North. Equation (6.190) is valid if ϑ is independent of γ_k. This assumption may be violated if safety measures with respect to the course are regularly taken by a ship's master in severe weather conditions. Considering the example of a containership on North Atlantic shipping routes, the stochastic model of $p(\gamma_k + \vartheta)$ is well approximated by a Bernoulli distribution with

$$p(\gamma_k + \vartheta = 90^\circ\, E) = p(\gamma_k + \vartheta = 90^\circ\, W) = 0.5. \tag{6.191}$$

With $K = 8$ angles γ_k used for the summation in (6.190), $p(\gamma_k)$ may be taken directly from [82]. Averaging again over the four North Atlantic observation areas 15/16/24/25 gives the probability $p(\vartheta)$ of Table 6.2.

The probabilities of the seaway from any direction deviate only slightly from the average value 0.125. Thus, the stochastic model for the angle between the main

Table 6.2. Probability distributions of angles of encounter

k	N	NE	E	SE	S	SW	W	NW
$p(\gamma_k)$	0.115	0.084	0.076	0.151	0.152	0.165	0.135	0.122
$p(\vartheta)$	0.106	0.125	0.134	0.137	0.106	0.125	0.134	0.137

direction of the seaway and the orientation of the ship is well enough approximated by a uniform distribution

$$F(\vartheta) = \frac{1}{2\pi}\vartheta \quad 0 \leqslant \vartheta < 2\pi. \tag{6.192}$$

Deterministic values used for the velocity and the loading condition yielding still-water stresses $\bar{\Sigma}$ can be found in Table 6.3.

Material parameter data should be modelled as random variables due to significant scatter of data obtained through tests. Basic parameters such as fracture toughness K_{Ic}, yield strength Σ_y and ultimate strength Σ_u are commonly modelled by a Weibull or log-normal distribution, while Σ_y and Σ_u are positively correlated with a coefficient of correlation $\rho_{yu} = 0.9$. If relevant data are available, other distribution types which fit the data should be used. The initial crack length A_0 is significantly related to the specific material, method of fabrication (cold-worked, surface-processed), type of connection (welded or bolted) and form of the component under consideration. In practice, the choice of the parameters of the distribution of the initial crack length and its type is one of the most difficult questions as it also involves the techniques for quality control and inspection and the definition of the termination of the crack initiation time. Due to the lack of specific information on such influences, a reasonable choice is presented in Table 6.3, as well as for the exponent $M = m$ in the Paris–Erdogan law (6.179), assumed to be deterministic. The factor C of the crack growth law (6.179) is modelled stochastically by a log-normal distribution (see Table 6.3 for the data).

Table 6.3. Stochastic model and parameter assumptions

Parameter	Stochastic model	Location parameter	Dispersion parameter
A_0	Rayl.	2 mm	—
m	—	4	—
C	Logn.	$5 \cdot 10^{-17}$	20%
K_{Ic}	Weib	100 MNm$^{3/2}$	15%
Y_c	—	1.12	—
Σ_y	Logn.	400 MPa	7%
Σ_u	Logn.	550 MPa	7%
H_V	Weib.	see (6.185)	see (6.185)
$T_V|H_V$	Weib.	see (6 187)	see (6.187)
ϑ	Unif	0–2π	—
$\bar{\Sigma}$	Unif.	80–100 MPa	—

6.3.3.5 Numerical Estimation of Fatigue Failure Probability as a Function of Time

The above outlined steps of analysis (Sect. 6.3.3.1 to 6.3.3.3) will now be illustrated using the stochastic models and data of Sect. 6.3.3.4 valid for the example of a modern container ship. The ship's geometry has already been represented by the finite element idealization in Fig. 6.42, and its main dimensions are collected in Table 6.4.

Stresses at highly loaded locations (hot spots) of the vessel's structure were calculated by the finite-element method. About 7000 nodal points and about 18 000 elements have been introduced to describe all strength relevant components of the structure. Plain stress elements (about 14 000) and truss elements (about 4000) have been used. A few beam elements were needed to represent, for instance, shipboard cranes and the rudder. Six support elements were needed for free support suppressing residual free body motions, which are due to numerical inaccuracies, but do not induce any relevant residual stresses in the ship structure. The finite element code used is a special version of SAP IV [81]. The global analysis was, however, too clumsy for accurately computing the stress concentrations in the considered hatch corner, so a detailed local FE model had to be developed to represent the neighbourhood of the hot spot in the hatch corner.

Stresses have been calculated for 15 different elementary wave periods T between 0.75 s and 40.0 s for unit wave amplitudes yielding the transfer functions $H_{\Sigma Z}(\omega, \phi)$. With the transfer functions the stress spectra $S_{\Sigma\Sigma}(\omega, \vartheta, H_V, T_V)$ were established according to (6.183) for all values of (H_V, T_V) of interest. Thus the spectral moments $m_{0/2/4}(\vartheta, H_V, T_V)$ required for the reliability calculations were determined according to (6.184) (Fig. 6.45).

Figure 6.45 may be regarded as the basic data set for the definition of response surfaces where the functions are approximations to cuts through these response surfaces for fixed angles ϑ. As explained in Sect. 6.3.3.3, response surfaces allow decoupling of the mechanical and the reliability analysis. Thus, in the next step, it is possible to evaluate (6.175) by advanced methods of reliability analysis making use of these response surfaces as quasi-analytical representations of the spectral moments m_i being functions of H_V, T_V and ϑ.

In the example, two service routes for the container ship were considered, which differ significantly with respect to their wave climate: the North Atlantic, as discussed above, and the Pacific. For the latter the stochastic models and data were obtained in a similar manner as already described in Sect. 6.3.3.4 for the North Atlantic. In Table 6.5 the fatigue failure probability $P_f(t)$ is given implicitly as the reliability index $\beta(t) = \Phi^{-1}(P_f(t))$ at different service times t, with Φ the standard normal distribution function (see (6.115)). The higher negative β-values represent the lower P_f-values.

Table 6.4. Main data of the container ship

L(m)	B(m)	H(m)	D(m)	Δ(t)	V(m/s) (av.)
157	23	11.8	8.9	20 514	7 0

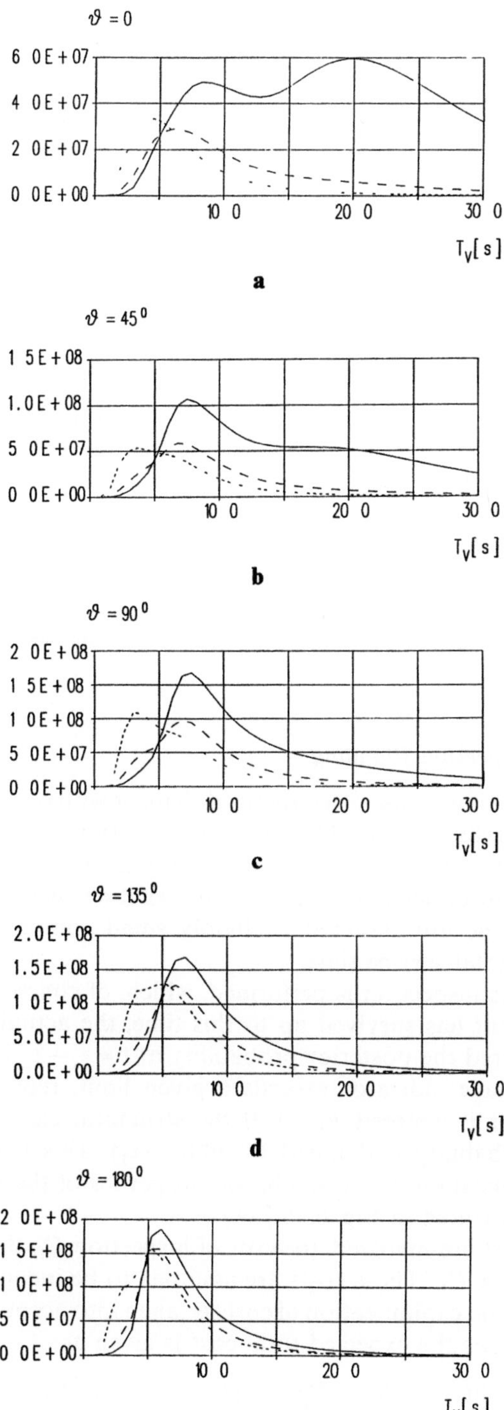

Fig. 6.45a–e. Spectral moments m_0/H_V^2 (solid line), m_2/H_V^2 (broken line) and m_4/H_V^2 (dotted line) as functions of the characteristic wave period T_V and five angles ϑ of encounter between the main direction of the seaway and the orientation of the structure.

Table 6.5. Reliability index β for different sea routes and service times

Route	Service Time	Days at sea	Location parameter H_V(m)	Dispersion parameter H_V(m)	$-\beta$
North	4y	800	2.71	1.54	6.02
Atlantic	8y	1600	2.71	1.54	5.97
	12y	2400	2.71	1.54	5 79
	15y	3000	2.71	1.54	3.89
Pacific	4y	800	2.12	2.01	7.62
	8y	1600	2.12	2.01	7.60
	12y	2400	2.12	2.01	7.47
	15y	3000	2.12	2 01	6 30

Table 6.5 indicates that there is a significant difference between the fatigue failure probability on different shipping routes. Thus, inclusion of statistical information on the wave climate as experienced by the ship on its actual route is one major concern of studies on adaptive inspection planning as described in the following section.

6.3.3.6 Adaptive Inspection Planning

We consider first the simple case of a structural element with one planned inspection in a pre-selected service time t_s. The time to inspection must be defined and a decision made whether or not to repair on a minimal cost basis. We assume that after repair the structural element has properties as if it was new. Further, inspection as well as repair time are assumed negligibly small compared to the time to inspection and the total service time.

If at time $t = t_1$ an inspection is performed, which, of course, is only possible if the structural element has survived up to this time, the actual state is observed. If no repair is required the posterior probability at $t = t_s - t_1$ is $P_f''(t_s - t_1)$. If this value or some damage indicator exceeds a given limit, repair is required. The probability of repair is denoted $P_R(t_1)$. If the structural element is repaired the residual failure probability is denoted by $P_f'''(t_s - t_1)$. This failure probability is based on a new realization of the stochastic properties of the structural element. The situation is illustrated in Fig. 6.46 [84].

Three types of cost are involved: the cost of inspection C_I, the cost of repair C_R and the cost of failure C_F. These costs are assumed to be independent of the time when they occur, i.e. no capitalization of costs is taken into account for convenience. Under such conditions the expected total cost is given as

$$C(t_1, t_s) = P_f'(t_1)C_f + [1 - P_f'(t_1)] \cdot [C_I + \{P_R(t_1)\} \cdot \{C_R + P_f'''(t_s - t_1)C_f\}$$
$$+ \{1 - P_R(t_1)\} \cdot \{P_f''(t_s - t_1)C_F\}]. \tag{6.193}$$

Because $P_f'(t_1)$ is increasing with t_1 as well as $P_R(t_1)$, and $P_f''(t_s - t_1)$ as well as $P_f'''(t_s - t_1)$ are decreasing with t_1 there must be a time t_1^* at which the total cost

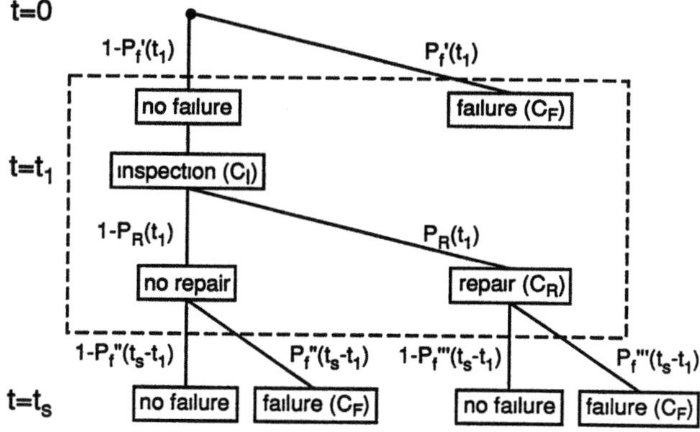

Fig. 6.46. Decision tree for inspection planning.

has a local minimum. Thus, an optimal value for the time to inspection can be determined.

The structural element is inspected at time t_1^* and actual observations become available upon which a decision about repair is taken according to a prespecified rule. After repair at time t_1^* the structural element is treated as new. In case of no repair the crack state observed during inspection (with or without measurement error) is taken as the initial crack state for the analysis of the interval to a second inspection. Proceeding in this manner establishes an adaptive scheme for planning of times to inspection, which takes account of all available information.

The failure probability $P_f''(t_s - t_1)$ can be calculated by modifying $P_f'(t_s)$ under the condition that the structural element has survived up to time t_1

$$P_f''(t_s - t_1) = P_f'(t_s) - P_f'(t_1). \tag{6.194}$$

This probability function of the time t_1 to the first repair is not updated, because the crack length at time t_1 is yet unknown. If, however, inspection is performed at time t_1 a new failure probability function $P_{f1}'(t - t_1)$ for time $t \geqslant t_1$ has to be calculated. Let $E(t_1)$ be the inspection event at time t_1, then the probability of failure at time of the first inspection and before any further inspection is calculated as

$$P_{f1}'(t - t_1) = P[F(t - t_1)|E(t_1)] = \frac{P[F(t - t_1) \cap E(t_1)]}{P[E(t_1)]}. \tag{6.195}$$

Nominator and denominator of the right hand side of this equation can be separately evaluated using modern reliability analysis software (e.g. [85]), which includes most advanced theoretical and numerical developments of the background on reliability analysis presented in Sect. 6.3.1.2. We mention in passing that the failure probability in the nominator is that of a 2-element parallel system discussed in Sect. 6.3.2.1.

According to [86], it is useful to distinguish between two types of crack observations. Let $\hat{a}_1$ be an observed crack length and $A(t|r, q)$ the crack length at time t according to the calculation model (6.180). (Compare (6.174) or (6.175) for the meaning of '$|r, q$' as realizations of '$|\mathscr{R}, \mathscr{Q}$'.) The observation events are then formulated as equality and inequality constraints

$$E_a(t_1) = \{(A(t_1|r, q) - \hat{a}_i = \varepsilon_i)\}. \tag{6.196}$$

where ε_i is the measurement error in the observation, or

$$E_{na}(t_1) = \{(A(t_1|r, q) - \hat{a}_b \leqslant \varepsilon_b)\}. \tag{6.197}$$

if no crack is observed. Asymptotic first- and second-order results for the evaluation of probabilities of events with equality constraints (6.196) were outlined for the first time in [87], and studied more intensively in [77]. Modern reliability software provides numerical solution schemes for this particular problem. In (6.197) $\hat{a}_b$ is the lowest detectable crack size and ε_b is the corresponding measurement error depending on the particular inspection method, as before.

Whether or not a structural element is repaired after inspection depends on the result of the actual observation. The necessary decision rule can, in principle, be based on the potential gain in residual reliability. A simpler and probably more practical criterion uses the measured crack length directly. An example of such a crack-length based decision rule D is given as

$$D(t_1) = \begin{cases} \text{repair} & \text{if } A(t_1|r, q) \geqslant a_{c,rep} \\ \text{no repair} & \text{if } A(t_1|r, q) < a_{c,rep} \end{cases}. \tag{6.198}$$

The critical repair crack length $a_{c,rep}$ needs to be predefined. Since the event of repair depends on the decision rule, the probability of repair is

$$P_R(t_1) = P[R(t_1)]. \tag{6.199}$$

$$R(t_1) = \{a_{c,rep} - A(t_1|r, q) \leqslant 0\}. \tag{6.200}$$

To clarify further the effect of repair (or replacement), the simplest model is to assume independence of the properties between the previous and the repaired condition of the structural element. The probability of failure after repair is then

$$P_f'''(t_s - t_1) = P_f(t_s - t_1). \tag{6.201}$$

Note that other repair strategies are possible and can also be incorporated.

Based on the results obtained for the fatigue failure probabilities of a hatch corner of a container ship (see Table 6.5) of the previous section, in Figs. 6.47 and 6.48 calculated probability functions are given for the North Atlantic and Pacific shipping routes, respectively. Abbreviations are chosen as in Fig. 6.46, with $P_f'(t_1) \equiv P_f(t = t_1)$ from the evaluation of (6.175): $P_f'(t_1)$ is the failure probability during $[0, t_1]$ of the structural element without inspection.

$P_f'(t_1)$ and $P_R(t_1)$ increase continuously with t, while $P_f''(t_1)$ and $P_f'''(t_1)$ tend to zero for $t_1 = t_s = 12$ years. Any other service time t_s would be a valid assumption, e.g. $t_s = 15$ years. Since greater service times had no significant effect on the optimal time to the first inspection t_1 but the appropriate graphical presentation of results became difficult, we used $t_s = 12$ years in the example.

Repair is performed if the crack size measured at time of inspection exceeds

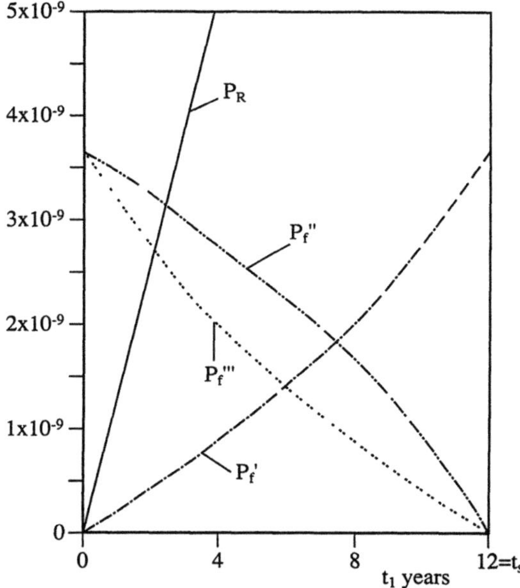

Fig. 6.47. Probability functions for the North Atlantic shipping route.

$a_{c,rep} = 15$ mm. Figure 6.48 indicates that there is always a small probability of repair P_R (i.e. there is a probability of a too large crack length according to (6.199) and (6.200) even for a new ship with time to inspection $t_1 = 0$).

To evaluate (6.193) the following cost assumptions were made in [84]:

$$C_I = 0 \quad C_R = 10^6 \quad C_F = 10^8.$$

The inspection cost C_I is constant with time and comparatively low. C_I was set to zero to make possible a clear graphical presentation of probable cost functions in Figs. 6.49 and 6.50, where the following abbreviations are used:

— $C' = P_f' \cdot C_F$ (probable cost of failure before inspection)
— $C'' = [1 - P_f'(t_1)] \cdot [\{1 - P_R(t_1)\} \cdot \{P_F''(t_s - t_1)C_F\}]$ (probable cost of no failure before inspection, but no repair after inspection and failure)
— $C''' = [1 - P_f'(t_1)] \cdot [\{P_R(t_1)\} \cdot \{C_R + P_f'''(t_s - t_1)C_F\}]$ (probable cost of no failure before inspection, but repair after inspection and failure)
— $C = C' + C'' + C''' + [1 - P_f(t_1)] \cdot C_I$ (expected total cost, i.e. sum of all probable costs, including probable inspection cost at t_1).

The expected total cost C increases with $t_1 > t_1^*$, which must be attributed to the strongly increasing probable cost functions $C'(t_1)$ and $C'''(t_1)$. Increasing the probable repair cost moves the minimum to the left, which strongly contributes to the result that the optimal inspection time for the North Atlantic shipping route is shorter than for the Pacific. Increasing the repair threshold moves the minimum to the right because large cracks can only be observed at comparatively late times. Decreasing the inspection and repair cost also moves the minimum to the right.

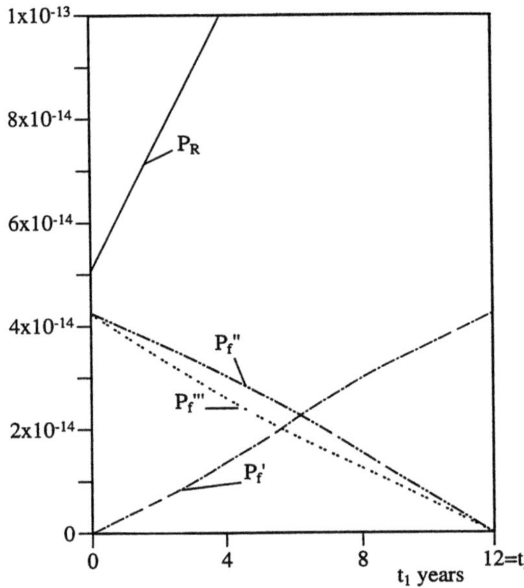

Fig. 6.48. Probability functions for the Pacific shipping route (Australia–USA).

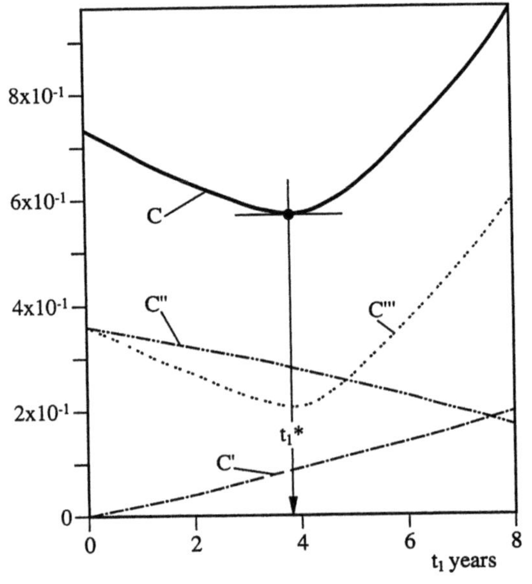

Fig. 6.49. Probable cost functions for the North Atlantic shipping route.

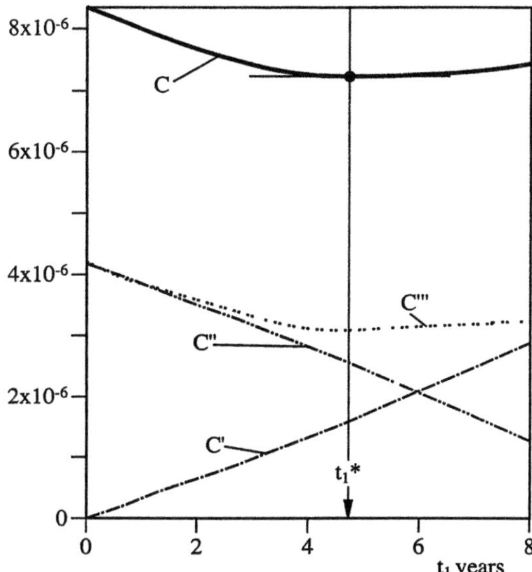

Fig. 6.50. Probable cost functions for the Pacific shipping route (Australia–USA)

According to Figs. 6.49 and 6.50 first inspections would be optimal at $t_1^* = 3.8$ or 4.7 years, respectively. This result, if generalized, leads to the conclusion that first optimal inspection of the hatch corner could be delayed by nearly one year for continuous operation in the Pacific compared to continuous operation in the North Atlantic. We thus conclude that adaptive inspection planning for offshore structures is as sensible as for ships [75].

After the first inspection, various actions should be based on the decision concept that we recalled from [74] in the previous chapter. These details were extensively discussed in [74], and will not be repeated here.

This concept of adaptive inspection and repair planning is based on a minimal cost principle. It aims at rational decisions on inspection intervals and on the amount of inspections and repairs during a ship's service time. Such rationalization of inspection strategies with application of modern methods of reliability analysis will probably be adopted in the future for effective control of operations of marine structures (see [88]) for alternative inspection strategies and further applications to offshore structures.

6.4 List of Symbols

The list of symbols includes some symbols which are used conventionally in the specialist literature or which are given additional clarification by the definition used here. Derived symbols (e.g. $E[A]$, m_A etc.) are after X (e.g. $E[X]$, m_X, etc.).

A	stochastic stress parameter (demand)
A_0	stochastic parameter of initial crack length

$A(t)$	stochastic parameter of the envelope of a seaway process at time t
$A(t)$	stochastic parameter of crack length at time t
$\mathrm{Cov}[X_1, X_2]$	covariance of random parameter X_1 and X_2
C	covariance matrix
C	Paris–Erdogan parameter
C	cost of failure
D	random parameter with realization d_i for indicating discrete states
D	diameter
D	dynamic amplification factor
D	damage indicator
D_R	damage indicator in reference time period T_R
D_{Rij}	damage in seaway class (H_{vi}, T_{vj}) within T_R
D_s	damage indicator at stress range level s
$E[X]$	expectation of random parameter X, $E[X] = m_X$
$E[X]$	vector of expectations
F	failure indicator
F	stochastic excitation force
$F_X(x)$	probability distribution function: $P[X \leqslant x]$
$F^{(2)}$	stochastic second-order excitation force
G	stochastic parameter of structure function $g(X)$
G	weight (platform)
H	wave height
H_v	observed wave height
H_{vi}	observed wave height representative of i-th seaway class
$H_{XY}(\omega)$	complex transfer function between the excitation process $\{Y(t)\}$ and the response process $\{X(t)\}$
$H_{1/3}$	significant wave height
$H_{1/3i}$	significant wave height representative of the ith seaway class
I	maximum value of i, e.g. number of seaway classes $H_{1/3i}$ or H_{vi}
J	maximum value of j, e.g. number of seaway classes T_{1j} or T_{vj}
K	maximum value of k, e.g. number of stochastic parameters X_k
K	stress concentration factor
K_{Ic}	fracture toughness
L	maximum value of l, e.g. number of stress ranges
L	stochastic sectional load parameter
L	characteristic (reference) length of a structure
M	maximum value of m, e.g. number of plastic hinges
M	Paris–Erdogan parameter
N	maximum value of n, e.g. number of degrees of freedom
N	integer random parameter (occurrence number)
$N(s)$	number of endurable stress ranges of magnitude s, $N(s) = N_s$
P_{SE}	up-crossing probability of x_S in T_E
P	matrix of transition probabilities p_{ij} from state d_i to d_j
$P[X]$	probability $P[x < X \leqslant x + dx] = f_X(x)dx$
$P[X_i]$	probability $P[x \in X_i]$
P_i	probability $P[H_{vi}]$ for seaway observations from the seaway class with the parameter H_{vi}

P_{ij}	probability $P[H_{Vi}, T_{Vj}]$ for seaway observations from the seaway class with the parameters H_{Vi} and T_{Vj}
Q	risk (index S: series system, P: parallel system, index R: redundant system)
$Q(s)$	mooring force characteristic of mooring system
R	reliability
$R_{XX}(\tau)$	autocorrelation function of stationary random process $\{X(t)\}$
$R_{XY}(\tau)$	cross-correlation function of stationary random processes $\{X(t)\}$ and $\{Y(t)\}$
$R_{ZZ}(\tau)$	autocorrelation function of seaway process $\{Z(t)\}$
$R_{A2A2}(\tau)$	autocorrelation function of random process $\{A^2(t)\}$
$R_{Z2Z2}(\tau)$	autocorrelation function of random process $\{Z^2(t)\}$
S	stochastic motion parameter
S	stochastic stress range parameter
$S_{XX}(\omega)$	autospectrum (spectrum) of stationary random process $\{X(t)\}$
$S_{XY}(\omega)$	cross-spectrum of stationary random processes $\{X(t)\}$ and $\{Y(t)\}$
$S_{ZZ}\{\omega\}$	seaway spectrum (autospectrum of stationary random process $\{Z(t)\}$)
$S_{A2A2}(\omega)$	autospectrum of random process $\{A^2(t)\}$
$S_{Z2Z1}(\omega)$	autospectrum of random process $\{Z^2(t)\}$
T	time interval (index n or k for number of years)
T	wave period, with harmonic oscillation $2\pi/\omega$
T	'transposition', when in exponent of vectors or matrices
T_B	design time (order of magnitude: years or decades)
T_E	service time ($< T_B$)
T_R	reference time (normally 1 year)
T_S	safety time ($> T_B$)
T_V	observed wave period
T_{Vj}	observed wave period representative of j-th seaway class
T_1	mean wave period ($= 2\pi/\omega_1$)
T_{1j}	mean wave period T_1 representative of j-th seaway class
T_0	mean time between successive zero up-crossings in the seaway process $\{Z(t)\}$
U	standardized random parameter $(X - m_X)/\sigma_X$
U	stochastic parameter of wave particle velocity
V	failure domain
V	stochastic parameter of generalized (modal) excitation forces
$\mathrm{Var}[X]$	variance (scatter) of random parameters $X(= \sigma_X^2)$
W	stochastic strength parameter (resistance)
W	stochastic parameter of generalized (modal) motion
X	random parameter, with derived values in this list of symbols also representative of $A, E, L, U, V, W, Y, Z, \Sigma$
X	random vector: $(X_1, X_2, \ldots, X_K)^T$
$\{X(t)\}$	random process of random value X
Y	random parameter representing general loads F, S
$Y(t)$	geometry factor
Y_c	constant geometry factor

Z	stochastic safety margin $Z = W - A$
Z	stochastic wave elevation (deviation from level of still water)
$\{Z(t)\}$	seaway process (random process of wave elevation)
a	crack length (index i: start, index f: end)
a	realization of A
$a(t)$	envelope of a random sample $\zeta(t)$
b	wave climate parameter of Weibull distribution of H
d	damping coefficient
d	water depth
f	excitation force
f_{WS}	sum of wind and current force
$f_X(x)$	probability density function of random value X
$f^{(2)}$	second-order excitation force
g	acceleration due to gravity, $g = 9.81\,\mathrm{m/s^2}$
g	structure (strength) parameter
$g(X)$	structure function in the X-domain (index L: linearized)
h	realization of stochastic value H (e.g. wave height)
$h(U)$	structure function in the U-domain (index L: linearized)
i	sequential index ($i = 1, 2, \ldots, I$), e.g. class index for wave heights $H_{1/3_i}$ or H_{V_i}
i	$\sqrt{-1}$, imaginary number
j	sequential index ($j = 1, 2, \ldots, J$), e.g. class index for wave periods T_{1j} or T_{V_j}
k	sequential index ($k = 1, 2, \ldots, K$), e.g. basic parameters X_k
k	S–N curve parameter
k	stiffness coefficient
l	sequential index ($l = 1, 2, \ldots, L$), e.g. class index for stress ranges s_1
l	sectional load
m	designation index ($m = 1, 2, \ldots, M$), e.g. for plastic hinges
m	mass
m	S–N curve parameter
m_{H}	hydrodynamic mass
m_X	mean value of random parameter X, $m_X = E[X]$
m_i	ith moment of spectrum
$\mathbf{m_X}$	vector of mean values: $(m_{X1}, m_{X2}, \ldots, m_{XK})^T$
n	sequential index ($n = 1, 2, \ldots, N$), e.g to indicate the n-th eigen mode or the n-th degree of freedom
n_{R}	number of waves in T_{R}
n^+	short form of $n_X^+(x)$
$n_X^+(x)$	number of up-crossings of value x (additional indices B, E, S for $x = x_{\mathrm{B}}; x_{\mathrm{E}}; x_{\mathrm{S}}$)
$n(s)$	number of stress ranges of magnitude $s = n_s$
p_{ij}	transition probability from state d_i to d_j
$q(s)$	mooring force characteristic of mooring element (chain or cable)
r_i	relative motion in direction i
s	motion values, $\dot{s} = \partial s/\partial t$, $\ddot{s} = \partial^2 s/\partial t^2$

s	stress range (index R for reference value of s in T_R)
$s_{1/3}$	significant motion amplitude
t	time variable
t_s	service time
u	realization of standardized random value U
u	velocity of wave particle, $\dot{u} = \partial u/\partial t$
v	generalized (modal) excitation force
w	realization of W (resistance)
w	normalized eigen frequency
w	generalized (modal) motion value
x	realization of random value X
x	horizonal coordinate axis, vertical to y, z
x_B	design value of X (exceeded once in T_B)
x_E	service value of X (exceeded once in T_E)
x_N	nominal value of X (code value)
x_S	safety value of X (exceeded once in T_S)
$x_{1/3}$	significant value of X (mean of 1/3 highest values)
x_k^*	component of design point vector: $x^* = (x_1^*, x_2^*, \ldots, x_K^*)^T$
x	vector: $x = (x_1, x_2, \ldots, x_K)^T$
(x, y, z)	orthogonal coordinate system (right-hand system)
y	horizontal coordinate axis, vertical to x, z
y	horizontal deviation (deflection)
z	vertical coordinate axis, vertical to x, y
z	realization of Z (safety margin, wave elevation)
Γ	stochastic safety factor, $\Gamma = W/A$
$\Gamma\{\cdot\}$	gamma function
Δ	damage indicator for failure
Φ	standard normal distribution function
Σ_E	random parameter of strength threshold
Σ_p	random parameter of peak stress
Σ_u	random parameter of ultimate stress
Σ_y	random parameter of yield stress
$\bar{\Sigma}$	random parameter of mean stress level
Σ	random stress parameter
α	width parameter of the spectrum
α	sensitivity factor
$\alpha^2(\omega)$	drift force coefficient for wave excitation with angular frequency ω
α_0^2	spectral mean of $\alpha^2(\omega)$
β	reliability index
γ	realization of γ (safety factor $\gamma = w/a$)
δ	damping ratio
δ	Dirac δ function
ε	relative change in length
ζ	wave elevation (deviation of wave contour from level of still water)
$\zeta(t)$	random sample of seaway process $\{Z(t)\}$
ζ_B	design value of wave elevation (exceeded once in T_B)

ζ_S	safety value of wave elevation (exceeded once in T_S)
$\zeta_a(\omega)$	real amplitude of an elementary wave of frequency ω
θ	time interval
ϑ	angle between structure and main direction of seaway
λ	eigenvalue
λ	rain-flow correction
v	event rate (index n for reference time interval T_n)
$v_X^+(x)$	up-crossing rate of value x in random process $\{X(t)\}$
$v_X^+(0)$	zero up-crossing rate in random process $\{X(t)\}$
v_{0ij}^+	zero up-crossing rate ($= v_0^+$) in seaway class with the parameters H_{Vi} and T_{Vj}
π	circumference of circle of unit diameter ($= 3.1415926535\ldots$)
$\pi_x(i)$	probability of reaching a state d_i at time $t = x$ (state probability)
π_x	vector of state probabilities $\pi_x(i)$
ρ	density
ρ	correlation coefficient
σ	normal stress (index B: bending stress, index N: tensile/compression stress, index BN: bending and tensile/compression stress)
σ_X	standard deviation of random process $\{X(t)\}$
ς	characteristic wave steepness
τ	time interval
ϕ	elementary wave direction
$\varphi(u)$	standard normal distribution density (standardized Gaussian normal distribution density)
ω	angular frequency, with harmonic oscillation $2\pi/T$
ω_0	natural angular frequency of an oscillating system, reference frequency for a pure amplitude modulation, $2\pi/T_0$
ω_1	angular frequency of centre of gravity of a seaway spectrum
∇	differential operator ('Nabla')

References

1   Clauss GF. Stability and dynamics of semisubmersibles after accidental damage. OTC 4729 (1984): 159–171

2   Grim O 'Sea Troll' Grenzarbeitsbedingungen eines Kranschiffes im Seegang (Limiting operating conditions for a crane vessel in the seaway) Inst. Schiffbau Rep. No 355 (1977)

3   Schellin TE. Motions and mooring study of a 3500 t crane vessel Germanischer Lloyd Report STB-1261 (1985)

4   Scharrer M, Schellin TE Pipe laying barge, motions and accelerations in waves Germanischer Lloyd Report STB-1480 (1986)

5   Östergaard C, Payer H. Rationale Beurteilung der Festigkeit von Halbtauchern (Rational evaluation of the strength of semisubmersibles) Proc Schiffbautech. Ges 76 (1973) 141–159

6   Tagaki M, Arai S-I A comparison of methods for calculating the motion of a semi-submersible. Tech. Res Inst. Rep., Hitachi Zosen Corp (1984)

7   Östergaard C, Schellin TE. Comparison of experimental and theoretical wave actions on floating and compliant offshore structures Appl. Ocean Res (1987) 9: 192–213

8   Ogilvie TF. Second order hydrodynamic effects on ocean platforms Proc Int Workshop on Ship and Ocean Platform Motion, Berkeley (1983) 205–265

9 Maruo H. The drift of a body floating in waves. J. Ship Res. (1960) 4: 1–10

10 Newman JN. Second order slowly varying forces on vessels in irregular waves. Proc. Int. Symp. Dynamics of Marine Vehicles and Structures in Waves, Univ. College London (1974) 182–186

11 Schellin TE, Scharrer M, Mathies G. Analysis of vessels moored in shallow, unprotected waters. Offshore Technol. Conf., OTC 4243 (1982) 161–176

12 Bracewell RN. The Fourier Transform and its application. 2nd ed. New York: McGraw-Hill (1978)

13 Papoulis A. Probability, random variables, and stochastic processes. 2nd ed. New York: McGraw-Hill (1984)

14 Pinkster JA. Low frequency phenomena associated with vessels moored at sea. Soc. Petrol. Eng. 4837 (1974)

15 Schellin TE, Scharrer M. Auslegungsprinzipien für Ankersysteme (Design principles for anchoring systems). Hansa No. 118 (1981) 432–438

16 Clauss GF, Schellin TE. Slow drift forces and motions on a barge type structure comparing model tests with calculated results. Offshore Technol. Conf, OTC 4435 (1982) 677–692

17 Guedes Soares C, Moan T. On the uncertainties related to the extreme hydrodynamic loading of a cylindrical pile. Reliability theory and its application in structural & soil mechanics. The Hague: Martinus Nijhoff (1987) 351–364

18 Borgman LE. Spectral analysis of ocean wave forces on piling. J. Waterways Harbors Div., Proc. ASCE (1967) 129–156

19 Madsen HO, Krenk S, Lind NC Methods of structural safety. New Jersey: Prentice-Hall (1986)

20 Barltrop NDP, Adams AJ Dynamics of fixed marine structures. Oxford: Butterworth-Heinemann (1991)

21 Malhotra AK, Penzien J. Response of offshore structures to random wave forces J. Struct. Div., Proc. ASCE (190) 2155–2173

22 Karadeniz H Spectral analysis and stochastic fatigue reliability calculation of offshore steel structures. Offshore Struct. Anal. Rep., Delft Univ. of Techn., Dept. C.E. (1983)

23 Hutchinson B Risk and operability analysis in the marine environment. Proc. Soc. Nav. Arch. Mar. Eng. 3 (1981) 127–149

24 Östergaard C, Schellin TE The use of risk analysis to evaluate the seaworthiness of offshore structures. Proc. Brasil Offshore (1979) (1st Reprint. Appl. Ocean Res. 2 (1980) 113–117; 2nd Reprint: Probabilistic Offshore Mechanics Southampton. CML Publ. (1985) 92–96)

25 Östergaard C, Schellin TE. On safety of offshore structures in the sea. ASCE Proc. Civ. Eng in the Ocean IV 2 (1979) 918–931

26 Östergaard C, Schellin TE. An approach to assess the seaworthiness of offshore structures using risk analysis. Schiff und Hafen (1979)12: 1070–1074

27 Östergaard C, Schellin TE. Berechnung der Seegangseigenschaften des schwimmenden LNG-Verflussigungs- und Lagersystems SEAGAS (Calculation of seaway characteristics of the floating LNG liquefaction and storage system SEAGAS). Germanischer Lloyd Report STB 564 (1978)

28 Östergaard C, Schellin TE, Sükan M. Zur Sicherheit von Seebauwerken – Hydrodynamische Berechnungen fur kompakte Strukturen (Safety of offshore structures – hydrodynamic calculations for compact structures). Schiff und Hafen (1979) No. 1 71–76

29 Pflugbeil C, Schafer P, Walden H. Wave observations by ship-borne German weather stations in the North Sea 1957 to 1966. Deutscher Wetterdienst No. 75 (1971)

30 Hogben N, Lumb FE Ocean wave statistics. Nat. Phys. Lab., HMSO (1976)

31 Veneziano D, Grigoriu M, Cornell CA. Vector-process models for system reliability. J. Eng. Mech. Div. (1977) (EM3) 103: 441–460

32 Rackwitz R, Fiessler B. Structural reliability under combined random load sequences. Computers and Structures (1978) 9. 489–494

33 Breitung K, Rackwitz R. Nonlinear combination of load processes. J. Struct. Mech. (1982) 10. 145–166

34 Miner MA. Cumulative damage in fatigue. ASME J Appl. Mech (1945) 12: A159–A164

35 Paris PC, Erdogan F. A critical analysis of crack propagation laws. ASME J. Basic Eng. (1963) 85· 528–534

36 Paris PC, Sih GC. Stress analysis of cracks. Fracture toughness testing and its applications. ASTM STP (1965) 381: 30–82

37   Gupta A. Singh RP. Fatigue behaviour of offshore structures. Berlin: Springer-Verlag (1986)

38   Crandall SH, Mark WD, Khabbaz GR. The variance in Palmgren–Miner damage due to random vibration. Proc. 4th U.S. Nt. Congr Appl. Mech. (1962) 119–126

39   Guers F, Rackwitz R. On the calculation of upcrossing rates for narrow-band Gaussian processes related to structural fatigue. Reports on Reliability of Structures 79. Munchen: Laboratorium fur den konstruktiven Ingenieurbau (1986)

40   Guers F, Rackwitz R Time-variant reliability of structural systems subject to fatigue. Proc. ICASP (1987) 497–505

41   Bogdanoff JL, Kozin F. Probabilistic models of cumulative damage. New York: Wiley (1985)

42   Marshall PW. Tubular joint design. Planning and design of fixed offshore platforms New York. Van Nostrand Reinhold (1986) Chap. 18

43   Marshall PW. Steel selection for fracture control. Planning and design of fixed offshore platforms. New York: Van Nostrand Reinhold (1986) Chap. 19

44   Wirsching PH. Shehata AM. Fatigue under wide band random stresses using the rain-flow method. ASME J. Eng. Mat. Techn. (1977) 2340–2356

45   Wirsching PH. Fatigue reliability for offshore structures Journ. Struct. Eng. (1984) No. 2340–2356

46   Kan DKY, Petrouskas C. Hybrid time-frequency domain fatigue analysis for deep water platforms Offshore Technol. Conf., OTC 3965 (1981) 127–143

47   Nolte KG, Hansford JE. Closed-form expressions for determining the fatigue damage of structures due to ocean waves. Offshore Technol. Conf, OTC 2606 (1976) 861–872

48   Luyties WH, Geyer JF. The development of allowable fatigue stresses in API RP2A. Offshore Technol Conf., OTC 5555 (1987) 47–57

49   Ditlevsen O. Uncertainty modelling. New York McGraw-Hill (1981)

50   Schuéller GI. Einfuhrung in die Sicherheit und Zuverlassigkeit von Tragwerken (Introduction to safety and reliability to load-bearing structures). Munchen: Ernst & Sohn (1981)

51   Thoft-Christensen P, Baker MJ. Structural reliability theory and its applications New York· Springer-Verlag (1982)

52   Augusti G, Baratta A, Casciati F Probabilistic methods in structural engineering. London· Chapman & Hall (1984)

53   Östergaard C. Zuverlässigkeit von Konstruktionen (Reliability of structures). In: Handbuch der Werften, Bd. XVI and XVII. Hamburg: Hansa Part I (1982) and Part II (1984)

54   Ang AH-S, Tang WH. Probability concepts in engineering planning and design. New York· Wiley Vol. I (1975), Vol. II (1984)

55   Thoft-Christensen P, Murotsu Y. Application of structural systems reliability theory. New York Springer-Verlag (1986)

56   Fujita M, Schall G, Rackwitz R. Time variant component reliabilities by FORM-SORM and updating by importance sampling. Rel. & Risk Analysis in Civil Eng. 1, Ed. Lind NC, Inst. Risk Res. Univ. of Waterloo (1987) 520–527

57   Fießler B Das Programmsystem FORM zur Berechnung der Versagenswahrscheinlichkeit von Komponenten von Tragsystemen (The FORM program system for the calculation of failure probability of load-bearing structures). Ber. zur Zurverlassigkeitstheorie der Bauwerke 43. Munchen· Laboratorium fur den konstruktiven Ingenieurbau (1979)

58   Parkinson DB Computer solution for the reliability index. Eng. Struct. (1980) 2: 57–62

59   Hohenbichler M, Rackwitz R. Non-normal dependent vectors in structural safety J. Eng. Mech Div., ASCE (EM6) (1981) 107· 1227–1238

60   Rackwitz R Analytische Verfahren 4 (Baukonstruktionen) (Analytical methods 4 (structural design)) In· Peters OH, Meyna A (eds.) Handbuch der Sicherheitstechnik 1 (Handbook of safety technology). Munchen Hanser (1985)

61   Chen Y. On the reliability analysis of structural systems. Inst Mechanik Univ. Innsbruck. Internal Working Rep. 1 (1986)

62   Frangopol DM, Nakib R. Accuracy of methods for structural system reliability evaluation. Eng. Comp. (1987) 4: 90–103

63   Stahl B, Geyer JF. Ultimate strength reliability of tension leg platform tendon systems. Offshore Technol. Conf., OTC 4857 (1985) 151–160

64   Watwood VB. Mechanism generation for limit state analysis of frames. J. Struct Div, ASCE 1090, (1979) ST1 1–15

65   Murotsu Y Combinatorial properties of identifying dominant failure paths in structural systems
     Bull. Univ. Osaka Pref, (1983) Ser A. 32: 107–116

66   Ditlevsen O. Generalized second moment reliability index. J Struct Mech. (1979) 7· 435–451

67   Östergaad C Anwendung der Zuverlassigkeitstechnik auf tragende Konstruktionen (Application
     of reliability techniques to load-bearing structures). Continuing Education Course 22, Institute of
     Naval Architecture (IFS), Hamburg (1986)

68   Koch T, Östergaard C. Methodik und Bestimmung dominanter Versagensformen der Rahmen-
     konstruktion von Großtankern bei Strandung auf annahernd ebenem Meeresboden (Methodology
     and determination of dominant failure mechanisms of frame structures of large tankers during
     stranding on an approximately level sea bed). Germanischer Lloyd (1985)

69   Guenard YF. Application of system reliability analysis to offshore structures. PhD Thesis. The J.
     Blume Earthq. Eng. Center. Stanford Univ Rep. No. 71 (1984)

70   Zuverlássigkeit der Bauwerke (Reliability of Structures). Final Colloquium SFB 96. Reports on
     Reliability of Structures 81 (in German). München: Laboratorium für den konstruktiven
     Ingenieurbau (1986)

71   Ximenes MCC. Fatigue reliability and inspection of TLP tendon system. Marine Technology
     (1991) 28(2): 99–110

72   Leadbetter MR, Lindgren G, Rootzen H. Extremes and related properties of random sequences
     and processes. New York. Springer-Verlag (1983)

73   Schall G, Faber M, Rackwitz R. Investigation of the ergodicity assumption for sea states in the
     reliability assessment of offshore structures. Proc OMAE, Houston (1990)

74   Fujita M, Schall G, Rackwitz R. Adaptive reliability-based inspection strategies for structures
     subject to fatigue. Proc. ICOSSAR, San Francisco (1987)

75   Bryla P, Faber MH, Rackwitz R. Second order methods in time variant reliability problems. Proc.
     OMAE, Stavanger (1991)

76   Schall G, Scharrer M, Östergaard C, Rackwitz R. Fatigue reliability investigation for marine
     structures using a response surface method. Proc. OMAE, Stavanger (1991)

77   Schall G, Gollwitzer S, Rackwitz R. Integration of multi-normal densities on surfaces Proc. 2nd
     Working Conference on Optimization of Structural Systems, London (1986)

78   Milne I, Ainsworth RA, Dowling AR, Stewart AT Assessment of the integrity of structures
     containing defects. Int. J. Pressure Vessels Piping (1988) 32 (188) 3–104

79   Abdo T, Rackwitz R Discussion to Kam, JCP and Dover WD. Fast fatigue assessment procedure
     for offshore structures under random stress history Proc. ICE (1989) 2(87): 645–649

80   Papanikolaou AS, Schellin TS, Zaraphonitis G 3-D method to evaluate motion and loads of
     ships with forward speed in waves. Proc. 5th. Congress on Marine Technology, Athens (1990)

81   Bathe KH, Peterson F, Wilson E SAP IV, a structural analysis program for static and dynamic
     response of linear systems (EERC 73-11). University of California, Berkeley (1974)

82   Global Wave Statistics. Feltham: British Maritime Technology (BMT) (1985)

83   Houmb OG, Overvik T. Parameterization of wave spectra and long term joint distribution of wave
     height and period. Proc. 1st Int. Conf on the Behaviour of Offshore Structures BOSS, Trondheim
     (1976)

84   Schall, G. Östergaard C Planning of inspection and repair for ship operation. Proc. SSC/SNAME
     Marine Structural Inspection, Maintenance, and Monitoring Symposium, Washington (1991)

85   NN. STRUREL; A Structural Reliability Analysis Program (Manual) Munchen: Reliability,
     Consulting, Programs (1991)

86   Madsen HO. Deterministic and probabilistic models for damage accumulation due to time varying
     loading. Danmarks Ing. Academy (DIALOG 5-82), Lyngby (1982)

87   Madsen HO. Random fatigue crack growth and inspection. Proc ICOSSAR, Kobe (1985).

88   Madsen HO, Torhaug R, Cramer EH. Probability-based cost benefit analysis of fatigue design,
     inspection and maintenance. Proc SSC/SNAME Marine Structural Inspection, Maintenance, and
     Monitoring Symposium Washington (1991)

89   Department of Energy: Guidance on the design and construction of offshore installations. Part
     II Sect. 4, (1974/77).

7. Dimensioning of Marine Steel Structures

Dimensioning involves, on the one hand, a multiplicity of regulations, guidelines and standards, and on the other, the application of rational analysis and evaluation procedures, which have been dealt with in detail in Chap. 3–6. The method of dimensioning on the basis of regulations, often used in design practice mainly to simplify the official approval process, is in many ways founded on rational analysis and evaluation procedures. Generally, it differs from rational procedures not in its principle, but in its form, which is that of a greatly simplified, and therefore at times confusing or even incomprehensible analysis procedure. For this reason, in this chapter we wish to discuss some regulations which are relevant to the dimensioning of marine structures, using selected examples from dimensioning practice, to demonstrate the relationship with the analysis and evaluation procedures already discussed.

In particular, questions of manufacture, materials selection and welding prove to be extremely complex. We can only touch on this subject briefly in Sect. 7.1, but in Sect. 7.2 we deal in greater detail with static dimensioning of typical components of marine structures according to regulations. Without claiming to be comprehensive, we consider the regulations of the international classification societies in Europe, namely Det norske Veritas (DnV) in Norway and Germanischer Lloyd (GL) in the Federal Republic of Germany, as well as the European Convention for Constructional Steelwork (ECCS), and from North America, Design Codes of the American Petroleum Institute (API) and the American Society of Mechanical Engineering (ASME).

In Sect. 7.3 we deal with questions of design for fatigue strength and fatigue strength analysis on the basis of regulations, using Design Codes GL and API–RP2A as examples to show to what extent the application of rational analysis and evaluation methods is accepted in fairly modern regulations for marine structures. In Sect. 7.4 we discuss trends in modern development of regulations, with special emphasis on integrating reliability techniques (described briefly in Sect. 6.3), and some aspects of general safety theory for load-bearing structures, discussed primarily in civil engineering as quality assurance and specified more narrowly, e.g. in the Joint Committee of Structural Safety (JCSS). Finally, in Sect. 7.5, examples of structural design are discussed to highlight present-day dimensioning problems.

7.1 Fabrication and Materials

Fabrication of marine structures from steel takes place in traditional shipyards or where a location offers advantages, in special construction and assembly yards.

Equipment for preparing plates and profiles, including numerically-controlled flame cutters and profile-bending machines in a shipyard, is also suitable for manufacturing marine structures. This includes, among other tools, rolling mills for straightening, presses and high-speed metal saws. Preliminary assembly of structures which are sometimes very complex, and which may be very large, is extremely important. Preliminary assembly can often only be undertaken outdoors, which involves considerable difficulties. For example, the assembly of semisubmersibles in a floating state is subject to particular problems. As marine structures are often made of high-tensile steels and must meet high static and dynamic strength requirements, constructional inaccuracies such as out-of-round tubes and misalignments, etc., must be avoided when manufacturing components and during assembly. Special equipment is necessary for checking the joints of large components before assembly. Theodolites, plus laser and ultrasonic measuring instruments, are used for these purposes.

In contrast to shipbuilding, a large number of different steels are used in marine structures. Although this is not always necessary for structural reasons, it has come about due to the fact that up to now the regulations of the various authorities and classification societies have not been harmonized. As a rule, high-strength carbon–manganese steels are used, which are smelted in a killed or fully-killed state. A deep secondary metallography with micro-alloying additions gives a fine-grain steel, with in some cases very high yield point combined with adequate toughness. The steels are supplied normalized, and often thermomechanically (TM) rolled. Particular importance is attached to satisfactory thickness characteristics (Z steels) to prevent lamellar tearing. The yield points are in the region of $355\,\text{N/mm}^2$ (EH36 shipbuilding steel) and $685\,\text{N/mm}^2$ (HY – 100). Table 7.1 on page 314 [1] gives a list of the most important steels corresponding to various international quality standards.

In the case of welded structures which are subjected to predominantly dynamic loading, strength is influenced by the notch effect of the welding process and is thus practically independent of the strength of the basic material. Therefore, we often cannot utilize the high yield points of many steels for building marine structures, so higher-tensile steels with yield points of 355 to $420\,\text{N/mm}^2$ are most frequently used. Conventional shipbuilding mild steel is used for less highly stressed components.

Welding technology [1] is very important in marine structures; so the quality control must be of a very high standard. Welding processes used include the inert gas methods (MAG, WIG), as well as manual electrode welding and submerged-arc welding. Welding of components causes a number of notches due to penetration grooves, seam reinforcement, root recoil, edge misalignment and a number of other internal and external faults, such as pore formation, etc. These effects can impair fatigue strength considerably and must be kept to a minimum. With high-strength steels, welding causes microstructural changes in the basic material near the heat-affected zone. This hardness increase (martensite formation) caused by fast cooling can lead to cracks. In addition, a certain amount of coarse grain formation with a reduction in the strength of the material is to be expected.

For these reasons, preheating is used to prevent excessively fast cooling during welding. Modern techniques with temperature control are necessary for this. Distortions and residual stresses can also be caused by welding, and can influence fatigue strength and produce lamellar tearing. Geometric, thick-walled structures such as tube joints are particularly susceptible to damage from residual stresses.

Careful planning of the welding process is therefore necessary in marine structures. This includes painstaking design of all structural details, including seam preparation. In all, marine steel structures require a higher standard of construction than conventional shipbuilding.

7.2 Dimensioning in Accordance with Regulations

Marine structures are rated according to the rules of the classification societies and other authorities. We will show, with the aid of some examples, how the theoretical analysis methods described in earlier chapters have influenced the regulations. Only a few examples can be described, and they are not very comprehensive. A full commentary on individual regulations would be very difficult, not only because of the scope, but also because regulations change quickly. The reason for this is that in the case of marine structures, there are always new structures for which regulations must be developed and adapted. Many marine structures are subject to extreme operating conditions, so new findings from practical experience, research and development must become available quickly to dimensioning practice in the form of regulations. Of course, such regulations often require modification.

7.2.1 Dimensioning of Slender Stiffeners for Plates Subject to Pressure

For the dimensioning of plate reinforcements subject to pressure, Det norske Veritas (DnV) requires three different types of failure to be examined:

1. Buckling failure, initiated by buckling failure of the plate (plate-included failure),
2. Buckling failure, initiated by buckling failure of the stiffener's flange (stiffener-induced failure),
3. Torsional buckling.

According to DnV,

$$\sigma_{Xd} \leqslant \frac{\psi f_K}{\varkappa \gamma_m} \frac{A_{te}}{A_t} \frac{1 + \delta + \eta - \sqrt{(1 + \delta + \eta)^2 - 4\delta}}{2\delta} \tag{7.1}$$

should apply to such structures under pressure. σ_{Xd} is the compression stress in the direction of the stiffener and ψ is a factor which takes into account the post-buckling behaviour of the structural element concerned. For components which have a marked post-buckling behaviour such as plates, it should be assumed that

$\psi = 1.05$, while for shells and struts with a less marked post-buckling behaviour, $\psi = 0.9$ can be used. The material coefficient γ_m has a value of 1.15, and f_K should generally be set equal to the yield point of the material used, provided torsional buckling can be excluded.

We can identify the last part of (7.1) as the Perry–Robertson formula (see (4.188)), with δ corresponding to the slenderness ratio $(\bar{\lambda}_K^2)$, for

$$\delta = \frac{f_K}{f_i}; \quad \bar{\lambda}_K^2 = \frac{f_y}{f_i},$$

if we substitute the yield point f_y for f_K. The factor f_i corresponds to the critical buckling stress of a simply supported column

$$f_i = \frac{\pi^2 E r^2}{a^2}, \quad r^2 = \frac{I}{A_t},$$

of stiffener length a, moment of inertia of the stiffener including plate I and stiffener cross-section including plate A_t and modulus of elasticity E.

Constructional inaccuracies and deformations, generally termed *imperfections*, have a considerable effect on the load-bearing capacity of slender structures. This effect is smaller in the case of plates than of shells, and is taken into account by the factor $\varkappa$. For slenderness ratios up to $\bar{\lambda}_K = 0.5$, this factor should be $\varkappa = 1.0$. For slenderness ratios $\bar{\lambda}_K > 1$ it should be $\varkappa = 1.1$ for plates, and 1.3 for shells. Linear interpolation can be used for slenderness ratios between $\bar{\lambda}_K = 0.5$ and $\bar{\lambda}_K = 1$. The parameter η describes the effect of initial deformation and eccentricity of the load

$$\eta = \begin{cases} \dfrac{Z_p \omega}{r^2} & \text{Plate-induced failure} \\[2em] \omega = 0.0015a + Z_p\left(\dfrac{A_t}{A_{te}} - 1\right) \\[2em] 0.0015\dfrac{Z_t}{r^2}a & \text{Stiffener-induced failure.} \end{cases}$$

The definitions of Z_p and Z_t can be found in Fig. 7.1. A_{te} is the effective flange face taking into account the post-buckling behaviour of the plate

$$A_{te} = b_e t_p$$

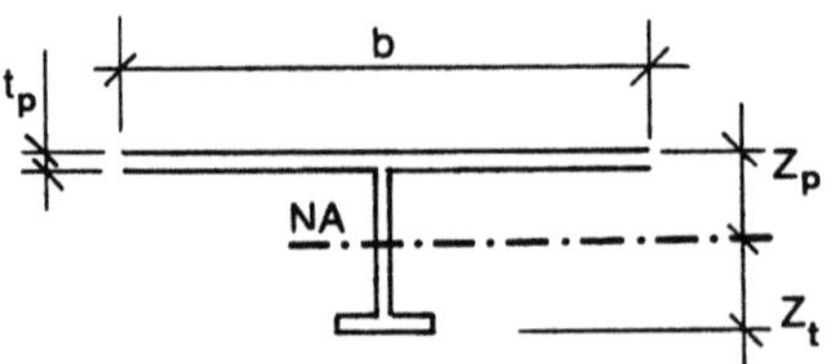

Fig. 7.1. Cross-section of a reinforced plate (NA: neutral axis).

with

$$b_e = bc_x c_y c_\tau.$$

The correction factor c_x stands for the effective breadth of plates in the range of postbuckling. The distance between stiffeners b is usually very much smaller than the stiffener length a, so the effective breadth for a plate with a large aspect ratio must be used here. For this DnV uses:

$$c_x = \begin{cases} \dfrac{1.8}{\beta} - \dfrac{0.8}{\beta^2} & \text{if } \beta \geq 1 \\[2mm] & \quad\quad\quad\quad \text{for plate failure} \\[1mm] 0.9 & \text{if } \beta < 1 \\[2mm] 1.1 - 0.1\beta & \quad\quad\quad\quad \text{for stiffener failure.} \end{cases}$$

β is the slenderness ratio $\beta = b/t_p \cdot \sqrt{f_y/E}$. The factors c_y and c_τ account for transverse and axial loading, and t_p is the plate thickness [2]

These formulae are based on the studies of various authors [3–5].

7.2.2 Dimensioning of Cylindrical Structures

Circular cylindrical structures with and without reinforcement can often be found in marine structures today. The popularity of such structures is due to the fact that of all imaginable types of structure, shell structures possess the best relationship between load-bearing capacity and weight, as in the case of shell structures the load is dissipated by membrane forces and not, as with plane structures, by bending moments. Let us visualize this situation with reference to the load-bearing behaviour of a cylindrical circular or square tube: the stresses in a thin-walled circular tube under external pressure p are exclusively membrane stresses of magnitude

$$\sigma^m = \frac{p}{2} \cdot \frac{2r}{h}. \tag{7.2}$$

We obtain the same membrane stress state for the square tube, to which we must add a bending stress component at the corners of magnitude

$$\sigma^b = \frac{p}{2} \cdot \left(\frac{2r}{h}\right)^2. \tag{7.3}$$

That means that the bending stress component is greater than the membrane stress by a factor given as the diameter/wall thickness ratio of $2r/h$. The $2r/h$ ratio is typically very large (100 to 500), so it is easy to see why shell structures are generally preferred to all other plane structures.

These advantages are countered by the fact that the stability of shell structures must be considered as well as the strength. Here it is found that the actual elastic buckling load of shell structures used in marine structures is often considerably lower than indicated by theoretical strength evaluations, so the advantage of a shell often cannot be fully utilized.

While the theoretical buckling loads of plates and columns can easily be determined experimentally, we know that the experimentally determined buckling loads of shells are often considerably lower than theoretically predicted. The reason for this is that many shell structures react extremely sensitively to initial deformation, local disturbance by reinforcements and unplanned or planned eccentric load application. So in dimensioning practice, the theoretical buckling load must be reduced with knock-down factors which are established in tests.

A prime example of this is the axially-loaded cylindrical shell. Initially, we restrict ourselves to elastic buckling. The component under consideration is dealt with in all important standards, such as the European Convention for Constructional Steelwork (ECCS), Recommendation R.4.6 [6], ASME Boiler and pressure vessel code N 284 [7], the API Recommendation for fixed offshore platforms RP2A [8], Det norske Veritas Rules for the design, construction and inspection of offshore structures [9]. All standards are based on the theoretical elastic buckling load in (4.126)

$$\sigma_{K_1} = \frac{E}{\sqrt{3(1 - v^2)}} \frac{h}{r}.$$

The actual ultimate (elastic buckling) load is then said to be

$$\sigma_e = \alpha \sigma_{K_1}.$$

The knock down factor α is defined differently in the various standards. ECCS R.4.6 uses

$$\alpha = \frac{0.7 \cdots 0.83}{\sqrt{1 + 0.01 r/h}}$$

for thin-walled cylinders, while ASME Code N 284 uses

$$\alpha = 1.52 - 0.473 \log_{10}\left(\frac{r}{h}\right)$$

and DnV uses

$$\alpha = 0.35 - 0.0002 \frac{r}{h}.$$

API recommends a reduction of $\alpha = 0.50$ which is independent of the diameter/wall thickness ratio $2r/h$.

Figure 7.2 shows these knock-down factors α. The differences can be explained partly by the fact that different safety factors are explicitly required, i.e. in some regulations extra safety factors are concealed in the α values.

The maximum permissible stress is defined with the aid of a safety factor as part of the actual ultimate load. Here the stipulations of the individual regulations differ. ECCS R.4.6 gives the ultimate load as

$$\alpha_K = \frac{\alpha \sigma_{Ki}}{\gamma}.$$

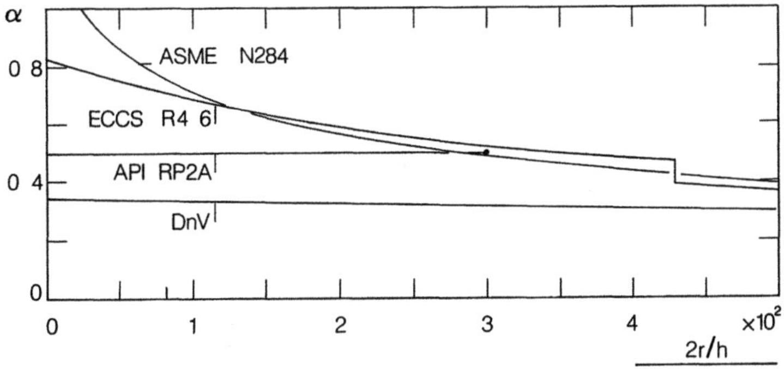

Fig. 7.2. Knock-down factors α for axially-compressed circular cylindrical shells.

The factor γ is given as 4/3, a general safety factor as 1.5. In the elastic domain this gives a total safety factor of 2.0 compared with σ_e.

DnV requires the safety format

$$\gamma_{f_1} S_d \leqslant R_d$$

to be fulfilled, with R_d as the characteristic resistance and S_d as the design stress

$$R_d = \frac{R_K \psi}{\gamma_m \varkappa},$$

where the individual values of ψ, γ_m and $\varkappa$ are given as $\psi = 0.9$, $\gamma_m = 1.15$ and $\varkappa$ equals

$$
\begin{array}{lll}
1.0 & \text{for} & \bar{\lambda}_K \leqslant 0.5 \\
0.7 + 0.6\bar{\lambda}_K & \text{for } 0.5 < \bar{\lambda}_K \leqslant 1.0 \\
1.3 & \text{for } 1.0 < \bar{\lambda}_K
\end{array}
$$

for the case being considered. The slenderness ratio related to the yield stress is

$$\bar{\lambda}_K^2 = \frac{f_y}{\sigma \alpha_{K_1}},$$

and f_y is the yield stress. In addition, a load amplification factor γ_{f_1} is defined which is set at a maximum of 1.3, depending on the load category. The ultimate load related to the yield stress is defined as

$$R_K = \frac{f_y}{\sqrt{1 + \bar{\lambda}_K^4}}.$$

Buckling in the elastoplastic domain is considerably more problematic than in the elastic domain. Today no theory suitable for practical dimensioning purposes is available. Therefore we generally use the following concept: buckling will not occur in thick-walled cylindrical shells. As the shell is in a constant state of compression stress and we can assume a material law with perfectly elastic–plastic behaviour without hardening, the ultimate load of a thick-walled cylindrical shell is defined at the yield point of the material.

Most regulations define an ultimate load curve which, on the basis of the behaviour of thick shells, has an empirical variation which changes to the ultimate load curve for elastic buckling in the case of thin shells. This transition can be continuous as for instance, in the above DnV formula, or as in ECCS R.4.6, only in the elastoplastic domain, where

$$\sigma_K = \sigma_F(1 - 0.4123\bar{\lambda}_K^{1.2}).$$

Here the yield stress is indicated by σ_F.

For the elastoplastic zone the API–RP2A Code uses the following formula for the load stress

$$F_{xc} = F_y\left(1.64 - 0.2735\left(\frac{r}{h}\right)^{1/4}\right),$$

if

$$F_{xc} \leqslant F_{xe}.$$

For small $2r/h$ ratios ($r/h \leqslant 30$) $F_{xe} = F_y$ is used. F_y is the yield stress and F_{xe} corresponds to the ideal buckling stress σ_{K_1}. The formula for elastic buckling becomes effective only with yield stresses over $600\,\text{N/mm}^2$. The above formula is restricted to tubes with a maximum of $r/h = 150$. The value of F_{xc} in the API–RP2A Code corresponds, therefore, to the ultimate load σ_K in ECCS R.4.6.

Definition of the safety factors in the API–RP2A Code is complicated, and appears to have little connection with column buckling. If the latter is disregarded, we obtain a minimum safety factor of 1.67.

ASME Code N 284 [7] applies to $r/h \leqslant 1000$. The knock-down factors α are very large for small r/h ratios, as Fig. 7.2 shows. They are therefore limited by an additional condition that α cannot be greater than

$$\alpha = 300\frac{\sigma_F}{E} - 0.033$$

(converted to metric units).

For St 52–3 steel we thus get a limit of $\alpha = 0.47$, which means that $\alpha = 0.47$ up to a $2r/h$ ratio of about 330. The ultimate load is determined with respect to the factor

$$\frac{\sigma_\phi \text{FS}}{\sigma_y},$$

where σ_ϕ is the membrane stress to be borne, FS is the safety factor, here assumed to be 2, and σ_y is the yield stress of the material. The reduction of the actual ultimate load F_{xc} or σ_{Ki} is

$$
\eta = \begin{cases}
1.0 & \text{for} \quad \dfrac{\sigma_\phi \text{FS}}{\sigma_y} < 0.55 \\[3ex]
\dfrac{0.18}{1 - 0.45\dfrac{\sigma_y}{\sigma_\phi \text{FS}}} & \text{for} \quad 0.55 < \dfrac{\sigma_\phi \text{FS}}{\sigma_y} < 0.738, \\[3ex]
1.31 - 1.15\dfrac{\sigma_\phi \text{FS}}{\sigma_y} & \text{for } 0.738 < \dfrac{\sigma_\phi \text{FS}}{\sigma_y} < 1.0.
\end{cases}
$$

To permit comparison with the other standards, the related ultimate stress is converted with

$$
\frac{\sigma_K}{\sigma_F} = \frac{\sigma_\phi \text{FS}}{\sigma_y} = \frac{\sigma_{Ki}\alpha\eta}{\sigma_F}.
$$

We then find

$$
\frac{\sigma_K}{\sigma_F} = \frac{\eta}{\bar{\lambda}_K^2}.
$$

Thus the related ultimate load becomes

$$
\frac{\sigma_K}{\sigma_F} = \begin{cases}
1.0 & \text{for} \quad \bar{\lambda}_K \leqslant 0.4, \\[2ex]
\dfrac{1.31}{1.15 + \bar{\lambda}_K^2} & \text{for} \quad 0.4 < \bar{\lambda}_K \leqslant 0.79, \\[2ex]
0.45 + 0.18/\bar{\lambda}_K^2 & \text{for } 0.79 < \bar{\lambda}_K \leqslant 1.35, \\[1ex]
1/\bar{\lambda}_K^2 & \text{for } 1.35 < \bar{\lambda}_K.
\end{cases}
$$

Figure 7.3 shows the ultimate loads, dimensionless with the yield stress, as related ultimate stresses against the ratio of diameter to wall thickness. We can see the sometimes considerable differences between the regulations. We must reiterate that the regulations provide different safety coefficients, so finally we obtain another

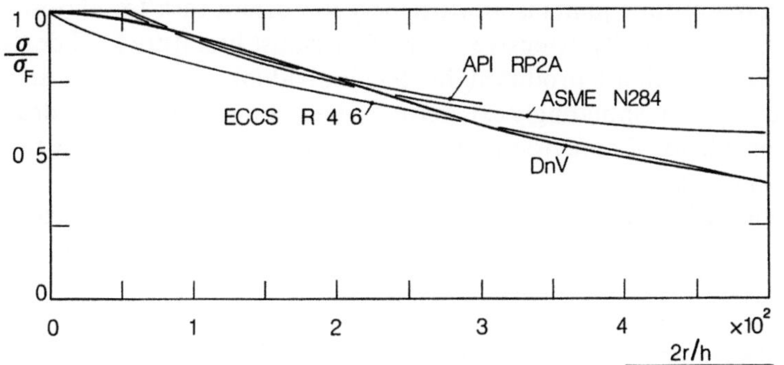

Fig. 7.3. Ultimate stress σ related to yield stress σ_F for axially-compressed circular cylindrical shells.

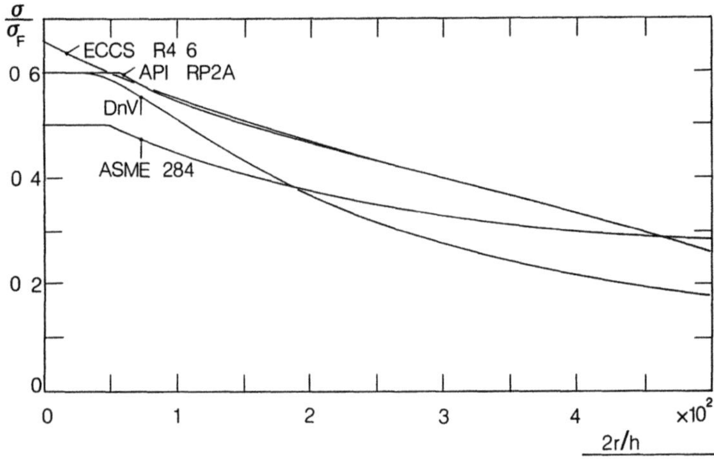

Fig. 7.4. Allowable stress σ related to yield stress σ_F for axially-compressed circular cylindrical shells

representation of the permitted stresses (see Fig. 7.4). We can see that the differences between the individual regulations can be up to 10% of the yield stress. All in all, this is a relatively small difference, but the difference between the minimum yield stress to be guaranteed by the steel manufacturer and the mean value of the actual yield stress found is often greater than 10%. Reference [10] deals with a comparison of no less than 18 standards or proposed standards for axially-loaded cylindrical shells. For reasons of space we are unable to give further information here on dimensioning guidelines for cylindrical tubes with potential stability problems. Reference [11] is particularly recommended, as it contains a detailed comparison of the individual standards for regular cylindrical shells subject to external pressure, which are a particularly important loading case for marine structures.

7.2.3 Tube Joints

Dimensioning of tube joints is an extremely important task for almost all marine structures. An unusually extensive work of specialist literature [12] deals with this subject. The guidelines produced by the individual classification societies and authorities are correspondingly extensive. Special congresses [13, 14] deal primarily with the problems of dimensioning welded seams which are subject to static and dynamic loading.

In the case of welded tube joints, a large number of welds are concentrated in the area of the junctions of the individual tubes, which is technologically extremely unsatisfactory [15], since apart from the difficulty of quality assurance, the heat application in such a small area can cause very high residual welding stresses. It is true that such residual stresses are not important for dimensioning tube joints for the limit of load-bearing capacity under static loads, but they are important for fatigue strength of such structures under dynamic loading.

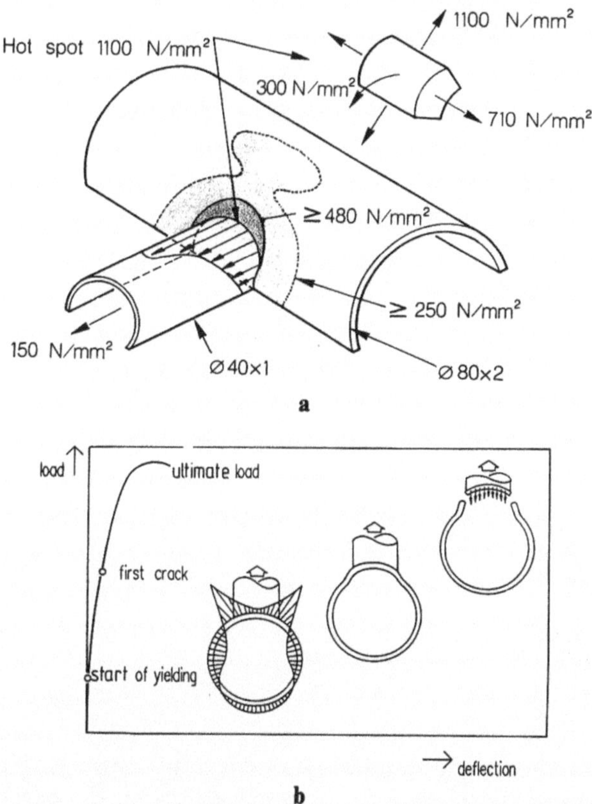

Fig. 7.5. a. Stress distribution in the region between brace and chord [14]; **b.** Typical load deformation distribution for failure between brace and chord

The API–RP2A recommendations and AWS structural welding code ASW D.1 of the American Welding Society are the definitive regulations for the dimensioning of tube joints. The works of Marshall (which can be found in [14], among others), are especially instructive.

The normal forces applied at a tube joint play an important part in static dimensioning. As the connection of tubes is geometrically complicated, very high peak stresses with increases in stress up to ten times the level of undisturbed stresses can occur (see Fig. 7.5a [14]). This means that even at very small load stresses (i.e. without a residual stress component), the yield stress can be reached locally without the load-bearing behaviour of the overall structure being noticeably affected. Only after a considerable increase in load (see Fig. 7.5b) do the first cracks occur, and the structure fails after repeated distinct load increases. This means that a purely elastic dimensioning process would result in extremely uneconomical structures. But there are a large number of test results available which make it possible to determine the actual loading limits for tube joints, with reference to the most important tube parameters (see Fig. 7.6), and to define acceptable stresses

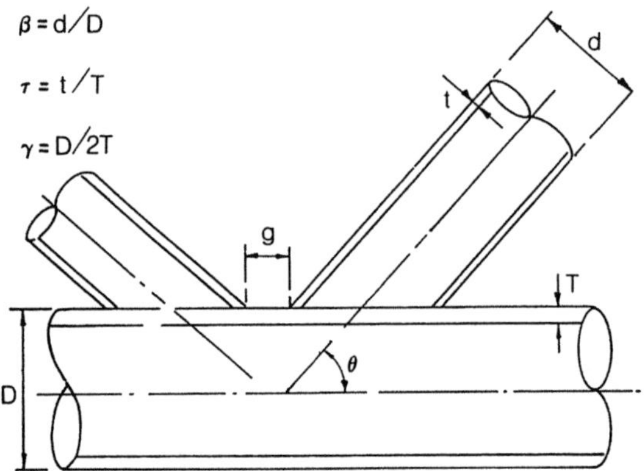

Fig. 7.6. Tube joint parameters according to API–RP2A.

after introducing safety coefficients. So API–RP2A gives

$$v_p = \tau \cdot f \cdot \sin \theta.$$

for the effective axial stress. The acceptable axial stress is

$$v_{pa} = Q_q \frac{F_y}{0.6\gamma}$$

(we have considered only the simple case of constant tensile loading $f = f_a$ here). F_y is the yield stress and Q_q is a factor which describes the effect of the type of load and geometry of the connection (see Fig. 7.6)

$$Q_q = \left(1.1 + \frac{0.2}{\beta}\right) Q_g,$$

$$Q_g = \begin{cases} 1.8 - 0.1\dfrac{g}{T} & \text{for } \gamma \leqslant 20, \\[3mm] 1.8 - 4\dfrac{g}{D} & \text{for } \gamma > 20. \end{cases}$$

We will assume an actual case

$$\gamma = 2.0; \ \tau = 0.5; \ \theta = 90°; \ \frac{g}{T} = 5.$$

Thus we find

$$v_{pa} = 0.108 F_y$$

i.e. a maximum acceptable axial stress is recommended at which only insignificant local yielding occurs. This is also advisable because the more frequent causes of damage are fatigue cracks, due to dynamic loading starting in areas with stress peaks (hot spot stress) [16].

7.3 Fatigue Strength Evaluation on the Basis of Regulations

The development of a regulation for evaluation of the fatigue strength of marine structures in the Federal Republic of Germany was influenced by the crane construction standard DIN 15018. Due to the special factors associated with this, in Sect. 7.3.1 we first look closely at the regulations of the internationally active German classification society, Germanischer Lloyd (GL), which are based on this crane construction standard [17].

The development of other international regulations for fixed steel platforms, in particular the development of basic rules for demonstrating their fatigue strength, has been strongly influenced up to now by the American Petroleum Institute (API) through Recommendations RP2A [8]. These distinguish between design for fatigue strength (Fatigue Design) and the demonstration of fatigue strength (Fatigue Analysis). The API–RP2A recommendations for fatigue design are based on the principles of developing a design stress range, which we have dealt with in detail in Sect. 6.2.3.4. Fatigue analysis is conducted on the basis of the Miner Rule, however, which we described briefly in Sect. 6.2.2.1, and according to which the method most suitable for solving the problem out of those described in Sect. 6.2.3.1–6.2.3.3 – and variations of it – is to be considered.

We will continue to examine the two analysis methods of API–RP2A, of which the second (fatigue analysis) is found in a similar form in many other international rules, in Sects. 7.3.2 and 7.3.3 to illustrate some particular features of dimensioning practice and practical safety analysis in fatigue strength questions which are relevant to regulations.

7.3.1 Fatigue Design on the Basis of GL Requirements

Design for fatigue strength (fatigue design) is based on S–N curves defined by GL to represent sectionwise linear relationships between the logarithm of the applied stress range $\Delta\sigma$ and the logarithm of the related number N of endured stress range cycles

$$\log N = 6.699 + m \cdot \Delta \tag{7.4}$$

with

$$\Delta = \log(\Delta\sigma_R) - \log(\Delta\sigma) - 0.1326$$

and m the inverse slope of the S–N curve (cf. (6.79) and Fig. 6.19), especially defined as $m = 3$ for negative Δ-values, $m = 5$ for $\Delta \leqslant 0.3204$ and $m = \infty$ for $\Delta > 0.3204$. The term $\Delta\sigma_R$ is a reference stress range at $2 \cdot 10^6$ stress range cycles, referred to as the 'detail category number', which is defined in relation to structural details in tables given in the GL regulations [17]. (In this section we use the nomenclature of the referenced codes to avoid confusion.)

Based on such S–N curves, where $\Delta\sigma_R$ has to be corrected for inadequate corrosion protection, plate thickness effects and possible weld improvements (giving $\Delta\sigma_{Rc}$), the calculation of the damage ratio D (defined in (6.75) as the Miner rule with abbreviation s instead of $\Delta\sigma$ for stress ranges) is now possible.

GL require the stress range $\Delta\sigma_i$ of any considered ith stress range block i to be multiplied by a safety factor γ given in relation to structural details, and to either a 'fail safe' or a 'non-fail safe' philosophy, respectively. For design, the number $n_i = n(\Delta\sigma_i)$ of stress cycles of the ith stress range block may be calculated using a standard Weibull distribution

$$\Delta\sigma_i = \Delta\sigma_0 \cdot [1 - \log n_i / \log n_0]^{1/h},$$

where h is a shape parameter (cf. b in (5.109)), given in tables as a function of $\Delta\sigma_{Rc}$, the corrected fatigue strength reference value of the S–N curve at $2 \cdot 10^6$ stress range cycles, and of the total number n_0 of service-life stress cycles of the considered structural element. The term $\Delta\sigma_0$ is the maximum stress range being exceeded once within n_0 stress cycles.

(Fatigue analysis is carried out analogously taking into account all stress fluctuations relevant to fatigue (normally wave induced stresses) during the planned life of the structure (see Sects. 6.2.3.1–6.2.3.3 for details of such computations).

This procedure has the advantage of relative simplicity combined with the possibility of influencing the result by constructional measures. However, we must remember that a simple method always implies a great degree of additional safety, and for this reason does not allow the most effective constructional solution. We therefore consider an alternative method of examining rule-based fatigue design, which promises improvements compared with the methods described above.

7.3.2 Fatigue Design on the Basis of API–RP2A Recommendations

In 1983 the API committee for fixed structures organized a special group to study fatigue strength questions, and to recommend fatigue design methods which would be compatible with stochastic evaluation methods, but which would be simple and realistic enough to be used in design guidelines.

The result was produced in the 17th and 18th editions of API–RP2A of 1987 and 1989 [8]. It is based on the analysis method, which we have already dealt with in Sect. 6.2.3.4 (Chap. 6, [47, 48]). A prime aim was to calibrate the parameter g introduced in (6.100) as a strength parameter for global understanding of the behaviour of a structure under wave loading, i.e. to derive generally valid values for g by matching the results of (6.105) – in an alternative form – with the results of fatigue analyses of a series of existing platforms. The parameter g was therefore treated as a calibration parameter, and all other parameters, especially the S–N curve, the long-term up-crossing frequency distribution of the wave height, and the stress concentration factors had to agree in the comparative calculations. The conditions under which this method can be used are given in Sect. 6.2.3.4.

The Stokes wave theory described in Sect. 3.2.2 of Vol. I was used to determine the strength parameter g for tubular joints of typical shallow water ductile steel platforms, the (design) wave periods T_B with hypothetical wave gradients of 1/16 and 1/12 being determined from the relevant (design) wave height H_B. With the design waves defined for a reference time T_R, the associated stress ranges $s_R = s((H_B, T_B)| T_R)$ at the monitoring points of the structure under consideration were determined as the difference between the maximum value σ_{max} and the minimum value σ_{min} of stresses $\sigma(t)$, $0 < 2\pi t/T_B \leqslant 2\pi$, in a complete wave cycle. Thus, the associated strength parameter g could be determined from (6.105), after $n_R = n(T_R)$ had been determined from the fatigue analysis conducted in parallel, with the service life defined by the reference time T_R. This, in principle, logical application of the concept developed in Sect. 6.2.3.4 was simplified for practical reasons by considering only the peak stress σ_{max} during a cycle of the design wave. The calculation value designated as pseudo stress range s^* could be determined by

$$s^* = \sigma_{max}\cdot(1 - \varkappa^*), \quad \varkappa^* = Q_{min}/Q_{max}, \tag{7.5}$$

Q being the total base shear wave force acting on the structure at each point in time of the wave cycle. The platforms examined produced $\varkappa^*$ values of between -0.32 and -0.42, so with $s^* \simeq 1.4\sigma_{max}$ we can give a 'rule of thumb' for the design stress range for structural elements of jacket platforms, suitable for the first design stage.

Of course, the strength parameters g determined with the pseudo stress variation s^* and (6.105) vary greatly from element to element within a structure, as well as from structure to structure. As, according to (6.105), greater values of g mean greater values of design stress range, the value recommended in the 17th edition of API–RP2A had to be at the lower limit of the results. This limit was $g = 1.2$ for structural elements in the region of the water line, and otherwise $g = 1.3$.

After g had been determined by means of calibration, it was possible for the API–RP2A to use (6.105) – in an alternative form – to define a design stress σ_{max} for ensuring fatigue strength, for any reference time T_R, which, with reverse application of (7.5), would be suitable for regulations. The greatest $\varkappa^*$ value found, (-0.32) was used for this in API–RP2A to be on the safe side. The rules give the maximum allowable stress concentrations (allowable peak hot-spot stresses) thus determined for reference times T_R of 40 and 100 years. All steps of the calibration procedure were carried out for different depths of water, so the regulations correctly reflect the special effect of water depths between 0 and 122 m on fatigue strength (see Fig. 7.7, from Chap. 6 [48]). (1 ksi = 6.8906 N/mm² = 0.68906 kN/cm², 1 ft = 0.30488 m.)

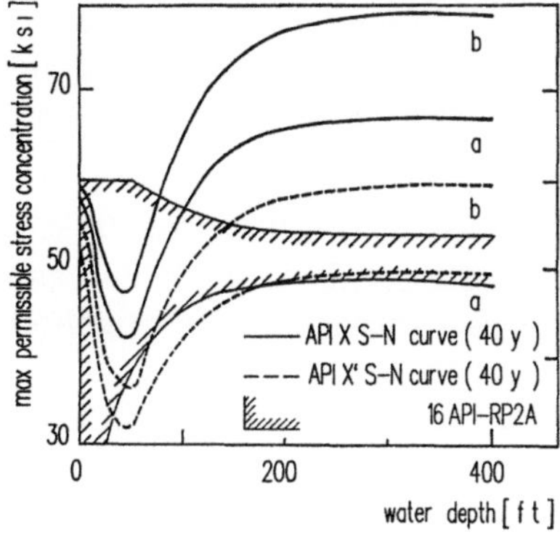

Fig. 7.7. Comparison of 16th and 17th API–RP2A from reference [41] (Chap. 6)

In Fig. 7.7 the parameter 'a' indicates structural elements close to the surface of the water, and parameter 'b' indicates all other elements. The designation 'X S–N curve' refers to welds which merge smoothly with the adjoining base metal, the designation 'X' S–N curve' refers to welds without such profile control, i.e. special manufacturing is also evaluated appropriately. From Fig. 7.7 we can see that the new API–RP2A allows greater maximum fatigue design stresses in deeper water for elements with API X S–N curves than the 16th edition of the rules.

We must point out, however, that the recommendations of API–RP2A developed with this method of fatigue design are only valid if the wave climate of the Gulf of Mexico is applicable. Wind and current effects are not taken into account, and the API reference water levels are used. Stress concentration factors of magnitude 3 were used for all structural elements.

7.3.3 Fatigue Analysis Based on API–RP2A Recommendations

The form of rule-based fatigue design described in Sect. 7.3.1 could only take into account the characteristics of the wave climate in the operating area of the structure in a general manner. The alternative form of rule-based fatigue design described in Sect. 7.3.2 is more advanced in this respect, but is currently applicable only to the Gulf of Mexico or areas with a similar wave climate.

Both rules recommend the performance of fatigue analysis of highly stressed structural elements according to the Miner rule if the conditions for simplified analysis are not met, or if the design has reached the detail stage. Nevertheless, the two rules and the regulations of the Norwegian classification society (Det norske Veritas (DnV)) demand safety factors depending on accessibility, structural significance, and position of the monitoring points [2, 9]. In API–RP2A and DnV the safety factor is defined as a parameter to be multiplied by the desired service life of the structure, i.e. when the Miner rule is applied, the service-life-related allowable damage indicator $D = 1$ must be divided by the safety factor (see (6.75) and (6.76)) while GL applies the safety factor to the considered stress range block Δs_i in the long-term distribution of stress range. A safety factor of magnitude 2 (i.e. $D = 0.5$ in the Miner rule) necessitates an increase in wall thickness of about 10% with the same material and the same production quality (see [8]). The magnitude of the safety factor thus has a decisive influence on material and fabrication costs. API–RP2A goes into great detail on the analysis stages of rule-based fatigue analysis associated with using the Miner rule. They accept two spectral analysis methods in principle:

1. Calculation of the damage indicator for each seaway class defined by H_V, T_V and also, by the main wave direction μ_V, and subsequent calculation of damage accumulation in the manner demonstrated by us as a spectral analysis method in (6.97) and Fig. 6.22.
2. Calculation of cumulative stress distributions for the reference time period in areas of highest stress concentration and their evaluation according to the concept which we described in Sect. 6.2.3.2 as an acceptable simplification of the spectral analysis method for static structures.

Also, a deterministic method as outlined in Sect. 6.2.3.1 is accepted under certain conditions.

Because we are generally concerned with evaluating fatigue strength at welded joints between structural elements, it is necessary to identify the areas of greatest stress concentration (hot spots) as there are many such joints in a jacket structure, and in the final analysis this can only be done iteratively by global stress analysis in conjunction with realistic values for the stress concentration factors. API–RP2A recommends examining the intersection between brace and chord centre line. The applied cyclic loads should be represented in such a way that the effects of load distribution along the member are included in the member end stresses. Distributed loads on brace members need to be considered only between intersection points. The choice of wave theory as well as mass and drag coefficients may be different from those used in strength analysis for design wave. Furthermore, the wave frequencies $\omega = 2\pi/T$ under consideration either must be very close together or must be carefully selected: API–RP2A first recommends examining the transfer functions of the total base shear wave force Q acting on the structure, and taking account of the frequencies of the relative maxima of Q in all subsequent calculations, especially the lower (elastic) natural frequencies in the case of dynamic structures, where a linear representation of the foundation may be used, provided the stiffness coefficients reflect the cyclic response for those sea states contributing significantly to fatigue damage.

As it is of significance for elements near the surface of the water that they are subjected to loading directly by the wave (inundation effect) as well as indirectly by deformation of the whole structure caused by wave loading, API–RP2A requires that the stress transfer function necessary for the analysis method be calculated using a carefully selected wave height H, even if linear wave theory and the spectral analysis method are applied. Wave height H can be found (as described above) from the wave periods $T = 2\pi/\omega$ in conjunction with an appropriate wave gradient H/L (1.20 to 1.25 for the Gulf of Mexico). (We demonstrated the calculation of the wave length L as a function of the wave period T in Chap. 3; we can use $L = 1.57 T^2$ as an estimate for deep water.)

If so defined waves of varying frequency ω pass through the structure in short time steps and, for each time step account is taken of all wave loads on tube joints lying below the actual wave contour, in order to calculate stresses at the considered hot spots, the inundation effect of the elements is considered correctly on average even for natural seaway conditions, as long as the maximum stress ranges calculated after a wave cycle are divided by the considered wave height to obtain the required transfer function value associated with each wave frequency (with inundation effect). We can then use this transfer function to determine the stress spectrum from the sea spectrum as in (5.54), and as described in Sect. 6.2.3.2 as the spectral analysis method.

If we want to make logical use of the advantage of linear wave theory, which, especially in the case of static structure, is that only two phases of the wave have to be analysed relative to the structure, to describe the overall effect of the wave cycle completely, we must first consider all wave loads on tube joints which lie under the still water surface. However, this generally underestimates the inundation effect on the stress range in the case of tube joints in the region of the wave contour above the still water surface (too small a stress range), and API–RP2A recommends subsequent correction of the results.

Effects of wind and current as well as of internal loads can often be disregarded in rule-based fatigue analysis of marine structures. Only the water depth at the highest astronomical tide should be determined as accurately as possible, e.g. with (5.134) or with adequate information, as water depth, in conjunction with the inundation effect just described, can have a decisive influence on fatigue strength (see Fig. 7.7). Reference water levels for the coast of the USA to be expected in conjunction with the 100 year design value of wave height are given in API–RP2A.

7.4 Development of Modern Regulations

In Sects. 7.2 and 7.3 we dealt with examples of rules in which the analysis methods we have called classical are combined with practical experience from about three decades of designing, building and monotoring marine structures to produce design regulations which have proved themselves repeatedly in the field. It has already been seen that safety formats, which describe the formalism according to which the stress and strength of a structure or structural element should be compared, only constitute a very small part of the regulation. Determining stress and strength makes up the greater part of the regulation, i.e. input data, and methods of analysis and evaluation are largely specified by regulations, as otherwise the safety factors used in the regulation's safety format would not be meaningful. Yet imponderables remain, and, of course, all engineers dealing with rule-making are interested in having as rational as possible a representation of precisely the safety factors and safety margins or alternative safety parameters in the format of the regulation.

For this purpose we have summarized the methods of evaluating the reliability of load-bearing structural elements in Sect. 6.3.1. More recent regulations already use this possibility (e.g. [18], but see also the summary in [19] and the application to ships in [20]), so in Sect. 7.4.1 we will make some comments on this to familiarize potential users of modern regulations with the principles on which they are based.

Independently of rule-based dimensioning, it is recognized during the course of developing increasingly larger marine structures that questions of safety are not answered merely by safe dimensioning of components, but just as much by permanently maintaining the planned status, i.e. by appropriate handling of the structure (cf. Sect. 6.3.3.6). In shipbuilding, both aspects of the safety problem are catered for by the concept of classification which has been developed and proved over more than a century. In marine technology, as in construction generally, a concept designated quality assurance, which is very similar to classification, has developed in the last decade. To give as complete as possible a view of dimensioning for marine structures, we give some indication in Sect. 7.4.2 of the current aims and possibilities of both quality assurance and classification, as both concepts play a decisive part in the dimensioning practice of marine structures.

7.4.1 Development of Regulations Based on Reliability Technology

During the first phase of applying the first-order FOSM-reliability methods (FORM/SORM) outlined in Sect. 6.3.1 for evaluating the reliability of structural elements, rationalizing conventional safety formats was seen by practising engineers as the most significant application. In civil engineering, from which the decisive impetus for developing reliability techniques had come, the initial and probably also most fundamental ideas were developed into new regulation formats, and tested and discussed in model codes. As familiarity with the FOSM methods grew rapidly, engineers in other disciplines, recently especially in marine technology, studied the same questions, so we devote a short section of this book to some answers obtained so far.

We are looking initially for ways of defining the safety factor γ on the basis of results of reliability analyses, or of clarifying reliability in the rule based safety format by means of safety factors. For this we will look first at Fig. 7.8.

To simplify our study, we concentrate initially on the global central safety format

$$\gamma_0 \cdot m_A \leqslant m_W. \tag{7.6}$$

(With non-skew probability distributions of stress A (load) and strength W (resistance), the mean values m_A and m_W are to be interpreted as the respective 50% fractiles of stress and strength, hence the term 'central'; the term 'global' is used as the opposite of 'partial': the difference is explained in more detail below in connection with the safety format (7.9).)

Of course, the definition of the global central safety factor γ_0 alone does not give a clear indication of the reliability achieved, because according to (6.114) and (6.115) the latter depends not only on the mean values m_A and m_W of stress and strength, but also on the respective standard deviations σ_A and σ_W. More recent regulations are starting to give, along with the safety factor, indications of the related values m and σ or associated values. It would naturally be more logical to summarize these findings in the reliability index. A formal relationship between the global central safety factor γ_0, the variation coefficients $V = \sigma/m$ and the

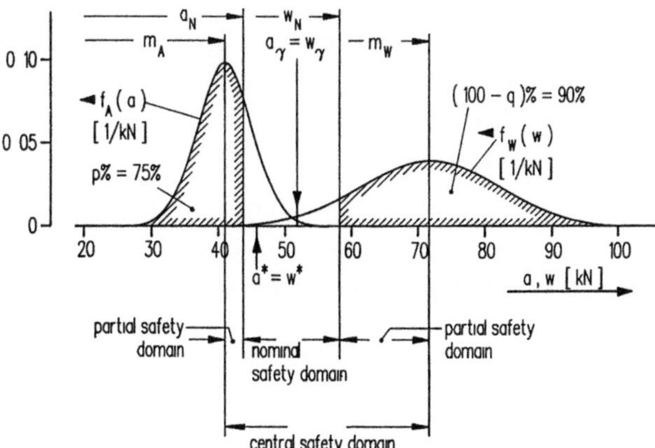

Fig. 7.8. Safety domains in an $A-W$ (stress strength) safety format.

reliability index β could be used, which can be obtained from the definition of the reliability index in (6.114) by using (7.6) in the limiting case $\gamma_0 \cdot m_A = m_W$

$$\beta = (1 - \gamma_0)/\sqrt{\gamma_0^2 V_W^2 + V_A^2}. \tag{7.7}$$

When solved for γ_0, (7.7) gives the relationship

$$\gamma_0 = (1 - \beta \cdot \sqrt{V_W^2 + V_A^2 - \beta^2 V_W^2 V_A^2})/(1 - \beta^2 V_W^2). \tag{7.8}$$

We can thus calculate the global central safety factor for a given reliability index, and with given variation coefficients of stress and strength it can be extended to nominal values (a_N, w_N) instead of mean values (m_A, m_W) [21].

In Fig. 7.8 we can identify partial safety domains for stress value A and strength value W. For the sake of simplicity, we make the nominal safety domain shown in Fig. 7.8 as zero, so the central safety domain consists of two partial safety domains for A and W, which may become adjacent at a value $a = w$. The safety format with partial central safety factors γ_{0A} and γ_{0W} is then

$$\gamma_{0A} \cdot m_A \leqslant m_W/\gamma_{0W}. \tag{7.9}$$

In the literature, we find the expression LRFD (Load and Resistance Factor Design) for structures designed on the basis of such a safety format with stress/strength values. With (7.6) and (7.9) we have indirectly defined the following relationship between global and partial central safety factors

$$\gamma_0 = \gamma_{0A} \cdot \gamma_{0W}, \tag{7.10}$$

but we have not decided in what way the safety defined jointly by the partial safety factors should be divided between the stress A and the resistance W, as (7.10) always gives the same product for an infinite number of factors γ_{0A} and γ_{0W}: γ_0. To achieve a suitable definition of partial central safety factors γ_{0A} and γ_{0W}, we must remind ourselves of the significance of the design point (a^*, w^*) in (6.123): in the case of failure, the design point for normally distributed physical basic parameters A and W is the point of most probable failure (see (6.125)). In this special case, the design point could be defined independently of the reliability index β and the sensitivity factors α. If we can agree on the design point as the boundary when making two partial central safety domains out of the central safety domain, we can calculate the partial central safety factors

$$\gamma_{0A} = a^*/m_A, \quad \gamma_{0W} = m_W/w^* \tag{7.11}$$

immediately with (6.125) in accordance with (7.9), if we know the mean values and standard deviations of stress and strength. It is also possible to extend this to nominal values (a_N, w_N) instead of mean values (m_A, m_W) [21].

If, instead of a^* and w^*, we use general design values of the stochastic basic parameters, which we designated in Sect. 6.3.1 as $x_k^*, k = 1, 2, \ldots, K$, then according to (6.138) (see also (6.123)) we find for each of K stochastically independent basic parameters X_k

$$x_k^* = m_{Xk} \cdot (1 + \beta \cdot \alpha_k \cdot V_k), \quad k = 1, 2, \ldots, K. \tag{7.12}$$

At the failure boundary the structure function must fulfil the condition

$$g(X|X=x^*) = g(x_k^*) = g(x_1^*, x_2^*, \ldots, x_K^*) = 0 \tag{7.13}$$

(see (6.132)). We should point out here that the structure function can of course be defined for different limit states which may be found in particular structures (e.g. for the limit states of failure under ultimate loads known as 'ultimate limit states' (ULS), or for the limit states of performance capability known as 'serviceability limit states' (SLS)). Different values (m, V) of the basic stochastic strength parameters can be considered depending on the limit state under consideration, but that does not affect in principle the formalism we are interested in here.

If we index all basic parameters X_k so that the most critical has the index $k = 1$, the next most critical the index $k = 2$, etc., we can use the following method to estimate conservatively the sensitivity factors α_k, which are (positive or negative) measures of the influence of the basic parameters on the safety of the components [22]. First we make

$$\alpha_1 = \pm 1. \tag{7.14}$$

(0.8 or -0.7 can also be used, e.g. [23] and [24].) The sign is negative for stress parameters $(X_k \in A)$ and positive for strength parameters $(X_k \in W)$, as according to (6.135) and (6.136) the sensitivity factors α_k contain the partial derivatives a_k of the structure function $g(X)$ depending on the relevant basic parameter. In (6.132) we defined $g(X)$ such that an increase in stress leads to a reduction in $g(X)$ (and conversely if the resistance rises). Furthermore, we make

$$\alpha_k = \pm [\sqrt{k} - \sqrt{(k-1)}], \quad k = 2, 3, \ldots, K. \tag{7.15}$$

We can use these estimates to calculate the multiplication factors ψ_k in (6.144). We now illustrate this procedure with the data for a truss element under tension

tensile force A	cross-section F	stress S
$m_A = 41\,\text{kN}$	$m_F = 175\,\text{mm}^2$	$m_S = 410\,\text{N/mm}^2$
$\sigma_A = 4.1\,\text{kN}$	$\sigma_F = 17.5\,\text{mm}^2$	$\sigma_S = 41\,\text{N/mm}^2$

The FOSM solution in Sect. 6.3.1 gives $\beta = -3.07$ and $a^* = 46.6\,\text{kN}$, $f^* = 144\,\text{mm}^2$, $s^* = 330\,\text{N/mm}^2$ and $\alpha_A = -0.448$, $\alpha_F = 0.632$, $\alpha_S = 0.632$, and of course (6.137) applies, i.e.

$$\alpha_A^2 + \alpha_F^2 + \alpha_S^2 = 1.$$

We now make $X_1 = S, X_2 = F, X_3 = A$, so according to (7.14) and (7.15) $\alpha_1 = 1$ (FOSM: 0.623), $\alpha_2 = 0.414$ (FOSM: 0.632), $\alpha_3 = -0.318$ (FOSM: -0.448), $K = 3$. Thus

$$\sum_{k=1}^{K} \alpha_k^2 = 1.273 > 1 \quad \text{(conservatively)}.$$

We can now evaluate (7.12) approximately. We obtain $x_1^* = 295\,\text{N/mm}^2$ (FOSM: $330\,\text{N/mm}^2$), $x_2^* = 154\,\text{mm}^2$ (FOSM: $144\,\text{mm}^2$), $x_3^* = 44.7\,\text{kN}$ (FOSM: $46.6\,\text{kN}$), and the structure function $g(x) = g(f, s, a) = f \cdot s - a$ of the truss element under consideration gives, with $\beta = -2.8$ (estimate)

$$g(x_k^*|\beta = -2.8) = [154 \cdot 0.295 - 44.7]\text{kN} = 0.73\,\text{kN} > 0,$$

i.e. the element is reliable within the framework prescribed by $\beta = -2.8$. With $\beta = -3.2$ (estimate) then

$$g(x_k^*|\beta = -3.2) = [152 \cdot 0.278 - 45.2]\text{kN} = -3\,\text{kN} < 0,$$

i.e. the element is not reliable within the framework prescribed by $\beta = -3.2$. We can see that this simple reliability test provides iteratively a good approximation for the reliability index and the design point, $-3.2 < \beta < -2.8$ (FOSM: -3.07), so if the mean values of the basic parameters are given, all associated partial central safety factors can be determined, including that of the load

$$\gamma_{0k} = x_k^*/m_{Xk} \quad (X_k \in A),$$
$$\gamma_{0k} = m_{Xk}/x_k^* \quad (X_k \in W). \tag{7.16}$$

(The same applies to the partial nominal safety factors with nominal and mean values.)

We have kept the structure functions of the truss element under tension as simple as possible to demonstrate the basic relationships. The principles developed here are directly applicable to studies of similar structural elements in marine technology (e.g. tendons of a tension leg platform or trusses of a jacket platform subjected to tensile loads). Of course, there is a whole range of appropriate simplifications or extensions, depending on the requirements of the particular structural element and its structure function (see [25, 26]).

The rational (reliability-oriented) determination of safety factors requires consideration of another problem, that of the appropriate combination of loads which are associated with different, simultaneously possible random processes.

We therefore take another look at a consequence of the Turkstra rule which is particularly important for practical applications. We consider the modified Turkstra rule, which is also called the Borges–Castanheta rule [27], instead of (6.65): (6.65):

$$A_{\max,\text{T}} \simeq \text{Max}\left[\underset{[0,T]}{\text{Max}\,[A_1]} + \underset{[t,t+\theta_1]}{\text{Max}\,[A_2]},\, A_1 + \underset{[0,T]}{\text{Max}\,[A_2]} \right], \tag{7.17}$$

$T/\theta_1 = N$ and N being an integer (see Fig. 6.18 with A instead of Y to symbolize the load in the rule format). Often the $p\%$ fractile ($F^{-1}(p)$) of the superposed extremal loading $A_{\max,\text{T}}$ is adequately approximated by the equation [28]

$$F^{-1}_{A\max,\text{T}}(p) \simeq \text{Max}[F^{-1}_{A1\max,\text{T}}(p) + F^{-1}_{A2\max,\theta1}(0.5),$$
$$F^{-1}_{A1}(0.5) + F^{-1}_{A2\max,\text{T}}(p)]. \tag{7.18}$$

For our purposes, the only important factor here is that the notation in (7.18) indicates an appropriate way of designing the safety formats used in regulations with regard to the combination of loads using partial safety and load combination factors, $\gamma_{Nk} = \gamma_N(A_k)$ or $\Psi_{kj} = \Psi(A_k, A_j)$, for reliability purposes. Using (7.18) for comparison we can develop meaningful safety formats of the form

$$\gamma_{n1} \cdot a_{n1} + \gamma_{n2} \cdot \Psi_{12} \cdot a_{n2} \leqslant w_n / \gamma_{nW} \tag{7.19}$$

or

$$\gamma_{n2} \cdot a_{n2} + \gamma_{n1} \cdot \Psi_{21} \cdot a_{n1} \leqslant w_n / \gamma_{nW} \tag{7.20}$$

with appropriate values for the partial load safety factors γ_{nk} or load combination factors Ψ_{kj}, and always decide in favour of the format with the greatest left-hand side.

In practical code development, load combination factors are seldom used explicitly. They are, however, included implicitly in the derivation of partial safety factors using so-called repetition numbers when deriving the maximum distribution of a load. We briefly exemplify this with a safety format developed for vertical bending loads on container ships [20]. In this case, random processes of wave (index s) and still water (index g) loads had to be superposed. The simplified safety format with partial safety factors is then

$$\gamma_{mg} \cdot m_{ng} + \gamma_{ms} \cdot m_{ns} \leqslant w_n / \gamma_W, \tag{7.21}$$

with m_{ng} and m_{ns} being nominal code values of still water and wave bending moments, and w_n the related nominal value of w.

For reliability analysis, a structural state may be defined by the structure function

$$G = W - M_g - M_s.$$

$G > 0$ indicates safe states of the structure, $G < 0$ indicates failure states as defined by the meaning of W, e.g. strength as a function of an upper working stress limit σ_P of linear elastic behaviour. W and M are random numbers of vertical bending moments with realizations w and m respectively. According to (7.17) – second alternative – we consider

$$M = M_g + M_s = M_g + \text{Max}(M_{sR})$$

or, with $N = 1$ and $M = 20$ in (6.64),

$$F_M(m) = \int_0^\infty [F_{MsR}(m - m_g)]^{20} \cdot f_{Mg}(m_g) \, dm_g. \tag{7.22}$$

$[F_{MsR}(\cdot)]^{20}$ is the extreme value distribution of the maximum of 20 independent random samples of M_s, having a stochastic model $F_{MsR}(\cdot)$, where index R refers to one voyage of, for example, 10 days on the North Atlantic shipping route. Thus, (7.22) defines the probability that $M = M_g + \text{Max}(M_{sR})$ is less than m for 20 voyages. Including harbour stops, this is about one year of operation. Hence, the estimated risk obtained in a reliability analysis may be interpreted either as the annual risk of $G < 0$ for one ship, or as the probability that one of 20 identical ships, which operate on North Atlantic shipping routes at the same time but independently of another, experiences the condition $G < 0$. This condition does not describe an ultimate limit state of the structure, but a serviceability limit state given as the exceeding of the nominal value of the strength against bending loads in a section close to mid-ship, i.e. basically exceeding σ_P somewhere in a structural element. This risk is used here as a reference value to base the definition of partial safety factors on a risk or reliability measure that is equal for all ships independent of beam B and length L.

The stochastic models $F_{Mg}(m_g)$, $F_{Ms}(m_s)$ and $F_W(w)$ were derived for the three random numbers M_g, M_s or W in [20]. Thus, three design values m_g^*, m_s^* and w^* could be defined with a standard reliability algorithm described in Chap. 6 [85], as those points on the failure boundary $g(m_g, m_s, w) = w - m_g - m_s = 0$ that have maximum probability.

Hence, partial safety factors were derived as

$$\gamma_{mg} = m_g^*/m_{ng}, \quad \gamma_{ms} = m_s^*/m_{ns}, \quad \gamma_w = w_n/w^*. \tag{7.23}$$

It is a characteristic of the reliability algorithm to estimate the reliability R or the reliability index β together with the design points (m_g^*, m_s^*, w^*) of the basic parameters (M_g, M_s, W), i.e. partial safety factors $(\gamma_{mg}, \gamma_{ms}, \gamma_w)$ were defined together with the related β value. In order to obtain the same reliability for all ships designed with these partial safety factors, the sample ship's scantlings were influenced in such a way that always the same target β value was obtained ($\beta = -2.88$). The partial safety factors related to this target β value were then used in a safety format, with the partial safety factors of strength included in the partial safety factors for the loads, i.e. a safety format analogous to (7.21) was introduced as

$$\gamma_{mg} \cdot \gamma_w \cdot m_{ng} + \gamma_{ms} \cdot \gamma_w \cdot m_{ns} \leq w_n.$$

Non-linear two-dimensional regression for beam B and length L of the sample container ships finally leads to the following prediction formulae for the resulting products of partial safety factors:

$$\gamma_{mg} \cdot \gamma_w = 71258 + 2.631E - 6 \cdot L^{2.2} - 71257 \cdot (B/L)^{(2/L^3)}$$

$$\gamma_{ms} \cdot \gamma_w = 5.6 - 0.014 \cdot L^{1.05} - 2270 \cdot (B/L^2)^{1.075},$$

at 95% confidence limits. These partial safety factors imply load combination factors (cf. (7.22)), but do not give them explicitly. With respect to the Borges–Castanheta Rule, we must say that these results are valid only under the assumption that in this example the second alternative in (7.17) gives higher design loads than the first.

Similar to (7.21) is the (limit state) load and resistance factor design (LFRD) format of the API–RP2A published for comment in 1989 [29], being extended to the North Sea environment (wave climate) in 1991 by a joint industry project [30].

7.4.2 Principles of Quality Assurance and Classification

The Joint Committee on Structural Safety (JCSS) has developed a clear presentation of a recent quality assurance concept [31]. In connection with this, we summarize some selected aspects of quality assurance (QA) in order to define the concept axiomatically, as it is not always used in the same way.

QA is always based on a number of structure-related requirements (performance requirements) to be met by the performance of the structure in service (i.e. on fixed engineering rules), in particular recommendations or regulations for the construction and operation of structures which must satisfy essentially three quality criteria: safety, serviceability and durability. The term *safety* implies here the requirement that the likelihood of material damage or loss of human life should be acceptably small. The term *serviceability* implies here only a recommendation that the likelihood of the structure becoming unserviceable should be acceptably small. The term *durability* means that the requirement of safety and the recommendation for serviceability should apply for the planned service life of the structure. What may constitute an acceptably small likelihood is not specified by the QA concept: this concept defines relevant measures to provide quality, but not the acceptable damage risk level itself. The following measures are considered relevant to quality assurance.

The QA concept should take account of all foreseeable hazard scenarios including all foreseeable accident situations and all foreseeable utilization scenarios, as well as possible combinations of these factors.

The QA concept should ensure that life-long critical technical support and control of the whole process between planning and scrapping is possible, i.e. during design, construction, operation and demolition, and that provision is made for maintenance and repair.

The QA concept should also ensure that all significant documents relating to the whole design process and service life of the structure should be accessible and continually updated.

Finally, the QA concept should ensure that adequate measures are taken to prevent human error.

We would like to compare these fundamental elements of the JCSS QA concept, which is described in great detail in [31], with the corresponding elements of a classification concept: we again make reference to the Germanischer Lloyd regulations on the construction and testing of marine structures [17], which are fairly clearly inclined to the classical concept of the classification of ships.

These regulations contain structure-related requirements for marine structures (performance requirements) in the form of dimensioning principles. Various load cases are defined for which specific loading values can be given in the regulation, and the stresses which can be calculated with them must be combined in the way prescribed by safety formats to give a comparative value of stress. If the components are designed appropriately, the comparative value of stress is smaller than or equal to the load-bearing capacity (strength). The various safety formats take adequate account of the different modes of failure, e.g. material yield, loss of stability of the component, fatigue of an element of the structure, etc. (see examples in Sects. 7.2 and 7.3). In addition, it is stipulated that certain component deformation limits are not exceeded, e.g. with the express requirement that the stiffness of the load-bearing parts of the structure is such that the normal operating functions are not impaired by deformations.

The classification regulation we have examined thus contains all quality criteria which are significant for the QA concept (safety, serviceability, durability), with the addition of an (implicit) specification of an acceptable damage risk, by using certain safety factors or comparable safety measures in the rule format, as, according to the relationships described in Sect. 7.4.1, there is a connection between the safety factor and the reliability index, the latter being a measure of the probability of safety (see text on (6.106) to (6.109)).

As far as the other relevant measures for quality assurance, mentioned in the *QA* concept, are concerned (control, documentation and human error), in the general survey and design review conditions of the *GL* regulation we find a list of requirements concerning scope and procedure of surveys and reviews in which questions of control (monitoring of construction and operation, inspections) and documentation (design documents) are dealt with in great detail, though with different points of emphasis compared with the JCSS *QA* concept. In this respect, there is no fundamental difference between quality assurance and classification.

With regard to the 'human risk factor', in the *GL* regulation examined there is an example of a qualification requirement for the welders working in steel construction (welding supervisors' and welders' qualification certificates), but in the much wider JCSS interpretation in [31] the possibility of human error as a controllable risk factor in the construction and operation of marine structures is not considered in the classification. The API–RP2A recommendation goes a little further in this regard [8], requirements being formulated for the qualifications of the inspection personnel, and significantly not only for the training of the inspectors, but also for their experience. In future there will probably be regulations which pay more attention to the risk of conscious or unconscious negligence, ambiguous responsibilities, inadequate working procedures, or similar risk factors within the sphere of influence of people engaged in the planning, design, construction and operation of marine structures. But these problems are beyond the scope of this book, which is primarily concerned with the specific aspects of the analysis and evaluation of marine structures.

7.5 Examples of Structural Components

Although dimensioning in marine construction differs both in range and method from normal practice in steelwork or shipbuilding, the structural solutions used are nevertheless quite similar. This is not surprising, since the great majority of marine structures are produced by naval architects or civil engineers, who tend to bring their own experience to bear on actual steelwork projects.

Such typical steel structures can be found in the two German platforms of the *Schwedeneck* oilfield in the Baltic (see Fig. 7.9). In each case, a two-stage steel platform stands on a concrete foundation containing intermediate storage tanks. Both platforms rest on four pillars which are fixed to a 'corbelring' at the transition from the pillars to the circular concrete structure. This corbelring consists of a massive cross structure, to the arms of which the pillars are fixed (see Fig. 7.10). The load is then transferred to the concrete structure via a circular frame. With a diameter of 20 m, this corbelring alone can be considered a large structural component. To ensure correct fitting, exceptionally small tolerances of a few millimetres were specified, and the manufacture therefore had to be carried out with extreme accuracy and special attention was paid to shrinkage and warping.

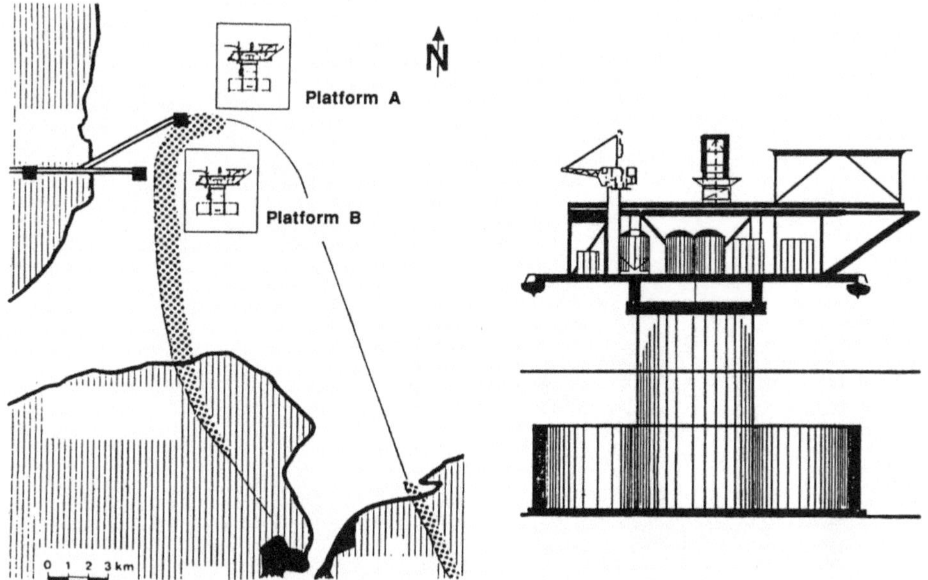

Fig. 7.9. Offshore platform Schwedeneck-See (Baltic Sea)

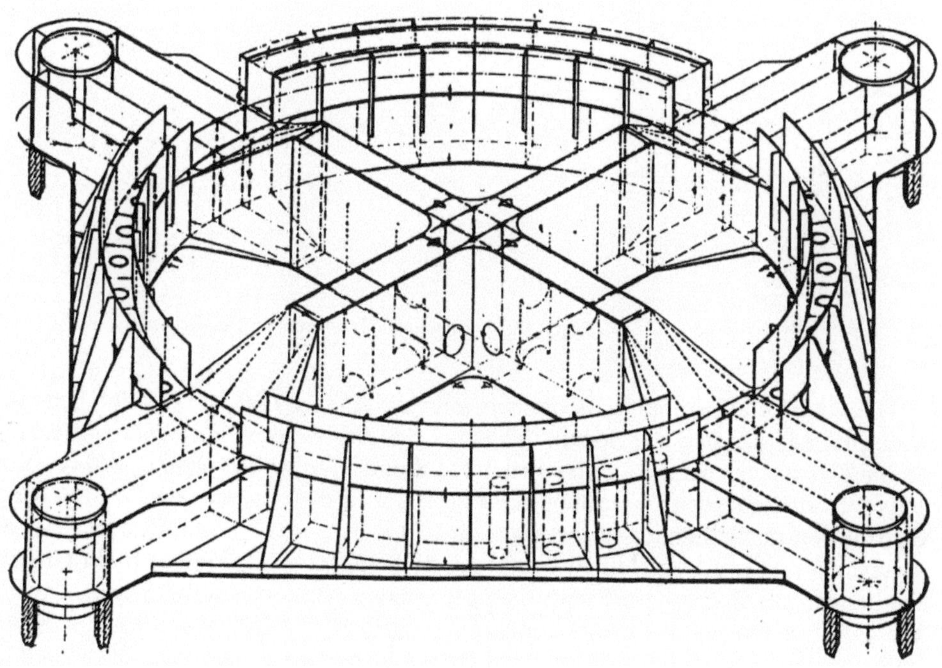

Fig. 7.10. Steel structure of a Corbelring of Schwedeneck-See platform types A and B, as built by HDW, Kiel.

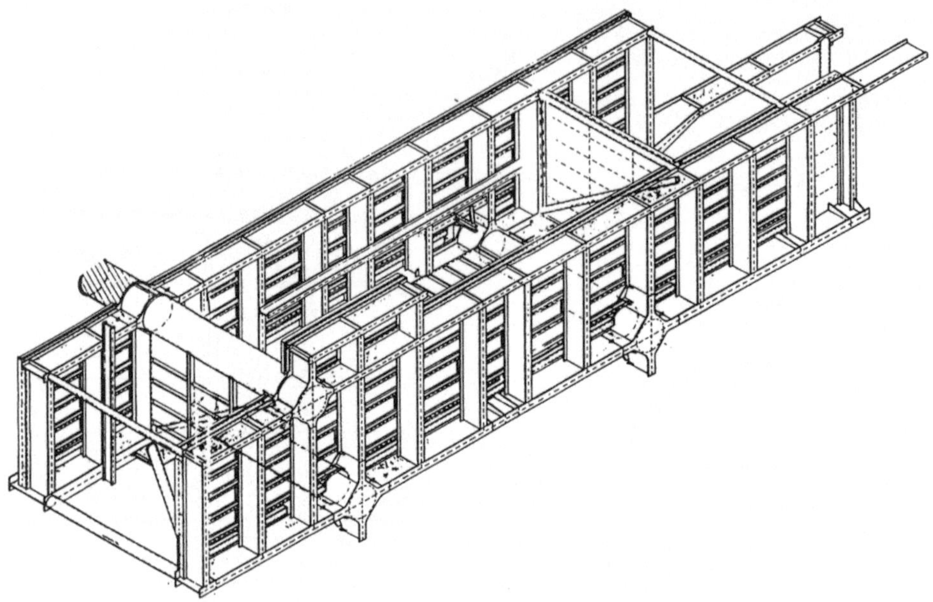

Fig. 7.11. Installation drawing of a part of Schwedeneck-See platform A.

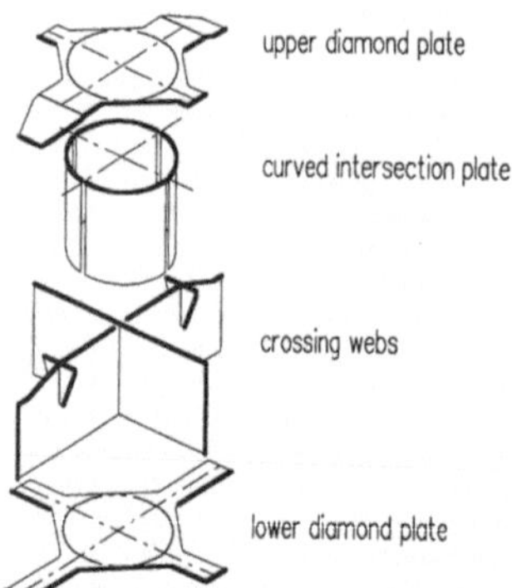

Fig. 7.12. Connection of columns and main girders

The actual platform consists of deep-web girders (see Fig. 7.11), which bear the deck-plating, assisted by local stiffeners. Special welding and manufacturing techniques are necessary for the structural detail involved in joining the support pillars to the load-bearing lattice (see Fig. 7.12). In order to ensure that the force distribution in the flanges of the main girders is as even as possible, and that the pressure forces are transformed continuously from the pillars to the main girders, the flanges are strengthened with diamond plates. This makes for manufacturing simplicity at the joint, since the pillars are welded via double-level grooves flat onto the diamond plates at both sides. Between the flanges the webs of the beams run crosswise along with four dished connecting plates, giving the impression that the pillars run straight through.

Large components are frequently constructed separately on land and then fitted together on site at sea, this still being more economic, despite the high costs of transport and on-site assembly of massive components. This is especially true of the North Sea and the Baltic, where weather windows are so narrow that assembly of many small components at sea is hardly possible. For this reason, such massive components must also be provided with suitable lifting eyes for the transport. Thus lifting eyes must be very carefully constructed, since they may have to bear weights up to several hundred tonnes. One should also note with regard to the dimensioning of structures that transport loadings can be of considerable significance, on a par with working load.

Living quarters form a special group of marine structures in the North Sea. They provide accommodation for drilling platform crew for whom daily transport back to land from their inhospitable workplaces is too costly. This accommodation in many cases consists of unused production platforms or semisubmersibles no longer needed for their original purpose. In some cases living quarters are also purpose built (see Fig. 7.13).

What they all have in common, however, is that they must provide a reasonable free-time environment for their inhabitants, with comfortable rest areas, common rooms, and sport and leisure facilities, while also conforming to increased safety requirements. People in these living quarters are in effect on holiday, and should therefore be able to expect a similar level of safety to that of comparable installations on land.

These structures are therefore subject to stringent construction standards. The actual living quarters are mostly block-type constructions, built as modules on land and shipped out to the site on crane-barges, where they are hoisted onto, for example, a prepared cradle. Floating units usually have living quarters installed directly at the building yard.

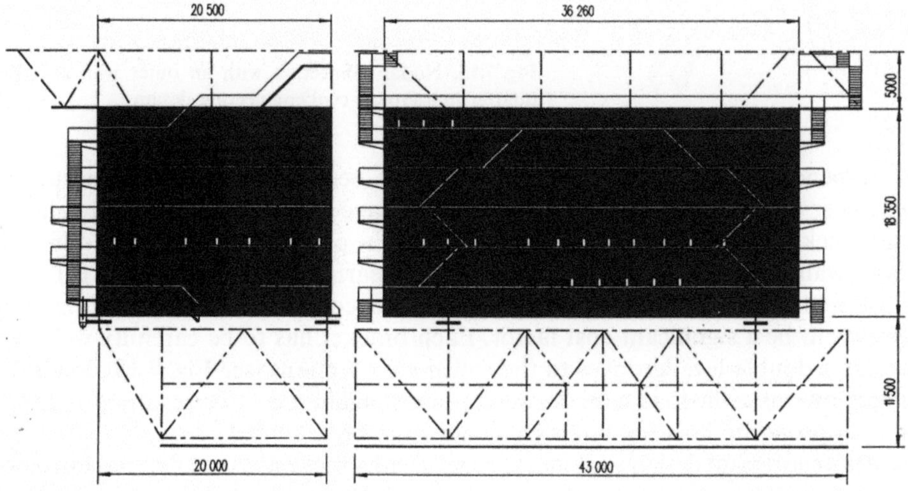

Fig. 7.13. View of living quarters of a platform, as built by Blohm & Voss, Hamburg.

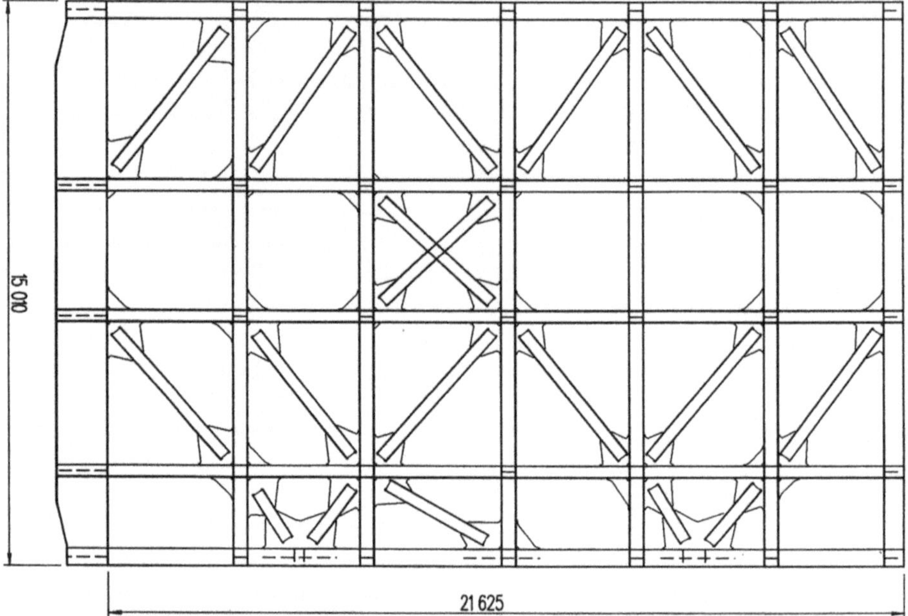

Fig. 7.14. Structure of an outer wall of a living quarter in a typical civil engineering design.

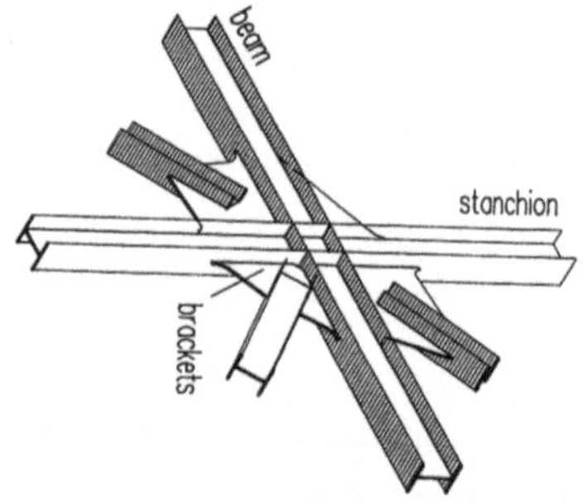

Fig. 7.15. Nodal connection with an outer wall of a living quarter in a typical civil engineering design

Depending on whether these modules have been initiated by a constructional engineer or a naval architect, they will either have a steel framework with cladding and decking (see Fig. 7.14), or, as with ships, a plate-stiffener structure in which both walls and decks form an integral load-bearing system (see Fig. 7.16).

The former structure requires a lot of end corrections with brackets, which proves to be a significant cost factor. Each bracket has to be carefully welded via suitable double-level grooves to the rolled-steel sections (see Fig. 7.15). With these living quarters, the load case during transport again has to be regarded as critical, and appropriate hoisting reinforcements must be provided.

Another typical design problem in marine construction is the local reinforcements to transmit the gravity forces of jack-up platforms from the pontoons to the legs. From a static viewpoint, the leg must be fixed as firmly as possible to the

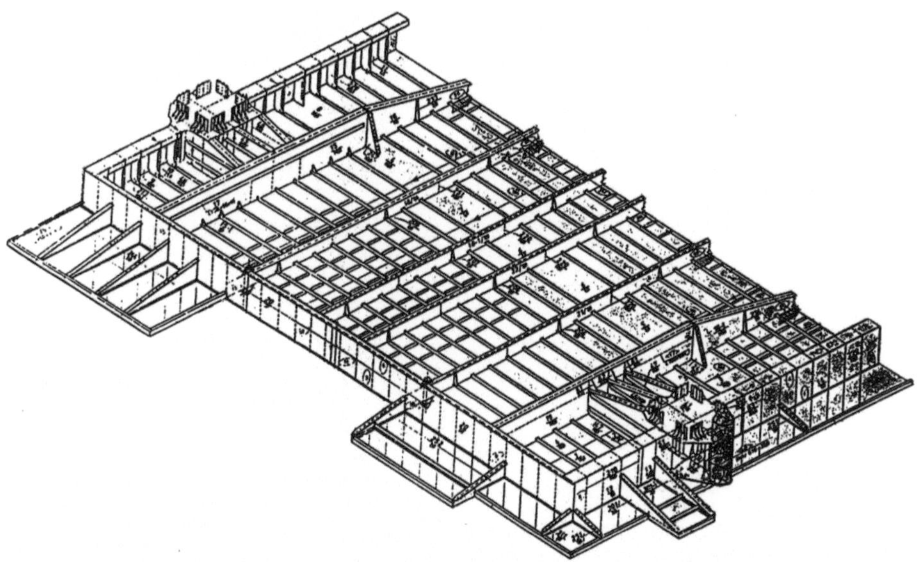

Fig. 7.16. Deck structure connection with walls of a living quarter in a typical ship building design.

pontoon. However, it can be seen that if, under lateral displacement, the leg remains vertical at the point where it enters the pontoon, then the 'buckling length' l of the leg is equal to $2H$. The critical buckling load is then given by (4.91)

$$N_{ki} = (EI^2\pi^2)/(4H^2).$$

Since the buckling length l appears as a quadratic term in this expression for the buckling load, it is clear that a small rotation of the leg in the pontoon will produce a significant increase in buckling length, and thus a sharp reduction in bucklings loads. Also, since the pontoon is generally stiffer than the leg, the buckling length depends in practice only on the local stiffness in the region of the joint between pontoon and leg, expressed in terms of the spring constant c_φ. By approximation, then, the buckling length is

$$l = 2H\sqrt{1 + (3EI)/(c_\varphi \cdot H)},$$

and the buckling load is

$$N_{ki} = (EI^2\pi^2)/[4H^2\{1 + (3EI)/c_\varphi \cdot H)\}].$$

In other words, it is desirable to ensure that the force-transmission structure at the joint between pontoon and leg is as stiff as possible. With tubular structures, a small amount of play is desirable, so that the horizontal force, P_H is distributed as evenly as possible in the tube, thus avoiding local buckling. Tubes are in general relatively stiff against localized loading.

It is rather different when lattice rather than tubular legs are used. Here, the horizontal forces which make up the restraining moment $P_H \cdot h$ give rise to considerable deformations in lattice-framework legs (see Fig. 7.17). In such cases,

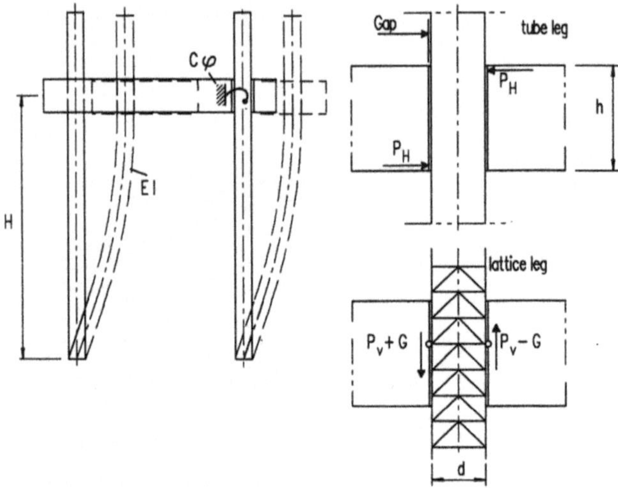

Fig. 7.17. Leg bearing in a jack-up structure.

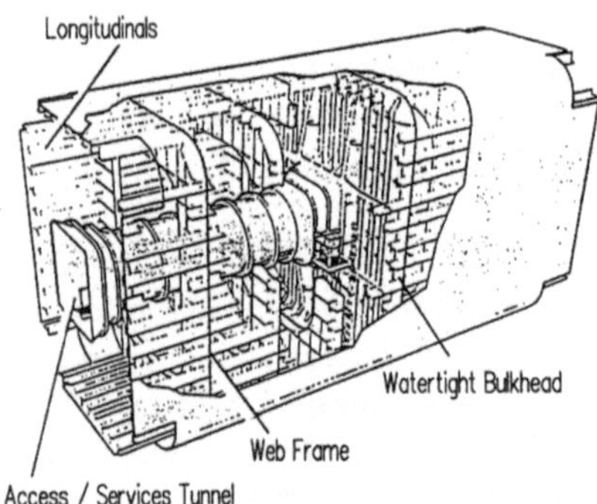

Fig. 7.18. Steel structure of a horizontal deck section of the Conoco platform, Ocean Industry, August 1984.

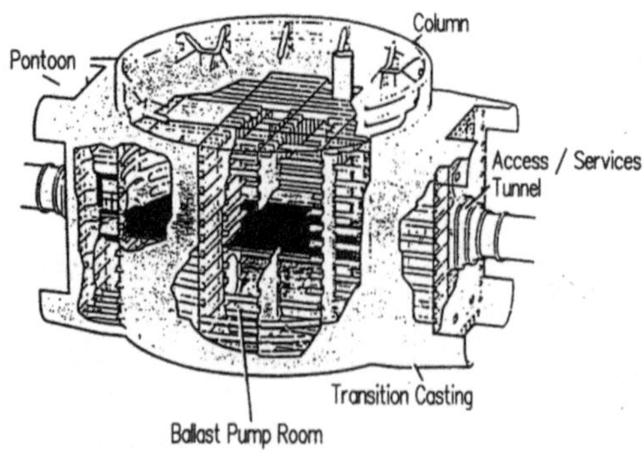

Fig. 7.19. Connection between columns and the horizontal deck section of the Conoco platform, Ocean Industry, August 1984.

it is preferable to have vertical force transmission through the strong vertical sections of the lattice. Now the necessary couple is produced by the moment $P_{\mathrm{v}} \cdot d$. With tubular legs, the weight forces of a pontoon are usually taken up by special lifting equipment placed above the pontoon, whereas with lattice-type legs, it is more practical to transfer the weight at the same points as P_{v}.

In order to obtain an impression of the constructional design of floating marine structures, the design details of a typical TLP are given in Fig. 7.18 [43], which shows the Conoco platform. The cut-out view of the pontoon structure shows a ship-like structure of longitudinals supported by strong transverse framing (Fig. 7.19). The connecting pieces between the rectangular pontoon sections and the vertical

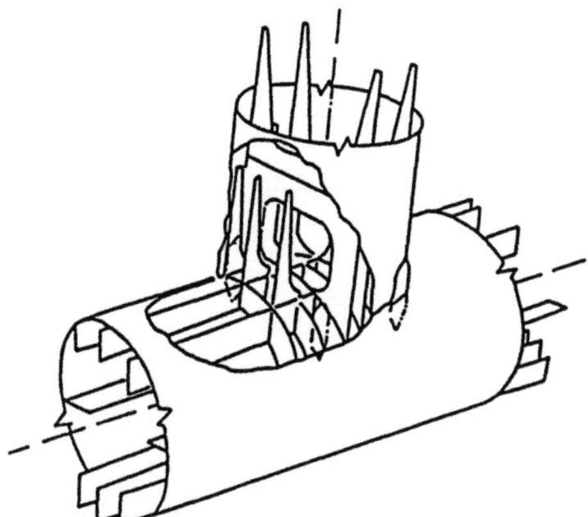

Fig. 7.20. Typical connection of stiffened cylindrical elements of an offshore structure.

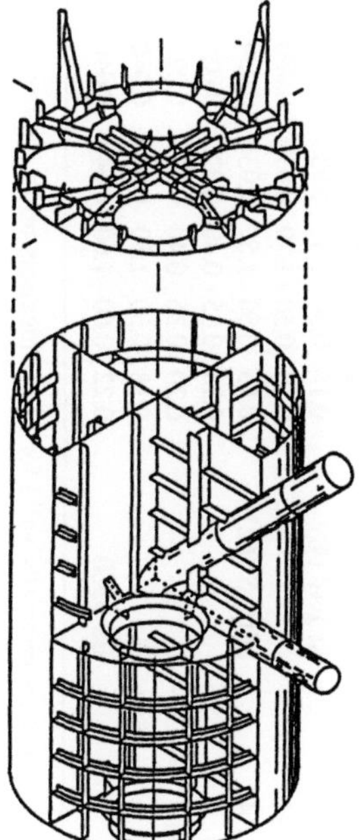

Fig. 7.21. Connection of unstiffened braces with stiffened chords.

Table 7.1. Materials for offshore engineering [1]

Steel type	Quality standard	C	Si	Mn	P	S	Cr	Cu	Mo	Ni	Nb V	Ti Zr	Yield stress N/mm²	Tensile stress N/mm²	Breaking elongation %	Notch impact energy °C	J	Other
D		0.21	0.35	1.40	0.040	0.040							235	400–490	22	−20	27	Al > 0.02%
E		0.18	0.35	1.50	0.040	0.040							235	400–490	22	−40	27	Al > 0.02%
DH 36		0.18	0.50	1.60	0.040	0.040	0.20	0.35	0.08	0.40	0.05 0.10		355	490–620	21	−20	34	Al 0.07%
EH 36	GL, LR, DnV	0.18	0.50	1.60	0.040	0.040	0.20	0.35	0.08	0.40	0.05 0.10		355	490–620	21	−40	34	
A 36		0.25		1.20	0.040	0.050		0.20					250	400–550	23			
A 514/A 517 Type A		0.21	0.80	1.10	0.035	0.040	0.80		0.28			0.13	689	793–931	16	−75	27	B 0.025%
A 514/A 517 Type G		0.21	0.90	1.10	0.035	0.040	0.90		0.60			0.13	689	793–931	16	−75	27	B 0.025%
A 514/A 517 Type F		0.20	0.35	1.00	0.035	0.040	0.65	0.50	0.60	1.00	0.08		689	793–931	16	−46	27	B 0.006%
A 533 Type C Cl.2		0.25	0.32	1.50	0.035	0.040			0.60	1.00			485	621–793	16			
A 537 Gr.A	ASTM	0.24	0.50	1.35	0.035	0.040							345	483–621	22	−50	27	
A 537 Gr.B		0.24	0.50	1.35	0.035	0.040							414	552–689	22	−60	27	
A 543 Type A Cl.1		0.23	0.35	0.40	0.035	0.040	2.00		0.60	3.25	0.03		586	724–862	14			
A 572 Gr.55		0.23	0.30	1.35	0.040	0.050							380	>485	20			
A 633 Gr.E(N)		0.22	0.50	1.50	0.040	0.050					0.11		415	522–690	23	−50	27	N 0.03%
A 633 Gr.E(QT)		0.22	0.50	1.50	0.040	0.050							520–670	<770	19	−50	27	
A 699		0.12	0.55	1.90	0.025	0.025			0.35		0.02		355	490–630	24	−75	27	
TT St E39	St EW 089	0.16	0.50	1.60	0.030	0.030					0.16		385	500–600	20	−50	27	
TT St E43		0.18	0.50	1.70	0.030	0.030				0.70	0.18		420	530–680	19	−50	27	
TT St E47	MIL-S 16216	0.15	0.50	1.50	0.030	0.030		0.70		0.70	0.18		460	560–730	17	−50	27	
HY 80		0.18	0.35	0.40	0.025	0.025	1.80		0.60	3.25			550		19	−85	67	
HY 100		0.20	0.35	0.40	0.025	0.025	1.80		0.60	3.50			685		17	−85	67	
50 D		0.18	0.50	1.50	0.040	0.040					0.10		355	500–620	20	−30	27	
55 E	BS 4360	0.22	0.60	1.60	0.040	0.040					0.10		450	550–700	17	−50	27	

Specified values are generally valid for plate thickness up to approx. 35 mm pillars, which pose difficult welding problems, are cast steel components. Individual sections are separated by watertight bulkheads typical of shipbuilding design.

Many floating marine structures require tubular structures of varying diameters to be interconnected, as shown in Fig. 7.20. The typical tubular nodes represent a special class of structures. Their treatment is a special subject, and is extensively covered in several monographs. Just as interesting from a design point of view are the connections between unstiffened tubes of smaller diameter with stiffened pillars (see Fig. 7.21). Here, quite complex steel structures are realized, including consideration of the widely varying functions of components, dimensioning criteria and assembly requirements, the latter being frequently decisive. For stiffening, rolled steel flats, flanged beams, and rolled girders are used. The erection of these partly geometrically developed structures is effected by specially made templates. To obtain as many suitable welding positions as possible, the components are frequently rotated during welding, which must again be considered in design.

7.6 List of Symbols

A	stress (random parameters)
$A_{\mathrm{max},T}$	extremal value of A in time period T (random parameters)
A_t	stiffener cross-section including plate
A_{te}	effective flange area (attention to post-buckling)
B	ship beam
E	modulus of elasticity
F	probability distribution function, e.g. $F_A(a_P) = P[A \leqslant a_P] = p$
F^{-1}	inverse function of F, e.g. $F_A^{-1}(p) = a_P$
F_{Mg}	distribution function for still water bending load
F_{Ms}	distribution function for wave bending load
F_{W}	distribution function for bending strength
FS	safety factor
F_y	yield stress
G	random structure function
H	wave height
H_{B}	design wave height
H_{V}	visually observed wave height
I	moment of inertia
L	wave length
L	ship length
M_g	random still water bending load
M_s	random wave bending load
P	probability
Q	total horizontal wave force acting on the structure
Q_{max}	maximum value of Q during a wave cycle
Q_{min}	minimum value of Q during a wave cycle
S	normal stress (random value)
T	wave period $2\pi/\omega$ or time interval $0 \leqslant t \leqslant T$

T_B	design wave period
T_R	reference time interval for fatigue strength
T_V	visually observed wave period
V	coefficient of variation (standard deviation σ/mean value m)
V_A	coefficient of variation of stress
V_W	coefficient of variation of strength
V_k	coefficient of variation of random parameter X_k
W	strength (random parameter)
X_k	random parameters, $k = 1, 2, \ldots, K$
a	realization of A
a	stiffener length
a_n	nominal value of A
a_p	$p\%$ fractile of A: $a_p = F_A^{-1}(p), p = p\%/100$
a^*	design value of A
b	stiffener spacing
f_{Mg}	probability density function of still water bending load
f_y	yield stress
g	strength parameter
$g(\cdot)$	structure function
h	wall thickness of regular cylinder
j	sequential index
k	sequential index
m	mean value
m_A	mean value of stress A
m_W	mean value of strength W
m_g^*	design value of still water bending load
m_s^*	design value of wave bending load
m_{ng}	nominal value of still water bending load
m_{ns}	nominal value of wave bending load
m_{Xk}	mean value of random parameter X_k
n_R	number of waves in T_R
p	pressure
p	measure of probability: $0 \leqslant p \leqslant 1$
$p\%$	measure of probability p in %: $0 \leqslant p\% \leqslant 100$
r	radius of inertia
r	radius of regular cylinder
s	stress range
s_R	design stress range for T_R
s^*	pseudo stress range $\sigma_{max} \cdot (1 - \varkappa^*)$
t	time variable
w_n	nominal value of strength W
w^*	design value of strength W
x_k^*	design value of X_k
Ψ_{kj}	load combination factor, k and j are load indices
α	sensitivity factor
α	knock-down factor (actual/theoretical buckling load)

α_k	sensitivity factor for X_k
β	reliability index
β	degree of slenderness
γ	safety factor
γ_{mg}	partial safety factor for still water bending load
γ_{ms}	partial safety factor for wave bending load
γ_{nk}	partial nominal safety factor for A_k
γ_{nW}	partial nominal safety factor for W
γ_W	partial safety factor for strength
γ_0	global central safety factor
γ_{0A}	partial central safety factor for A
γ_{0W}	partial central safety factor for W
δ	factor for degree of slenderness $(\bar{\lambda}^2)$
θ	time interval $T/N, N$ is integer
$\varkappa^*$	pseudo stress limit ratio Q_{min}/Q_{max}
$\bar{\lambda}$	degree of slenderness
μ_V	visually observed main direction of seaway
σ	normal stress
σ	standard deviation
σ_A	standard deviation of stress
σ_B	bending stress
σ_F	yield stress
σ_W	standard deviation of strength
σ_m	mean value of stress during a wave cycle
σ_{max}	maximum stress during a wave cycle
σ_{min}	minimum stress during a wave cycle
ψ_k	multiplication factor
ω	wave frequency $2\pi/T$

References

1 Schonfeld H. Fertigung und Stahlwerkstoffe (Production and steels). Handbuch der Werften. Hamburg: Hansa (1980)

2 Rules for the design, construction and inspection of offshore structures. Oslo· Det norske Veritas (1981)

3 Carlsen CA. Collapse of stiffened panels in compression. Det norske Veritas 79–306 (1976)

4 Carlsen CA. Simplified collapse analysis of stiffened plates. Norwegian Maritime Res. (1977) 4: 20–36

5 Faulkner D. Design against collapse for marine structures Int. Symp. Adv. Marine Techn., Trondheim (1979)

6 Europ. Rec. Steel Constr., 4.6: Buckling of Shells. New York: The Construction Press (1981)

7 Boiler and Pressure Vessel Code. Case N 284 Sect. III. Div. 1 Class MC. Washington: ASME (1987)

8 Recommended practice for planning, designing and constructing fixed offshore platforms. RP2A. Washington: Am. Petr. Inst. (1987 and 1989)

9 Rules for the construction and inspection of offshore structures. Oslo: Det norske Veritas (1977)

10 Bornscheuer B-F. Einheitliches Bemessungskonzept für gedrückte Schalen, Platten und Stäbe aus Baustahl (Uniform dimensioning concept for shells, plates and trusses under compression). Stuttgart: Forschungsberichte Inst. für Tragkonstr. und Entwerfen (1984)

11 Stracke M. Stabilität kurzer stählerner Kreiszylinderschalen unter Außendruck (Stability of short steel circular cylinders under external pressure). Dtsch. Verb. für Schweißtechnik, Forschungsbericht (1987) 12

12 UEG Offshore Res. Design of tubular joints for offshore structures. Norwich: Page Bros. Ltd. (1985)

13 Proc. Int. Conf. on Fatigue and Crack Growth in Offshore Structures, London (1986)

14 Int. Conf. on Welding in Offshore Constructions. Newcastle: Welding Inst. (1974)

15 Bainbridge CA. Fatigue analysis offshore structures. Trans. Inst. Marine Eng. London: Marine Management Ltd. (1986)

16 Dept. of Energy. Background to new fatigue design. Guidance for steel welded joints in offshore structures. Rev. Guidance Notes Drafting Panel. London: HMSO (1984)

17 Rules for classification and construction, offshore technology. Part 2 – Offshore Installations, 1990 Edition. Hamburg: Germanischer Lloyd (1990)

18 CEC Eurocode 4. Gemeinsame einheitliche Regeln für Verbundkonstruktionen aus Stahl und Beton (Community unified rules for composite structures of steel and concrete). Report EUR 9886 DE (1985)

19 Mansour AE, Ian HY, Zigelman CI, Chen YN, Harding SJ. Implementation of reliability methods to marine structures. Proc. SNAME (1985) 353–377

20 Östergaard C. Partial safety factors for bending loads on container-ships. J. Offshore Mechanics Arctic Eng. 114 (1992) 129–136

21 Östergaard C. Neuere Sicherheitsformate für Vorschriften, die sich an der Zuverlässigkeitstechnik orientieren (New safety formats for regulations based on the reliability concept). Continuing Education Course 22, Institute of Naval Architecture (IFS), Hamburg (1986)

22 Rackwitz R. Implementation of probabilistic safety concepts in design and organizational codes. In: Moan TE, Shinozuka M (Eds.). Structural safety and reliability. Amsterdam: Elsevier (1981) 593–614

23 NaBau. Grundlagen zur Festlegung von Sicherheitsanforderungen für bauliche Anlagen (Basics for establishing safety standards for structural installations). Dtsch. Inst. für Normung (DIN), Berlin: Beuth (1981)

24 Int. Organization for Standardization (ISO). General principles on reliability for structures. ISO Draft ISO/DIS 2394 (1984)

25 Ravindra MK, Galambos TV. Load and resistance factor design for steel. J. Struct. Div. ASCE 104, ST9 (1978) 1337–1353

26 Yura JA, Galambos TV, Ravindra MK. The bending resistance of steel beams. J. Struct. Div. ASCE 104, ST9 (1978) 1355–1370

27 Borges IF, Castanheta M. Structural safety, 2nd. ed. Lisbon: Nat. Civil Eng. Lab. (1971)

28 Turkstra CJ, Madsen HO. Load combinations in codified structural design. J. Struct. Div. ASCE 106 ST12 (1980)2527–2543

29 Draft Recommended practice for planning, designing and constructing fixed offshore platforms – load and resistance factory design. RP2A – LRFD. Washington: American Petroleum Institute (1989)

30 API RP2A – LRFD, its consequences for and adaption to North Sea Offshore design practice. Project background report on a joint industry project on application of API RP2A LRFD to the North Sea. Guildford (Great Britain): Advanced Mechanics & Engineering Ltd (1991)

31 Joint Committee of Structural Safety (JCSS): General principles on quality assurance for structures. Ber. der Arbeitskommissionen. Bd. 35 IABSE – AIPC – IVBH (1981)

Appendix

A1 Selected Principles of Probability Theory

We begin with the planning of a hypothetical experiment in which we want to measure the difference between the inlet and outlet temperatures in the coolant circuit of an engine. We expect that this temperature difference T will be in the range between 0 and 100 °C, and therefore, before the tests are carried out, we can divide this range into I temperature classes T_i of the extension Δt_i, $i = 1, 2, \ldots, I$. During the tests we want to count how frequently a measurement t can be assigned to each of these classes. The result after M measurements will be the frequency μ_i per temperature class T_i. If we define the term relative frequency by means of

$$v_i = \mu_i/M \tag{A1.1}$$

we know *a priori* that v_i must fulfil the conditions

$$0 \leqslant v_i \leqslant 1, \tag{A1.2}$$

$$\sum_{i=1}^{I} v_i = 1/M \cdot \sum_{i=1}^{I} \mu_i = 1/M \cdot M = 1, \tag{A1.3}$$

$$v_{ij} = (\mu_i + \mu_j)/M = v_i + v_j. \tag{A1.4}$$

We use v_{ij} to indicate the relative frequency of measurements which fall into the class T_i or T_j or both, and we call this the union of two classes. Henceforth we will abbreviate the union with the symbol $\cup$ 'or' taken from set theory: thus $T_i \cup T_j$ means T_i or T_j, or both.

When we have calculated v_i for the i-th class after M measurements, as in (A1.1), and ask ourselves how probable it is that the $(M + 1)$th measurement will bring about a temperature difference which will again fall into the i-th class T_i of the extension Δt_i, we would answer: the probability is v_i, if the results of all measurements can be regarded as random or stochastic. By the terms random or 'stochastic' we mean that we know of no rational reason for the course the experiment takes. So if we are looking for a mathematical formulation of the term probability, it should fulfil the same conditions as apply (*a priori*) to the relative frequency v_i.

In our hypothetical experiment we now want to use the random numbers X instead of the random temperature T, and instead of the sum of all I classes from 0 to 100 °C we define a number set (number class) X_S, the safe class or the safe domain of probability. (Instead of random number we also use the term *stochastic parameter* for X.) Furthermore, instead of I temperature classes T_i, we examine I subsets X_i, which have the extension Δx_i instead of Δt_i. A test result t is now called the realization or value x of the random number X, and we write $x \in X_i$, if this value belongs to the subset X_i; all subsets X_i, $i = 1, 2, \ldots, I$ form jointly the domain X_S of all possible realizations x, which we symbolize with $X_i \subset X_S$. None of these subsets X_i shall coincide with another. Then we can give three axioms, as with (A1.2) to (A1.4), with which we can define the concept of the probability that a realization x of the random number X will fall into the subset X_i, symbolized by $P[x \in X_i]$ (instead of v_i), as follows:

Axiom 1 (domain of probability):

$$0 \leqslant P[x \in X_i] \leqslant 1, \tag{A1.5}$$

Axiom 2:

$$P[x \in X_S] = \sum_{i=1}^{I} P[x \in X_i] = 1, \tag{A1.6}$$

Axiom 3:

$$P[x \in X_i \cup x \in X_j] = P[x \in X_i] + P[x \in X_j]. \tag{A1.7}$$

For any set $X_T \subset X_S$ consisting of a number J of subsets X_i, $1 < J \leqslant I$, we have, according to Axiom 3

$$P[x \in X_T] = P\left[\bigcup_{i=1}^{J} x \in X_i\right] = \sum_{i=1}^{J} P[x \in X_i]. \tag{A1.8}$$

We abbreviate the above notation somewhat and substitute

$$X_i := x \in X_i. \tag{A1.9}$$

In addition, it will be useful to show the probability of infinitesimally small intervals $\Delta x_i \to dx$ of an arbitrarily narrow subset $X_i \to X$. We than write

$$P[X] = P[x < X \leqslant x + dx] = f_X(x)dx, \tag{A1.10}$$

and we call the function $f_X(x)$ the probability density function, taking account of its dimension. Thus we define X as a random number (stochastic parameter), $P[X]$ being the probability that a realization x of the random number X lies in the interval $(x, x + dx]$.

Referring to Axiom 2, we than have

$$\int_{-\infty}^{\infty} f_X(x)dx = 1, \quad -\infty \leqslant x \leqslant \infty \tag{A1.11}$$

and, similarly to (A1.8) we write

$$P[x_i < X \leqslant x_2] = \int_{x_1}^{x_2} f_X(x)dx. \tag{A1.12}$$

In particular, we find that

$$P[X \leqslant x_{\mathrm{p}}] = F_{\mathrm{X}}(x_{\mathrm{p}}) = \int_{-\infty}^{x_{\mathrm{p}}} f_{\mathrm{X}}(x)\mathrm{d}x. \tag{A1.13}$$

Here $F_{\mathrm{X}}(x)$ is called a probability distribution function. Due to Axiom 2, we find for the complementary distribution function $\bar{F}_{\mathrm{X}}(x)$

$$\bar{F}_{\mathrm{X}}(x_{\mathrm{p}}) = \int_{x_{\mathrm{p}}}^{\infty} f_{\mathrm{X}}(x)\mathrm{d}x = 1 - F_{\mathrm{X}}(x_{\mathrm{p}}). \tag{A1.14}$$

The inverse function of (A1.13)

$$x_{\mathrm{p}} = F_{\mathrm{X}}^{-1}(P[X \leqslant x_{\mathrm{p}}])$$

gives a value x_{p}, designated as a $p\%$ fractile, for a probability $P[X \leqslant x_{\mathrm{p}}]$ specified to be between 0 and 1 (0 and 100%). As long as there is no possibility of confusion, $f_{\mathrm{X}}(x)$ and $F_{\mathrm{X}}(x)$ as well as $\bar{F}_{\mathrm{X}}(x)$ can all be called probability distributions. Table A1 summarizes a selection of density functions $f_{\mathrm{X}}(x)$ used in the main text.

By differentiating the integral in (A1.13) with respect to the upper limit $x = x_{\mathrm{p}}$ we also obtain

$$f_{\mathrm{X}}(x) = \mathrm{d}F_{\mathrm{X}}(x)/\mathrm{d}x. \tag{A1.15}$$

Thus we have used the three axiomatic demands on the term probability to develop a way of obtaining probabilities on the basis of distributions, if these distributions fulfil two prerequisites

1. They are acceptable in the sense of (A1.11)
2. They are acceptable for the problem under consideration.

The first condition is met by a whole range of distributions which are called stochastic models in the literature [1]. A selection of them as used in Chaps. 5 and 6 is given in Table A1 on page 330. The second condition, of course, cannot be dealt with generally. Statistical data and/or theoretical formulations, however, relate to a number of practical problems. From time to time we must rely on hypothetical assumptions which we believe will give relatively reliable results.

Up to now we have looked at classes (sets) X_i, $i = 1, 2, \ldots, I$, which must not coincide. Now we consider probability statements related to classes or sets which do overlap. For this purpose we define the joint probability of X_i and X_j by means of

$$\begin{aligned} P[X_i \cap X_j] &= P[X_i | X_j] \cdot P[X_j] \\ &= P[X_j | X_i] \cdot P[X_i]. \end{aligned} \tag{A1.16}$$

The term after the symbol '|' is to be understood as a specified condition: $|X_j$ means that the condition $x \in X_j$ already exists. We can see that $P[X_i | X_j] = 1$ when X_i and X_j overlap completely, and that $P[X_i | X_j] = 0$ when sets X_i and X_j do not coincide (i.e. if they are 'mutually exclusive'). The conditional probability $P[X_i | X_j]$ therefore indicates something about the probable extent of overlap (intersection) of sets X_i and X_k, while the joint probability $P[X_i \cap X_j] = P[X_i, X_j]$ indicates how probable it is that a value x lies in that overlapping subset formed jointly by X_i and X_k. (We use the symbol $\cap$ taken from set theory for the intersection of two sets, interpreted here as the degree of overlap of two sets X_i and X_j.) A generalization to two-dimensional sets is visualized with the Venn diagram (see Fig. 6.37).

If we now consider a set $X_T \subset X_S$ which we have divided into subsets $X_i \subset X_T$, $i = 1, 2, \ldots, J$, which do not overlap, because $X_i = X_T \cap X_i$, we get

$$P[X_T] = \sum_{i=1}^{J} P[X_T \cap X_i] = \sum_{i=1}^{J} P[X_T | X_i] \cdot P[X_i] \tag{A1.17}$$

using (A1.8) and (A1.16). This statement is designated as a total probability theorem. As the division of X_T into mutually exclusive subsets X_i is completely free, this threorem has achieved great practical importance. We must only take care that the subsets X_i defined by us completely fill the set X_T, and we can select the X_i as seems most appropriate. (We make frequent use of this possibility in Chaps. 5 and 6.) Of course, we can also work with infinitesimal sets $x < X \leqq x + dx$ or $y < Y \leqq y + dy$ instead of with finite sets X_i and X_j, i.e. using the form

$$f_{XY}(x, y) = f_{XY}(x|y) \cdot f_Y(y) = f_{XY}(y|x) \cdot f_X(x) \tag{A1.18}$$

instead of (A1.16), or using the form

$$f_X(x) = \int_{-\infty}^{\infty} f_{XY}(x, y) dy = \int_{-\infty}^{\infty} f_{XY}(x|y) \cdot f_Y(y) dy \tag{A1.19}$$

instead of (A1.17) ($-\infty \leqq y \leqq \infty$). With respect to the second term of the total probability theorem in (A1.19), we also call $f_X(x)$ marginal probability distribution density of the joint distribution $f_{XY}(x, y)$ of X and Y, and with respect to the third term we talk of making the distribution $f_X(x)$ conditional, $f_{XY}(x|y)$ being called a conditional probability density distribution.

If we find that

$$P[X|Y] = P[X] \text{ or } f_{XY}(x|y) = f_X(x) \tag{A1.20}$$

we designate X and Y as stochastically independent random numbers. With stochastic independence, according to (A1.16) and (A1.18),

$$P[X, Y] = P[X] \cdot P[Y] \quad \text{or} \quad f_{XY}(x, y) = f_X(x) \cdot f_Y(y). \tag{A1.21}$$

We now calculate the first moment about $x = 0$

$$m_X = \int_{-\infty}^{\infty} x \cdot f_X(x) dx. \tag{A1.22}$$

Taking account of (A1.11) we can also write

$$m_X = \int_{-\infty}^{\infty} x \cdot f_X(x) dx / \int_{-\infty}^{\infty} f_X(x) dx,$$

and this is a lever arm according to the laws of mechanics which indicates the centre of gravity of the area under $f_X(x)$, i.e. a mean value. We also call this mean value of the random number X the expectation $E[X]$, so we can use the expressions 'first moment', 'mean value' and 'expectation' for $m_X = E[X]$. We prefer the notation $E[X]$ when we want to demonstrate the mathematical operation of taking the mean, and m_X when we mean the result.

We will also define the second central moment

$$\sigma_X^2 = \int_{-\infty}^{\infty} (x - m_X)^2 \cdot f_X(x)dx. \tag{A1.23}$$

We can see that this is the mean quadratic deviation from the mean, which is also called scatter or variance Var

$$\mathrm{Var}[X] = E[(X - m_X)^2] = \sigma_X^2. \tag{A1.24}$$

Here σ_X is the standard deviation. We use the expression $\mathrm{Var}[X]$ when showing the mathematical operation, and the expression σ_X^2 when indicating the result. To avoid confusion with the stress symbol σ, we use the term $\sqrt{\mathrm{Var}[\Sigma]}$ for standard deviation when stress and standard deviation appear together (see (6.92)).

We now consider the functional dependence of two random numbers Y and X, which is determined by a continuously increasing or decreasing function of their realizations

$$y = g(x), x = g^{-1}(x)$$

the exponent -1 again symbolizing the inverse function. With $P[X] = f_X(x)dx$ and $P[Y] = f_Y(y)dy$, as in (A1.10), we get for $P[Y] = P[X]$

$$f_Y(y) = f_X(x) \cdot |dx/dy| = f_X(g^{-1}(y))/|dy/dx|. \tag{A1.25}$$

This relationship is called 'monotonic one-to-one transformation'. In principle, it is easy to extend this transformation to more than two random values (see also [1]).

With $U = (X - m_X)/\sigma_X$, i.e. $u = g(x) = (x - m_X)/\sigma_X$, (A1.25) gives the standard normal distribution $\varphi(u)$ for U if $y := u$, assuming that X follows the Gaussian normal distribution in (5.19).

We can now give formally the expectation of the expression $(X - m_X) \cdot (Y - m_Y)$, and thus define a parameter

$$\mathrm{Cov}[X, Y] = E[(X - m_X) \cdot (Y - m_Y)], \tag{A1.26}$$

which we call mixed central moment or covariance, using the terms introduced in connection with (A1.23) and (A1.24). With $X = Y$ we find the relationship

$$\mathrm{Cov}[X, X] = \mathrm{Var}[X], \tag{A1.27}$$

and if we multiply out the brackets in (A1.26), because $E[CX] = CE[X]$, we get

$$\mathrm{Cov}[X, Y] = E[X \cdot Y] - m_X \cdot m_Y. \tag{A1.28}$$

in accordance with (A1.22). Looking at (A1.21) and (A1.22) we can see that the covariance is zero if X and Y are stochastically independent. We cannot conclude conversely, however, that if the covariance disappears, they are stochastically independent: covariance and stochastic dependence are not identical concepts!

Positive covariance does, however, always indicate that if the parameters $x \in X$ increase, the parameters $y \in Y$ probably also increase, and with negative covariance the latter would probably decrease. To this extent we can regard the covariance of two random numbers X and Y as a measure of their tendential bond or (linear) correlation. For purely practical reasons, we need to standardize this measure defining the correlation coefficient ρ, with the latter assuming the following extreme values: $\rho = 1$: full correlation, $\rho = 0$: no correlation, $\rho = -1$: full negative correlation. We obtain these boundary values automatically with the following definition of the correlation coefficient

$$\varrho_{XY} = \text{Cov}[X, Y]/(\sigma_X \cdot \sigma_Y), \tag{A1.29}$$

as this expression has the characteristic $-1 \leqslant \rho \leqslant 1$ (see [48], Chap. 6, or [2]). The mean of standardized random parameters U with standard normal distribution is always zero, and the standard deviation one (see (5.18)), so with (A1.28) and (A1.29) we get

$$\varrho_{12} = \text{Cov}[U_1, U_2] = E[U_1 \cdot U_2]. \tag{A1.30}$$

A2 Selected Principles of Matrix Calculus

Matrix calculus serves to abbreviate linear algebra and thus makes it clearer; furthermore, computers are ideally suited to handling problems formulated using matrices. The principles of matrix calculus were laid in the last century (J. J. Sylvester, 1850, and A. Cayley, 1857), but its first real applications appeared only in the 1920s in the field of nuclear physics and, with the creation of more efficient computers in the 1950s, in the performance of complex structural analyses in aerospace engineering.

The term 'matrix' means an arrangement of values or elements in n columns and m rows. These elements can be individual numbers or matrices. Complete mathematical expressions, functions or statements are all allowed. If there are the same number of rows and columns, it is a quadratic matrix. If all elements a_{ij} fulfil the condition $a_{ij} = a_{ji}$, then the matrix is called *symmetrical*.

Often, the elements are structured so, for instance, for

$$a_{ij} = 0, \quad i \neq j; \quad a_{ij} \neq 0, \quad i = j$$

we get diagonal matrices, i.e. only on the main diagonal are elements which are not equal to zero.

If only a few diagonals have elements which are not equal to zero, the matrices are called *band matrices*. If all values are equal to zero below or above the main diagonals and not equal to zero above or below the main diagonals, it is an *upper* or *lower triangular matrix*.

The rules of determinant calculus (A. T. Vandermonde, 1771) are applied to the elements. The determinant of a matrix A with four elements gives

$$\det A = \begin{vmatrix} a_{11} & a_{12} \\ a_{21} & a_{22} \end{vmatrix} = a_{11}a_{22} - a_{21}a_{12}.$$

If we subtract two rows or columns of a determinant from one another, the value of the determinant does not change

$$\det A = \begin{vmatrix} a_{11} & a_{12} \\ a_{21} - a_{11} & a_{22} - a_{12} \end{vmatrix} = a_{11}a_{22} - a_{21}a_{12}.$$

This gives the easiest way of finding the determinant of large matrices. By appropriate addition we can achieve triangular decomposition of a matrix (Gauss, 1801). The determinant can then easily be calculated as the product of the main diagonal elements.

The general algebraic processes of addition and subtraction take place on the element level, i.e.

$$A \pm B = C \quad \text{corresponds to} \quad a_{ij} \pm b_{ij} = c_{ij}. \tag{A2.1}$$

Constant factors of a matrix are multiplied by each element.

Matrix multiplication is special. The sequence must be observed, for

$$A \cdot B \neq B \cdot A. \tag{A2.2}$$

This becomes obvious if we multiply the matrix A of the value $m \times n$ by the matrix B of the value $n \times o$. For the product of

$$A \cdot B = C \tag{A2.3}$$

we obtain a matrix of the value $m \times o$. The elements of the matrix C are

$$c_{ij} = \sum_{k=1}^{n} a_{ik} \cdot b_{kj}.$$

We can see that $B \cdot A \neq C$.

If we want to interchange rows and columns of a matrix, this is called *transposition* of a matrix, and is indicated with a superscript T. For a symmetrical matrix we then have, for instance,

$$A = A^T. \tag{A2.4}$$

The effect of sequence on multiplication indicated above is particularly marked in the multiplication of column by row vectors. If a and b are two column vectors, the product is

$$a^T \cdot b = c,$$

i.e. a scalar value. If, however, a is multiplied by the transposed column vector b^T

$$a \cdot b^T = C,$$

C is a quadratic matrix, which can also be called a *dyadic product*. In transposition of the product of several matrices, the sequence of the individually transposed matrices is inverted

$$(ABCD)^T = D^T C^T B^T A^T. \tag{A2.5}$$

When two symmetrical matrices are multiplied, the symmetry generally disappears, as is shown by the following example of forming the product of two 2×2 matrices A and B

$$A = \begin{bmatrix} a_{11} & a_{12} \\ a_{21} & a_{22} \end{bmatrix}; \quad B = \begin{bmatrix} b_{11} & b_{12} \\ b_{21} & b_{22} \end{bmatrix};$$

$$A \cdot B = \begin{bmatrix} a_{11}b_{11} + a_{12}b_{21} & a_{11}b_{12} + a_{12}b_{22} \\ a_{21}b_{11} + a_{22}b_{21} & a_{21}b_{12} + a_{22}b_{22} \end{bmatrix}.$$

If $a_{ij} = a_{ji}$ or $b_{ij} = b_{ji}$, then the product of the matrices A and B is

$$\begin{bmatrix} a_{11}b_{11} + a_{12}b_{12} & a_{11}b_{12} + a_{12}b_{22} \\ a_{12}b_{11} + a_{22}b_{12} & a_{12}b_{12} + a_{22}b_{22} \end{bmatrix}.$$

We can see that the product of symmetrical matrices is generally asymmetrical

$$a_{12}b_{11} + a_{22}b_{12} \neq a_{11}b_{12} + a_{12}b_{22}.$$

This also applies if matrix A or B is a diagonal matrix, which is occasionally overlooked.

A frequent task of matrix calculus is to answer the question of how to obtain the inverted matrix of A, so that

$$AA^{-1} = I. \tag{A2.6}$$

A^{-1} is the inverse of the matrix A, I is the unit matrix (diagonal matrix with elements 1). The inversion is formally necessary if, for example, we want to solve the linear system of equations

$$A \cdot x = r$$

with respect to x.

$$x = A^{-1} r$$

this inversion requires a large number of manipulations, and is therefore to be avoided if possible. There is a simple inversion rule for a 2×2 matrix. For

$$A = \begin{bmatrix} a_{11} & a_{12} \\ a_{21} & a_{22} \end{bmatrix}$$

we obtain

$$A^{-1} = \frac{1}{\det A} \begin{bmatrix} a_{22} & -a_{12} \\ -a_{21} & a_{11} \end{bmatrix}.$$

From this we can see that matrix A must have a determinant other than 0. If a matrix fulfils this requirement, we regard it as regular; if, however, the determinant of a matrix disappears, it is a singular matrix.

Quadratic forms occur with the derivation of stiffness matrices

$$S = x^T A x. \tag{A2.7}$$

S is a scalar value. If we carry out a non-singular transformation with

$$x = T \cdot y$$

we obtain

$$S = y^T C y$$

$$C = T^T A T.$$

If A is a symmetrical matrix, C is always symmetrical, whether T is symmetrical or not. These are similarity transformations. Derivations of S with respect to values x_i of the vector x are often required, e.g. for

$$S = \tfrac{1}{2} x^T A x$$

we obtain

$$\frac{\partial S}{\partial x} = A x. \tag{A2.8}$$

This result can be proved, for example, for a 2×2 matrix of A

$$S = \tfrac{1}{2} [x_1 x_2] \begin{bmatrix} a_{11} & a_{12} \\ a_{12} & a_{22} \end{bmatrix} \begin{bmatrix} x_1 \\ x_2 \end{bmatrix},$$

$$S = \tfrac{1}{2}(a_{11} x_1^2 + 2 a_{12} x_1 x_2 + a_{22} x_2^2),$$

$$\begin{bmatrix} \dfrac{\partial S}{\partial x_1} \\[2mm] \dfrac{\partial S}{\partial x_2} \end{bmatrix} = \begin{bmatrix} a_{11} & a_{12} \\ a_{12} & a_{22} \end{bmatrix} \cdot \begin{bmatrix} x_1 \\ x_2 \end{bmatrix}.$$

The inverse is then

$$\int A x \, dx = \tfrac{1}{2} x^T A x + \text{const.} \tag{A2.9}$$

We can test the assumption that the matrix A is symmetrical with the above example.

Eigenvalues, Eigenvectors, Modal Analysis

A linear system of equations is assumed

$$A \cdot x = r.$$

We are looking for a factor λ, which, when multiplied by the vector x, gives just the vector of the right-hand side r

$$A x = \lambda x.$$

We thus obtain a homogeneous system of equations

$$(A - \lambda I)x = 0. \tag{A2.10}$$

Such a system of equations has a non-trivial solution only if the coefficient matrix disappears, i.e.

$$|A - \lambda I| = 0. \tag{A2.11}$$

λ represents the eigenvalues we are looking for. For a 2×2 matrix A we obtain

$$\left| \begin{bmatrix} a_{11} & a_{12} \\ a_{21} & a_{22} \end{bmatrix} - \lambda \begin{bmatrix} 1 & \\ & 1 \end{bmatrix} \right| = 0.$$

The determinant gives the characteristic equation

$$(a_{11} - \lambda)(a_{22} - \lambda) - a_{21}a_{12} = 0.$$

The zeros of the characteristic equation give the eigenvalues

$$\lambda_{1,2} = \frac{a_{22} - a_{11}}{2} \pm \sqrt{\left(\frac{a_{22} - a_{11}}{2}\right)^2 + a_{12}^2}.$$

The product of the eigenvalues gives

$$\lambda_1 \cdot \lambda_2 = a_{11}a_{22} - a_{12}a_{21} = \det A$$

and the sum is

$$\lambda_1 + \lambda_2 = a_{11} + a_{22}.$$

This is generally true. The product of all n eigenvalues of an $n \times n$ matrix is equal to the determinant of A. The sum of the eigenvalues is equal to the sum of the main diagonal elements. From the first principle we can see that a non-singular matrix A, in which $\det A \neq 0$, always has eigenvalues which also must be $\neq 0$.

We determine the solution of the homogeneous equation $(A - \lambda I)x = 0$ by substituting the eigenvalues. We must select one unknown x_1, e.g. $x_2 = 1$, and can then determine the two eigenvectors $x = b_1$ or b_2

$$b_1 = \left\{ \begin{array}{c} \dfrac{-a_{12}}{a_{11} - \lambda_1} \\ 1 \end{array} \right\}, \quad b_2 = \left\{ \begin{array}{c} \dfrac{-a_{12}}{a_{11} - \lambda_2} \\ 1 \end{array} \right\}.$$

With the product

$$b_1^T \cdot b_2 = -\frac{a_{12}^2}{a_{12}a_{21}} + 1$$

we can see when the matrix A is symmetrical: in this case

$$b_1^T \cdot b_2 = 0.$$

This means the vectors are in a vertical relationship to one another, they form an orthogonal system. We call the matrix of the eigenvectors the modal matrix T.

Transformation of the matrix A with the aid of the scaled modal matrix gives the eigenvalues

$$\Omega = T^{-1}AT,$$

$$\Omega = \begin{bmatrix} \lambda_1 & & & \\ & \lambda_2 & & \\ & & \ddots & \\ & & & \lambda_n \end{bmatrix}. \tag{A2.12}$$

We can use the special feature of this modal transformation if we want to obtain the value

$$A^n$$

(n is any rational number) of a symmetrical matrix A with $\det A > 0$.

With the aid of (A2.12) we transform the matrix A into Ω. Then we produce Ω^n, which can be done directly, as there are values only on the main diagonals, and then by reverse transformation we get

$$A^n = T^T \Omega^n T. \tag{A2.13}$$

The frequently encountered problem of solving simultaneous equations of a system capable of oscillation with n degrees of freedom

$$M\ddot{s} + D\dot{s} + Ks = f(t) \tag{A2.14}$$

can now be dealt with by applying the features of modal matrices using n separate systems each with one degree of freedom. It is obvious that processing equations with one degree of freedom n times is considerably more convenient than dealing with the coupled system. A prerequisite is that it must be possible to represent the damping matrix D as a linear combination of the stiffness matrix K and the mass matrix M

$$D = \alpha M + \gamma K. \tag{A2.15}$$

Then the eignevalue problem can be solved without taking account of damping

$$|K - \omega^2 M| = 0.$$

From the eigenvalues ω_i^2, we obtain the eigenvectors b_i, $i = 1$ to n. This can be summarized as the modal matrix T

$$T = [b_1 b_2 \cdots b_n]. \tag{A2.16}$$

Applying the transformation to the equation of oscillation (A2.14) gives

$$T^T M T = I,$$
$$T^T K T = \Omega$$

and

$$\begin{aligned} T^T D T &= T^T(\alpha M + \gamma K)T \\ &= \alpha T^T M T + \gamma T^T K T \\ &= \alpha I + \gamma \Omega. \end{aligned}$$

Table A1. Probability density functions used in the main text

Type	Probability density function $fx(x)$	Domain of random number		Mean	Standard deviation	Reference to text
Exponential	$\lambda \cdot \exp\{-\lambda x\}$	$0 \leqslant x < \infty$	$\lambda > 0$	$\dfrac{1}{\lambda}$	$\dfrac{1}{\lambda}$	(5.29) with $x = t; \lambda = v$
Poisson	$\dfrac{\lambda^x}{x!} \cdot \exp\{-\lambda\}$	$x = 0, 1, 2, \ldots$	$\lambda > 0$	λ	λ	s. 5.1.3.1 with $\lambda = v \cdot t; x = n$
Standard normal	$\dfrac{1}{\sqrt{2\pi}} \cdot \exp\{-\tfrac{1}{2}x^2\}$	$-\infty < x < \infty$	—	0	1	(5.18) with $x = u$
Normal	$\dfrac{1}{\sqrt{2\pi} \cdot \sigma} \cdot \exp\left\{ -\dfrac{1}{2}\left(\dfrac{x-m}{\sigma}\right)^2 \right\}$	$-\infty < x < \infty$	$\sigma > 0$	m	σ	(5.19) with $m = m_x$; $\sigma = \sigma_x$
Rayleigh	$\dfrac{x}{\sigma^2} \cdot \exp\left\{ -\dfrac{1}{2}\left(\dfrac{x}{\sigma}\right)^2 \right\}$	$0 \leqslant x < \infty$	$\sigma > 0$	$\sigma \cdot \sqrt{\dfrac{\pi}{2}}$	$\sigma \cdot \sqrt{\dfrac{4-\pi}{2}}$	(5.23) with $x = a$; $\sigma = \sigma_x$
Weibull	$\dfrac{\lambda}{\sigma} \cdot y^{\lambda-1} \cdot \exp\{-y^\lambda\}; y = \dfrac{x-\mu}{\sigma}$	$\mu \leqslant x < \infty$	$\sigma > 0; \lambda > 0$	see [A1]	see [A1]	(5.109) with $x = h_{1/3}$ $\lambda = b; \sigma = d; \mu = h_0$
Gumbel	$\dfrac{1}{\sigma} \cdot \exp\{-y - \exp\{-y\}\}; y = \dfrac{x-\mu}{\sigma}$	$-\infty < x < \infty$	$\sigma > 0$	$\mu + 0.577 \cdot \sigma$	$1.282 \cdot \sigma$	(5.118) with $\delta = 1/\sigma$

We obtain the new, generalized or modal coordinates w from

$$s = Tw; \quad \dot{s} = T\dot{w}; \quad \ddot{s} = T\ddot{w}.$$

n uncoupled equations result

$$\ddot{w}_i + (\alpha + \gamma\omega_i^2)\dot{w}_i + \omega_i^2 w_i = b_i^T \cdot f(t).$$

These can now be dealt with individually by classical methods of vibration theory.

References

1   Hahn, G. Shapiro, SS. Statistical models in engineering. New York: Wiley (1967)
2   Ditlevsen, O. Uncertainty modeling. New York: McGraw-Hill (1981)

Subject Index